AUTOMATIC LANGUAGE TRANSLATION

HARVARD MONOGRAPHS IN APPLIED SCIENCE

These monographs are devoted primarily to reports of University research in the applied physical sciences, with especial emphasis on topics that involve intellectual borrowing among the academic disciplines.

1. Matrix Analysis of Electric Networks, P. Le Corbeiller

2. Earth Waves, L. Don Leet

3. Theory of Elasticity and Plasticity, H. M. Westergaard

4. Magnetic Cooling, C. G. B. Garrett

5. Electroacoustics, F. V. Hunt

6. Theoretical Elasticity, Carl E. Pearson

7. The Scattering and Diffraction of Waves, Ronold W. P. King and Tai Tsun Wu

HARVARD MONOGRAPHS IN APPLIED SCIENCE

NUMBER 8

AUTOMATIC LANGUAGE TRANSLATION

LEXICAL AND TECHNICAL ASPECTS, WITH
PARTICULAR REFERENCE TO RUSSIAN

ANTHONY G. OETTINGER

with a Foreword by Joshua Whatmough

HARVARD UNIVERSITY PRESS
Cambridge, Massachusetts, 1960

Library of Congress Catalog Card Number 60-7999

Printed in the United States of America

To
Mother and Father,
Anne, and Marilyn

PREFACE

This book presents the results of a study, begun in 1950 and carried on intermittently ever since, of the problems of automatic language translation. I emphasize the fundamental lexical and technical aspects of these problems, although it goes almost without saying that a satisfactory method of automatic translation also requires the solution of many profound syntactic and semantic problems. Indeed, the future value of research on automatic translation might well hinge more on its contributions to a fundamental understanding of all levels of language structure and of the nature of automatic information processing than on any resulting machine-produced translations.

The initial emphasis on lexical and technical problems stems from my conviction that an operating automatic dictionary is a necessary basis for effective, linguistically significant research on syntax and on semantics, as well as a necessary first approximation to an automatic translator. The reader may judge for himself the validity of this conviction: with the exception of the important work on affix interpretation described in Chapter 10 and only approaching completion at the time of writing, a proved system of dictionary compilation and operation is described in the full detail necessary to enable others to evaluate our results, to use them directly in their own research, to extend them to areas of discourse in Russian other than those for which our existing dictionary file is intended, or to adapt them to the study of the many pairs of languages different from Russian and English but sufficiently similar in structure for adaptation. My own attention is now turning to syntax with far greater hope for significant progress than I felt at the inception of this study.

My investigations have been confined to translation from Russian to English. There has been much speculation about general methods of translation, whether directly between pairs of languages or via some natural or artificial intermediate language. It is my belief that we are yet too ill-equipped for a frontal assault and that valid generalizations will emerge naturally from more modest and more promising investigations of individual language pairs. I have tried, however, to emphasize what I consider to be fundamental, in order to assist those who seek them in discerning universals once the time is ripe.

I did not intend to write a textbook or a treatise; but, because the investigation of automatic translation draws on the resources of several disciplines heretofore unrelated or related only vaguely, I do begin with some basic background material that sets the stage for the later descriptions of our investigations. In addition, several chapters are provided with bibliographic notes.

Working between conventional disciplines is a bit like having one foot on each side of a crevasse; I am deeply indebted to Professors Howard Aiken and Joshua Whatmough for sound guidance and inspiration over many years and for anchoring me sternly on both sides. I owe much also to Professor Roman Jakobson, who has given me freely of his time and knowledge both directly and through many of his students who collaborated with me. The friendship and wise counsel of Professor Philippe Le Corbeiller are warmly appreciated.

A. G. O.

The Computation Laboratory of
Harvard University
Cambridge, Massachusetts
June 1959

ACKNOWLEDGMENTS

I have been assisted in much of the work described in this book by my students and by members of the Staff of the Computation Laboratory. The Seminar in Mathematical Linguistics which I have conducted in collaboration with Dr. Lawrence G. Jones since 1954–55 has been a fertile source of ideas; fortunately, students' contributions to this seminar have been recorded by the students themselves, so that it is possible in most instances both to make a direct acknowledgment and to point to greater detail by bibliographic references in the text. Successive stages of work on automatic translation by members of the Staff have been described in a series of reports, so that acknowledgments and references are given in the text in this case also.

The investigation of automatic translation was first suggested to me by Professor Howard Aiken, who has since given me constant support and encouragement. The joint efforts of Professor Aiken as Director of the Computation Laboratory and of Professor Joshua Whatmough as Chairman of the Department of Linguistics have enabled me to combine the teaching of mathematical linguistics with research on automatic translation without many of the trying strains that frequently beset or prevent any interdisciplinary work. I owe my education in linguistics chiefly to Professor Whatmough and Professor Roman Jakobson.

I am much obliged to Dr. Warren Semon for his unfailing assistance in matters both intellectual and administrative, especially in connection with our use of the Univac installation at the Computation Laboratory. I have profited greatly from many discussions with Professor Kenneth Iverson, Professor Robert Ashenhurst, now of the University of Chicago, and Dr. Lawrence G. Jones. By assuming some of my responsibilities while this book was being written, Dr. Vincent Giuliano contributed greatly toward its completion.

The following people were kind enough to read and comment on the manuscript in whole or in part: Mr. William Foust, Dr. Vincent Giuliano, Mr. James Hannan, Dr. Lawrence G. Jones, Mr. Paul E. Jones, Jr., Mr. Roger Levien, Mr. Karol Magassy, Dr. Ladislav Matejka, Mr. Warren Plath, Dr. Richard See, Mr. Murray Sherry, Mr. Louis Tanner, Mr. Joseph Van Campen, Miss Stefanie von Susich, and Professor Joshua Whatmough; I thank them for saving me from some errors.

Every member of the secretarial staff of the Computation Laboratory contributed to the typing of the manuscript at some stage; the final draft was prepared by Mrs. Erma Bushek and Mrs. Eileen Seaward; my deep thanks go to Miss Jacquelin Sanborn, who coordinated the preparation of the whole manuscript, for raising once again the superb standards of editorial and secretarial craftsmanship she has set for herself over the years, and to Mr. Robert Burns for his assistance in many phases of preparation. I am glad of this opportunity to thank Mr. Paul Donaldson of Cruft Laboratory, who made the photographs, and Mr. William Minty, who prepared the drawings, for contributing of their fine craftsmanship.

The initial financial support for my investigations came from the Harvard Foundation for Advanced Study and Research and from the Ohio Oil Company in the form of very modest but greatly appreciated grants to which the present project owes its start. More recently, the work has been supported in part by the National Science Foundation and the Rome Air Development Center of the United States Air Force, and it is a pleasure to thank both organizations.

CONTENTS

FOREWORD

To write a good book is much harder than to produce a decent baby. The naming of either is not always apt to tell just what is inside. In Professor Oettinger's title Language is an important word, however you look at it. If I were a librarian I should strain one rule by buying more than one copy of his work, and break another by cataloguing them separately—one for the linguists, another for the communication engineers, and yet another (at least) for the computation laboratory.

But these interests are now running into one another. The trick of designating knowledge as humane (put language there), social (communication there), scientific (applied mathematics there), however generalizing, has not as yet actually proved educationally unifying. Despite all the lip service done to cross-fertilization between departments of human knowledge, few do anything more than preach, and those who talk the most perform the least.

Professor Oettinger performs. He combines a firsthand knowledge of three disciplines, which is better than being a committee, since committees do not have ideas. His book is quite a remarkable baby. If I mistake not, Linguistics has much to gain in rigor and method from it; those of the growing band of workers on automatic translation and related problems will be in his debt for the help that he gives them in coping with some thorny linguistic problems without setting up terminological roadblocks of the kind for which linguists have (and need) their own technical glossaries.

Sic uos non uobis.

JOSHUA WHATMOUGH

Department of Linguistics
Harvard University
8 September 1959

INTRODUCTION

Research on automatic language translation draws from and contributes to two major disciplines, linguistics and automatic information processing, one with firm roots in antiquity and strong traditions, the other newborn and not yet fully conscious of its identity. The use of automatic machines as tools for linguistic investigations and the use of linguistic theory as a guide in certain applications of automatic machines are only symptoms of a deeper and broader interest in an area where linguistics and mathematics meet, an area now commonly labeled *mathematical linguistics*. The concepts and techniques of modern algebra, statistics, and mathematical logic are contributing to the evolution of linguistics from a humanity into a science. Linguistics, in turn, is contributing to the solution of problems of language synthesis arising in studies of the foundations of mathematics and logic and in the design of automatic information-processing machines.

This book is concerned primarily with certain fundamental lexical and technical problems of automatic language translation. To provide a firm foundation not only for the solution of these problems but also for a more profound investigation of automatic language translation and for the study of mathematical linguistics in general, the first four chapters are devoted to certain basic questions of automatic information processing and linguistics.

It would be patently unreasonable to claim that amenability to automatic processing should be the only touchstone by which linguistic theories may be tested, but amenability to automatic processing is a criterion at least as reasonable and as significant as any of those conventionally used in linguistics. The venturesome linguist may find that scholarly investigations of automatic translation have raised a number of interesting linguistic problems never adequately treated by linguists, that the effective use of automatic machines imposes high standards of theoretical rigor and elegance, and that many tools and techniques of value to linguists have been and will continue to be invented in the course of these investigations.

The first chapter describes the essential features of automatic machines of the kind now employed in experimental work on automatic translation and outlines the problems facing the designers and the users of these machines. The reader who is unacquainted with automatic information-processing machines may find it best first to browse through the whole chapter, then to return to it once again after reading Chapter 2, and finally to refer to it as needed in later chapters. The presentation of Chapter 1 is new, and much of the material either has never been presented before or else can be found only in scattered journals or technical reports; Chapter 1 may therefore prove of interest even to the specialist in automatic information processing.

Several of the chapters, Chapter 1 among them, include examples and exercises, chiefly by way of answering the question "How does the machine do it?" in more detail than can be given in the body of the text. It is difficult to find a middle ground between the deceptively casual "Oh, you feed it in and the answers come out" and the labyrinthine detail of a machine program complete with operating instructions; but to give the one answer is to throw sand in the eyes of the inquirer, to give the other is to make him count the grains. The examples are simple and concrete illustrations of certain fundamental details; the exercises are intended to stimulate the reader to test his comprehension of these details or to generalize from them. Examples generally, and exercises always, may be skipped at first reading without break in the continuity of the text.

Chapter 2 is concerned with fundamental properties of certain classes of signs. In ordinary discourse in a familiar language one uses signs without much conscious thought about how and why. In discourse of which several natural and artificial languages are themselves the principal subjects, serious confusion inevitably results unless the signs *used* to carry on the discourse are carefully distinguished from those *mentioned* by it; logicians have long been aware of this problem and have devised various methods for coping with it; a new notational system, introduced informally in Sec. 1.2 and described in detail in Sec. 2.1, has been inspired by these methods. The new notation is richer than that normally required by logicians, for several reasons of which the following are the chief: first, it is necessary to relate the natural languages on which a translating machine is to operate to the artificial languages in which machines are programmed; second, the physical vehicles of signs, which are normally of little concern to linguists, logicians, or mathematicians, come to the forefront when machine design and applications are considered; finally, as should become apparent in Chapter 5, inflected languages have certain properties that can be described only by awkward and confusing circumlocution in the absence of an adequate notation.

Chapter 3 describes so-called automatic coding techniques that are proving extremely valuable in making automatic information-processing machines easier to use; it also presents flow-charting methods for the concise and easy description of the processes that machines are called upon to execute. The arts of flow charting and automatic coding are still in a rudimentary stage, and both linguistics and mathematical logic may be expected to contribute toward their development into more universal and rigorous techniques; conversely, the powerful, if not always complete and consistent, descriptive methods that have already been developed have interesting applications in those realms of linguistics and logic requiring the precise description of complex structures and processes.

The exploration of more than a few ramifications of the questions outlined in the first three chapters is beyond the scope of this book, not only on grounds of relevance to the main subject, but also because it is too early, in many cases, to discern fundamentals among the ideas bustling about in a rapidly growing literature. Bibliographic notes are appended to several chapters.

Translation, conventional and automatic, is the subject of Chapter 4, which discusses both what is expected of translation and what can be accomplished by way of meeting expectations. A brief introduction to the linguistic problems of translation is given.

The remaining chapters are devoted chiefly to the lexical aspects of automatic translation. Throughout these chapters, but especially in Chapter 10, a lexicon is regarded as the foundation on which syntax and semantics are to be built. The exposition is based upon a study of translation from Russian to English and upon extensive experimental work using a Univac I computer, but it emphasizes the fundamental problems common to other pairs of languages and to other information-processing machines. Chapters 5 through 7 describe dictionary compilation and Chapter 8 deals with the operation of an automatic dictionary; important problems slighted in Chapters 7 and 8 to avoid undue obscurity in exposition are treated in full in Chapter 9. Chapter 10 points the way to a transition from automatic dictionary to automatic translator.

CHAPTER 1

AUTOMATIC INFORMATION-PROCESSING MACHINES

Machines such as lathes are prized chiefly for their physical products; the value of the products of the machines with which we shall be concerned is not intrinsic, but derives from roles played by these products as vehicles for *signs*. Such machines are called *data-* or *information-processing machines* to distinguish them from the tools of production. Of the wide variety of automatic information-processing machines now in use, we shall consider only those of the class described as *large-scale automatically sequenced digital machines*. These machines were originally developed to do at high speed the extensive calculations necessary for solving many problems of science and engineering. Machines similar or related to the scientific calculators are widely applied to such commercial accounting activities as payroll preparation and customer billing. To the extent that the process of language translation is of a systematic but routine and tedious character, it is sufficiently similar to the processes of computing or accounting to have suggested itself to many as another area where automatic high-speed machines might be employed to advantage.

The problems of automatic translation cannot be approached without a clear understanding of what automatic information-processing machines are, and of what they can and cannot do; an introductory description of their relevant logical and physical characteristics is therefore given in this chapter. Emphasis on the description of existing machines is inevitable, since these machines are being used to help with the vast clerical tasks inherent in developing methods of translating. However, machines intended to serve the general purposes of scientific calculation or of commercial information processing are not necessarily ideal for translating. One of the objectives of research on automatic translation is the specification of design criteria for practical, economical, safe-failing procedures and equipment. (The *fail-safe* principle, a fundamental tenet of sound engineering, dictates that machines be built to act safely in the event of likely failures.) Whether these criteria can be met best by general- or by special-purpose equipment is an open question.

Describing existing machines and their use without recourse to some concrete examples would not be fruitful. While all machines share most of

1

the important basic features to be described, they differ enough in some details to make absolute generality impossible. Examples must therefore refer to specific machines, real or hypothetical. I have chosen to base my examples on the Univac I, which has been used extensively in connection with the translation studies described in this book. All basic results presented in later chapters are given, insofar as possible, in a form independent of the characteristics of any specific machine.

1.1. The Organization of Information-Processing Machines

Figure 1 illustrates a pattern of organization common in essence to most information-processing machines; it reflects accurately only the general logical structure of these machines, disregarding for the moment all physical aspects. Obviously, some information must be introduced into a machine, and processed information returned to the outside. The mechanisms for communication with the outside are represented in Fig. 1 by boxes labeled *input* and *output*. Storage facilities for information arriving from outside or generated within the machine in the course of operation are also provided. In Fig. 1, the storage or *memory* unit of the machine has been sketched as a set of *n* individual storage *registers*, each labeled with a numeral, its *address*. The address usually serves as a name for the corresponding register.

The lines connecting different elements of the diagram represent paths over which information may be transmitted. Any pair of units may be connected by the path labeled *information transfer bus*, provided that the proper *in-gates* and *out-gates* are closed, in other words, provided that these gates are so positioned as to establish a continuous path from one unit to the other. Information then can flow through the out-gate of one unit, over the transfer bus, and through the in-gate of the other unit in the direction indicated by the arrows. It is assumed that only one in-gate and one out-gate may be closed at any given instant, and that transfers of information between different pairs of units must therefore succeed one another in time. Some recent machines have facilities for simultaneous transfers under certain conditions.

Operations on information are initiated by transferring the operands from memory to a special unit. For example, in a machine with an addition operation, first the augend is transferred to the operation unit, then the addend. After the delay necessary for performing the addition, the sum may be transferred from the operation unit to any desired storage register or to the output. Addition is an example of *dyadic* operation, that is, one involving two operands; subtraction, multiplication, and division are likewise dyadic. An operation on k operands is called k-*adic*. Obtaining the letter next in alphabetic order to a given letter of the alphabet is an example of a monadic operation, and selecting the largest of three distinct numbers is an example of a triadic one. One k-adic operation unit is sketched in Fig. 1.

The actions of the machine are specified chiefly by prescribing a sequence of information transfers. The prescription of a single transfer is called an *order* or an *instruction*, and a sequence of instructions is a *program*. Fully automatic operation of the machine is enabled by providing a *control unit*

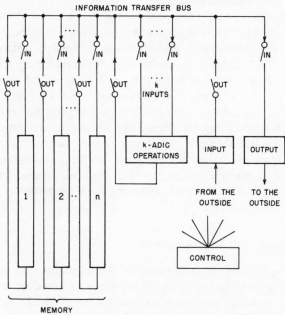

Fig. 1. The organization of an automatic information-processing machine.

capable of interpreting instructions and of opening and closing gates accordingly. To initiate automatic operation, a program is set into an appropriate portion of the memory. The control unit then takes over and executes the instructions in a prescribed sequence.

Although the sequence of instructions may be prescribed in many ways, one convenient and widely used system requires that consecutive instructions be placed in registers with consecutive addresses. In this system, once the control unit has executed the instruction in some register m, it usually executes next that in register $m+1$, then that in $m+2$, and so on.

An outstanding exception to the foregoing procedure provides the basis for the great flexibility and usefulness of automatic machines. Register m may contain a *control transfer* instruction which specifies some register other than

$m+1$ from which the control unit is to take the next instruction. *Conditional* control transfer, or *branch*, instructions direct a transfer of control *only if some specific auxiliary condition is also met*, namely, if some register p contains special information; otherwise, the next instruction is taken from register $m+1$ as usual. When branch instructions are present in a program, the machine can automatically select one of several alternative procedures depending upon what information was placed in p as a result of the operations preceding the execution of the branch instruction.

Instructions commonly comprise several characters that indicate the type of operation to be executed and name the affected register. Many machines with otherwise comparable properties use completely different instruction characters. For example, the instruction for transferring a sum from the addition unit to register m is $H\,m$ in the Univac, $19\,f\,m$ in the Cambridge University Edsac II, $10\,m\,11$ in the Harvard Mark IV, and $F\,m$ in the IBM 705. Each of these instructions mentions the register address explicitly, but the characters specifying the nature of the transfer are obviously different. Other more subtle differences exist among these instructions, and it is clear that, if concrete examples are to be given in a single notation, they can refer directly only to a single type of machine. We shall proceed with the description of a specific machine after examining some properties of physical media for storing information.

1.2. Signs and Their Representations

It will be helpful in understanding the methods used for storing and manipulating information in digital machines to introduce at this point the concepts of *sign, sign representation,* and *token*. A fuller explanation of these concepts and of such typographical devices as the joint use of italics and asterisks is given in Chapter 2, but something of their significance may be conveyed here by the following illustrative statements: the word *example** is a sign; conventionally, *example** has the printed representation "example", although other representations may be used, for instance, the Morse code ". _ · · _ · _ _ _ · _ _ · · _ · · ·"; the ink impressions on paper which we read as *example** are tokens for that word.

The notion of a token is essential to enable reference to multiple physical instances of a unique sign; that of representation enables us to speak of a conventional representation ("example") or a Morse code representation (". _ · · _ · _ _ _ · _ _ · · _ · · ·") of the unique word *example**.

Many physical configurations, including notches in clay or stone, ink marks on paper, and electromagnetic waves, have served mankind as tokens; each has special advantages and limitations. Most of the physical devices found suitable for storing information in automatic digital machines share the property of having only two stable configurations easily usable as tokens. Devices of this kind are called *bistable*; they include, among others, pieces of paper or cardboard, magnetic materials, vacuum tubes, and transistors. Thus, a given spot of cardboard may be perforated or not, a piece of magnetic material may or may not be magnetized, a vacuum tube may be conducting or nonconducting.

Table 1. A correspondence between integers and the configurations of three bistable devices.

Integers		Configurations of three bistable devices		
Decimal notation	Binary notation	1	2	3
0	000	0	0	0
1	001	0	0	1
2	010	0	1	0
3	011	0	1	1
4	100	1	0	0
5	101	1	0	1
6	110	1	1	0
7	111	1	1	1

A single bistable device in one of its stable configurations readily serves as a token for one sign, and in its other stable configuration, as a token for a second sign. More than two intermediate configurations are produced, maintained, or recognized by an automatic electronic machine only with excessive difficulty, in contrast with the ease with which some ink and a small spot of paper may be put in one of 26 or more stable configurations by striking a typewriter key. Consequently, the most natural number of characters on which the representation of signs in digital machines may be directly based is not 26 or more, but two. The mode of representing signs in digital machines therefore differs from the conventional mode, and requires further explanation.

Because the first large digital machines were calculators, it was natural directly to identify one configuration of a bistable device with the numeral 0^*, and the other with the numeral 1^*. In this way, a correspondence is readily established between numbers expressed in the binary notation and the configuration of a set of bistable devices. Table 1 illustrates how the integers from 0 to 7 may be placed in correspondence with the eight possible configurations of three bistable devices. In Table 1, "**0**" is used to denote a bistable device in one configuration, and "**1**" denotes a bistable device in the other configuration. Note that if more than eight integers are to be represented an additional bistable device is necessary and that, with its aid, sixteen integers may be represented. Groups of four devices are frequently used in so-called binary-coded decimal machines, where ten of the sixteen combinations are used to represent "0" through "9", and the remaining six do not occur except through malfunction. In general, n bistable devices have 2^n configurations which can be used as tokens.

It is natural, when dealing exclusively with numerical data, to ignore the distinction between "**0**" and "**1**", which name tokens, and "0" and "1", which name numbers. Thus, "101" is frequently used *homographically* to denote either the number 5 or a particular configuration of a set of three bistable devices. This practice is confusing in discourse about machines designed for or adapted to purposes other than numerical computation, especially when the relation between concrete machine tokens and abstract signs is itself the chief subject of discussion. In particular, it leads to the mistaken notion that every problem to be solved by an information-processing machine must somehow be "reduced to numbers".

This notion has a superficial validity: it is sometimes necessary, when using a computer for nonnumerical problems, to adopt a numerical representation for nonnumerical data. If we agree, for example, that "a" corresponds to "1", "b" to "2", "c" to "3", and so on, the word *cab**, normally represented by "cab", can be equally well represented by "312". In this context, the characters "3", "1", and "2" do not represent numerals at all, but letters. Using the same characters for several purposes is analogous to using the same word to express different meanings (homography or homonymy) and entails the same advantage of economy and disadvantage of confusion.

Although circumstances may force us to use a numerical representation of nonnumerical signs, it should be clear that this is *not* the same as "replacing words by numbers". It is senseless to add *cab** to *cab**, or to take its square root, even though the word may be represented by the string of characters "312". Computers have been used to do nonnumerical work only because such machines have been more widely available than machines specially designed for nonnumerical work. Their use is possible, not because arithmetic applies to nonnumerical signs, but because many of the *logical* processes *underlying* arithmetic are also fundamental in nonarithmetic systems.

1.3. The Organization of the Univac

The correspondence established in the Univac between common characters and sets of bistable devices is given in Table 2. Such a correspondence table establishes what is commonly called a *binary coding system* for the characters. Each character is represented by a token consisting of a configuration of six bistable devices. Tokens are therefore available for a maximum of 64 characters, and 63 are actually used. The configurations of the first two of the

Table 2. The binary coding system of the Univac.

	00	01	10	11
0000	i	r	t	Σ
0001	△	,	"	β
0010	—	.	|	:
0011	0	;	)	+
0100	1	A	J	/
0101	2	B	K	S
0110	3	C	L	T
0111	4	D	M	U
1000	5	E	N	V
1001	6	F	O	W
1010	7	G	P	X
1011	8	H	Q	Y
1100	9	I	R	Z
1101	'	#	$	%
1110	&	¢	*	=
1111	(	@	?	not used

six bistable devices are given at the top of the column in which a character is displayed, and those of the remaining four at the left of the row. For example, **010100** is a token for "A", and **111100** is a token for "Z". The space character ("△") is associated with the token **000001**. Note that the conventional (ink-on-paper) token for a space is simply blank paper. This is most inconvenient in discourse where every character must be carefully accounted for, since an intended space cannot be distinguished from accidentally blank paper. The token Δ is therefore adopted to replace blank paper as a token for a space in the ink-on-paper system.

A basic correspondence between machine tokens and the characters or signs they represent is established by fixing the design and construction of the input and output mechanisms of the machine. For example, after adopting the token **010100** as a token for the letter A^*, matters are arranged so that such a token will cause an output printer to print the character "A". When a set of signs, say the English alphabet, is ordered in a particular way, it is naturally desirable to define the correspondence between the members of the set of signs and their tokens in such a way that the machine will recognize, by simple physical operations on tokens, one token as "occurring after" some other, whenever this relation holds between the corresponding signs.

Characters or signs of certain classes are singled out in a machine whenever physical means are provided for handling their tokens as a group and in particular ways. For example, in some machines, a single bistable element can be isolated and used individually, whereas in others these elements must be handled in fixed groups. It is convenient to use special generic names for characters in the distinguished classes. A single bistable device is said to have a storage capacity of one *bit*. In the Univac, with a few highly technical exceptions, bistable devices are contained in groups of no fewer than six, which represent *characters* as described in Table 2. Sequences or *strings* of 12 characters are grouped in *words*, 10 words form a *blockette*, and 6 blockettes constitute a *block*. The characters in a word are numbered 1 to 12 from left to right; the words in a blockette are numbered 0 to 9; the words in a block are numbered 0 to 59. Similar groupings are used in most other machines. The name *word** is likely to be confusing when discussing linguistic matters, so *machine word** will be used instead when the occasion demands it.

A highly simplified diagram of the Univac system is given in Fig. 2; only salient characteristics of importance to a user of the machine have been shown. The portion of the diagram within dashed lines represents the *central computer*, whose functions roughly correspond to those of the machine sketched in Fig. 1. Information transfer paths are again represented by solid lines but, for simplicity, no gates have been shown. Information may flow in the directions indicated by the arrows, and simultaneous transfers over more than one path may occur under special circumstances.

The memory unit consists of 1000 registers with addresses ranging from *000* to *999*, each capable of storing one word. The machine word is singled out as the basic unit of transfer, although a few instructions provide for the transfer of word-pairs, blockettes, and even blocks as units. No explicit instructions are provided for the direct transfer of single characters but they can be synthesized by a series of special instructions (Example 1-2, Sec. 1.4). Such idiosyncrasies are usually justifiable on technical grounds, and similar ones appear in most machines.

The *central registers F, L, A*, and *X* each have a storage capacity of one word, and are used in conjunction with the operation unit. In the course of addition, for example, the operands are transferred from memory to registers

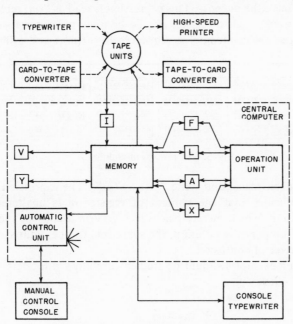

Fig. 2. The organization of the Univac.

A and *X*; their sum is then formed in the operation unit and returned to register *A* where it becomes available for transfer to any memory register. The central registers *V* and *Y* have storage capacities of two and ten words respectively; they serve as intermediate registers in the transfer of word-pairs and blockettes.

1.4. Elements of Programming

The basic technique of programming will be illustrated with a few simple examples, each of which could be a fragment of a larger program. The purpose of each instruction will be explained when it is first introduced, and a table of selected Univac instructions is provided for reference in the appendix to this chapter. No attempt will be made to fulfill all functions of a textbook on programming; the reader who is interested in mastering programming techniques will find references to helpful textbooks in the bibliographic notes of Sec. 1.8.

A shorthand notation for "the word contained in the register whose address is m" will be useful, and we shall use "(m)" for this purpose. Usually, "(m)" is read simply as "the content of m", or "the word in m". A notation expressing concisely the replacement of (n) by (m) will also be useful, and "$m \rightarrow n$" will serve the purpose. This expression is read as "(m) replaces (n)", or as "(m) is transferred to register n". To be quite precise, we should distinguish between the content of a register before $((m)_b)$ and after $((m)_a)$ a

		500	501	502	A	X
(1)	INITIAL	000012345678	000012121212	XXXXXXXXXXXX	XXXXXXXXXXXX	XXXXXXXXXXXX
(2)	B 500	000012345678	000012121212	XXXXXXXXXXXX	000012345678	000012345678
(3)	A 501	000012345678	000012121212	XXXXXXXXXXXX	000024466890	000012121212
(4)	C 502	000012345678	000012121212	000024466890	000000000000	000012121212

Fig. 3. The process of addition.

transfer has taken place, and write "$(m)_b \rightarrow (n)_a$". The expression "$(m)_a = (m)_b$" then indicates that the content of register m is unaltered by the transfer operation, which merely places in register n a copy of the word in register m. To simplify the notation, the subscripts will be omitted whenever there is no danger of confusion.

The following examples should be studied carefully, with reference to the appendix as needed.

Example 1–1.

Program (a) B 500
                    A 501
                    C 502

When the instruction B 500 is executed, $(500) \rightarrow (A)$ and also $(500) \rightarrow (X)$, that is, the word in register 500 is transferred to both register A and register X. The previous contents of A and X are destroyed as a result of this operation; but $(500)_a = (500)_b$, that is to say, the word originally in register 500 is merely copied, and not destroyed. The contents of the several registers, initially and after the execution of each instruction, are shown in Fig. 3 ("xxxxxxxxxxxx" in a box indicates that the content of the register is irrelevant). Executing A 501 brings (501) to X, again replacing the previous content of that register, and instructs the operation unit to form $(A) + (X)$ and to put this sum in A. The instruction C 502 has two effects: $(A) \rightarrow (502)$ and $0 \rightarrow (A)$, in that order; hence $(A)_a \neq (A)_b$. The original contents of registers 502, A, and X play no role in this operation, and are destroyed; had they been of importance, orders to transfer them to other registers should have preceded the addition program.

Program (b) B 500
                    A 501
                    H 502

This program is identical with (a) except that while $C\,m$ resets A to zero after transferring the sum to m, $H\,m$ does not alter (A).

$$\text{Program }(c)\quad B\,500$$
$$X$$
$$H\,502$$

When this program is executed, $(500) + (500) \rightarrow (502)$. The instruction X has this effect: $(A) + (X) \rightarrow (A)$. The final result follows from the fact that $(A) = (X) = (500)$ once $B\,500$ has been executed.

Programs (a) and (b) will be executed by the machine in approximately 1.5 ms (0.0015 sec) and (c) in 1.2 ms.

Exercise 1–1. Write programs analogous to (a) and (b) for placing the difference $(500) - (501)$ in register 502. Select your instructions from among those given in the table appended to this chapter.

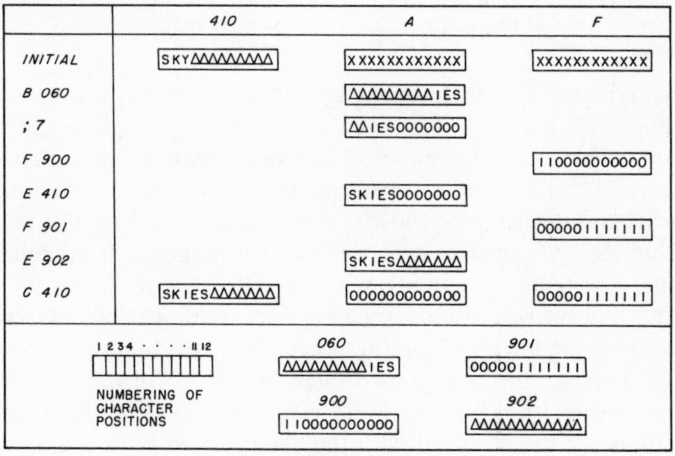

	410	A	F
INITIAL	SKY△△△△△△△△△	XXXXXXXXXXXX	XXXXXXXXXXXX
B 060		△△△△△△△△IES	
; 7		△△IES0000000	
F 900			110000000000
E 410		SKIES0000000	
F 901			00000IIIIIII
E 902		SKIES△△△△△△△	
C 410	SKIES△△△△△△△	000000000000	00000IIIIIII

NUMBERING OF CHARACTER POSITIONS

1 2 3 4 · · · · 11 12

060: △△△△△△△△IES
901: 00000IIIIIII
900: 110000000000
902: △△△△△△△△△△△△

Fig. 4. Extraction and shifting.

Example 1–2. Extraction and Shifting. Imagine that the English word *SKY** is stored in register *410.* Since "SKY" has only three characters and a register always holds 12 characters, other characters must be used as "fill". In this example, as shown in Fig. 4, spaces (each represented by " △ ") have been used. We should like to get the plural of the word by deleting "Y" and adding "IES". Initially, registers *410, 060, A,* and *F* are filled as indicated in Fig. 4, and certain useful data are stored in registers *900, 901,* and *902.* For convenience, the character positions in each register are numbered from 1 to 12, going from left to right. When all is done, "SKIES" are to occupy the first five positions of register *410,* and spaces are to fill the remaining positions.

The position of the characters of a word stored in register *A* can be changed by a *shift* instruction. The instruction with operation characters "*;n*" shifts (A) to the left *n* places, and that with character "*.n*", shifts (A) to the right *n* places. These instructions affect only register *A,* hence require no address. Characters "shifted out" of *A* are lost, and vacant positions are filled with zeros. The string "IES" is brought into proper position by executing *B 060* and *;7,* the first two instructions in the program.

We may now introduce the "SK" of "SKY" into register *A* by using the *extract* instruction *E 410.* The execution of an instruction *E m* requires having in register *F* an *extractor,* that is, a word of which each character is either decimal "0" or decimal "1". When $E\,m$ is executed, if there is a "1" in position p of register F, then the character in position p of register m is extracted and replaces that in position p of register A. Characters of (A) in positions where (F) has a "0" are not affected. The extractor necessary for our purpose is in register *900*, and the instruction $F\,900$ will transfer it to register F. Then, once $E\,410$ has been executed, the word "SKIES0000000" appears in register A. The zero-fill is replaced with space-fill by a second extract operation using (*901*) as the extractor, and finally the desired word is transferred to register *410*, and register A is reset to zero. The machine accomplishes the whole process in approximately 3 ms.

We note that the extract instruction enables the transfer of a variety of character groups, ranging in composition from a single character to combinations of up to 12 characters. The extractor itself may be a result of earlier processing, and hence provides a useful means of controlling the pattern of character transfers in accordance with these results. Thus, should we want to form plurals in several paradigms, we would have the means for selecting among several suffixes of varying length.

> *Exercise 1–2.* Can you write a program like that of Example 1–2, except that (*410*) is brought to register A first? What changes must be made if "SKY" initially occupies positions 10, 11, and 12 of register *410*?

Suppose we wish to form the sum of 200 numbers stored in registers *700* through *899*, and place it in register *950*. A very simple program capable of accomplishing this summation might begin with $B\,700$, followed by $A\,701$, then $A\,702$, and so on to $A\,899$, followed by $C\,950$, for a total of 201 instructions. In general $n+1$ instructions would be necessary to sum n numbers in this way. The prospect of writing such programs is appalling, but more subtle, less tedious methods are available. The principles on which these so-called *iterative* methods are based are quite general, but certain details depend on properties of Univac instructions which must be described first.

Six character positions are required to store a Univac instruction. Not all characters are significant in every instruction, and nonsignificant character positions are usually filled with the character "*0*", for example, "*A00950*", although the nonsignificant zeros are conveniently omitted in writing, for example, "*A 950*". Since a storage register can store 12 characters and an instruction consists of only 6, it is economical to store two instructions in each register, a *left* instruction in positions 1–6, a *right* instruction in positions 7–12. We can imagine the program of Fig. 4 to be held in storage as follows:

010	*B00060*	*;70000*
011	*F00900*	*E00410*
012	*F00901*	*E00902*
013	*C00410*	*000000*

The numbers in the left-hand column are the addresses of the registers storing the instructions. For convenience in reading, the left and right instructions are shown spaced apart. (Note the necessity of the "$\triangle$" convention to distinguish between this typographical convention and an indication that a space is to be stored in a register.) The null instruction "*000000*" has no effect; in this example, it is used merely as fill but it is sometimes deliberately introduced to provide gaps where significant instructions may be inserted to modify or to correct a program.

The control unit includes a special register, the *control counter*, containing the address of the next instruction-pair to be executed. In normal operation, the left instruction of a pair is executed, then the right instruction, and the address in the control counter is increased by unity. Instructions are therefore executed in the order in which they are stored in memory.

Control transfer instructions, however, cause the address given by their own address characters to be transferred into the control counter. The control unit then selects the next instruction-pair, not from the register immediately following the one storing the control transfer instruction, but from the register whose address has been placed in the control counter. If the transfer of a new address into the control counter takes place only when certain auxiliary conditions are met, the control transfer instruction is called conditional. Conditional control transfer instructions are frequently called *branch* instructions, since they enable a selection among alternative paths.

The Univac has three control transfer instructions. The instruction $U\,m$ unconditionally transfers control to the instruction in register m, while $Q\,m$ transfers control only if $(A) = (L)$ at the time of execution, and $T\,m$ only if $(A) > (L)$, that is, if (A) is numerically greater or alphabetically later than (L).

One final word about the operation of the addition unit will lead us to a concrete example. An instruction word is distinguished from other words in storage only when it is transferred into the control unit. Otherwise, it may be treated like any other word and may, for example, be transferred to register A. The addition unit is designed to perform the "addition" of a machine word containing alphabetic characters and of a plain numerical word in the following peculiar manner: an alphabetic character occurring in a given position of the "augend" will occur in the same position in the "sum". Numerical characters are treated in the usual way, subject to exceptions that need not concern us here. The significance of this mode of operation will become apparent in the following example.

Example 1–3. An Iterative Program. The program of Fig. 6 is equivalent in effect to the summation program with 201 instructions, but comprises only 15 instructions. The key to the structure of this program is the *iterative loop* (Fig. 5), a segment of program designed to perform the same function over and over again but each time on different signs. A few instructions are executed prior to the first entry into the loop to set up certain initial conditions. The instructions within the loop perform the basic function of the program, effect the necessary modifications prior to each new iteration, and keep track of the number of iterations or otherwise provide for leaving the loop. Following the iterative loop, some concluding instructions may be necessary to complete the process.

While tracing through the following explanation of the operation of the program shown in Fig. 6 the reader may find it useful to draw for himself a chart like that of Fig. 4. The first instruction-pair of the program transfers the content of register *100* to register *004* where it will serve as an instruction after one further modification. This procedure is known as "planting" an instruction. When writing a program, it is helpful to mark the holes for planting with square brackets, and to show within the brackets the instruction that will be executed the first time around the loop. Note that register *A* acts as a go-between in the transfer of a sign from one memory register to another. Such a method of transfer is used in most machines. The right member of the first instruction-pair causes the content of register *A* to be the number 0, which is then transferred to register *103* by *C 103*. Register *103* will hold the partial sums throughout the process. Setting initial conditions is completed once the instruction *L 101* has brought (101) to register *L* for eventual comparison with (A).

The instruction *B 004* is the first within the loop and initiates a test for the end of the iteration. At this stage

$$(004) = B00103A00699 = (A) \neq (L) = B00103A00899,$$

hence *no transfer of control takes place*, and the instruction-pair in register *003* is executed next; the execution of the *Q* instruction leaves (A) unchanged, hence the effect of the pair in register *003* is to plant the instruction

$$(A) + (102) = B00103A00699 + 000000000001$$
$$= B00103A00700$$

ENTRANCE

INITIAL SETTING

YES END OF ITERATION?

NO

LOOP

ITERATED OPERATION

CONCLUSION

EXIT

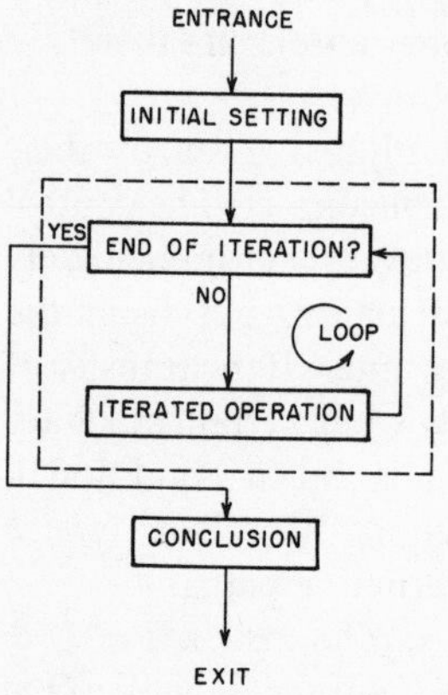

Fig. 5. The structure of an iterative program.

in register *004*. It is precisely for this purpose that the addition unit is designed to perform the "addition" of a word containing alphabetic characters and a plain numerical word. The proper instruction-pair is now planted in register *004*, and the execution of *B 103*, *A 700*, and *C 103* places the first partial sum, namely $0 + (700)$, in register *103*. Control is then transferred to the instructions in register *002*.

Although now $(004) = B00103A00700$, transfer of control by *Q 006* again cannot take place, and the loop is traversed once more. This time, the instruction-pair in register *004* will be *B00103A00701* and the second partial sum, $(700) + (701)$, will be placed in register *103*. Eventually (004) will be *B00103A00899*, and the complete sum will be in register *103* following the execution of the left instruction in register *005*. The instruction *U 002* will return control to the beginning of the loop. This time $(A) = (L)$ and control is transferred to the pair *B 103 C 950*, which sends the desired sum to register *950*. The next instruction stops the machine, and the process is complete. We shall consider this process again in Chapter 3, where means for a simpler description will be provided.

Exercise 1–3. A list of words is stored in registers $0, 2, 4, \ldots, n$, where n is even, and another list is stored in registers $1, 3, 5, \ldots, n + 1$. A word known to be in the first list is stored in register *998*. Write a program that will compare (998) successively with (0), (2), (4), and so on, until $(998) = (k)$ for some $k \leq n$. When k has been determined, $(k + 1)$ should be transferred to register *999* and the machine stopped. Given your program, what is the maximum allowable value of n? Can you think of applications for this program?

The advantages of the iterative-loop technique over the direct approach are manifold. There is, of course, the obvious and important reduction in the number of instructions that must be written and stored in memory. Moreover, the "straight-line" program can do only the specific job for which it is written, while it is easy to adapt the iterative program to sum numbers held in any group of consecutive storage registers. The reader can easily check that only the constants held in registers *100* and *101* need be changed to vary the group of registers whose contents are to be summed. Finally, the iterative loop is the natural technique for performing the important classes of mathematical and logical processes that have recursive definitions.

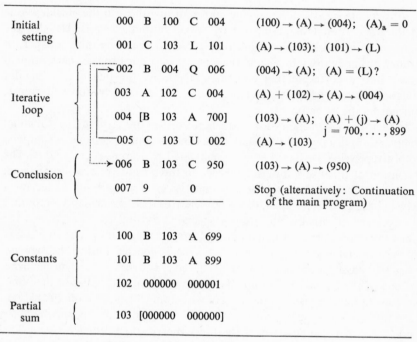

Initial setting		000 B 100 C 004	$(100) \rightarrow (A) \rightarrow (004);$ $(A)_a = 0$
		001 C 103 L 101	$(A) \rightarrow (103);$ $(101) \rightarrow (L)$
Iterative loop		→002 B 004 Q 006	$(004) \rightarrow (A);$ $(A) = (L)?$
		003 A 102 C 004	$(A) + (102) \rightarrow (A) \rightarrow (004)$
		004 [B 103 A 700]	$(103) \rightarrow (A);$ $(A) + (j) \rightarrow (A)$ $j = 700, \ldots, 899$
		005 C 103 U 002	$(A) \rightarrow (103)$
Conclusion		→006 B 103 C 950	$(103) \rightarrow (A) \rightarrow (950)$
		007 9 0	Stop (alternatively: Continuation of the main program)
Constants		100 B 103 A 699	
		101 B 103 A 899	
		102 000000 000001	
Partial sum		103 [000000 000000]	

Fig. 6. An iterative program for summation.

Against these advantages, a few disadvantages must be balanced. Iterative loops are trickier to write than straight-line programs. It is very easy to make a mistake in specifying the conditions for terminating the iteration, but the difference between exactly n and $n-1$ or $n+1$ iterations can be that between right and wrong results. Special care is therefore absolutely necessary. Straight-line programs are always executed faster than the equivalent iterative loops. In the present example, the execution times are roughly in the ratio 1 : 7. The reason for this is not difficult to see: several "bookkeeping" instructions must be executed each time around the loop; these are absent in the straight-line program. The effect is less pronounced in programs where the "useful" operations outnumber those necessary for bookkeeping.

In any case, to write compact programs in a reasonable time, it is nearly always necessary to use iterative loops. Without this technique, the usefulness of automatic information-processing machines would undoubtedly be negligible in many applications.

1.5. Information Input and Output

Introducing signs into a machine and returning them to the outside is one of the major functions performed by "input-output" devices. Usually, the signs to be introduced into a machine are represented by ink on paper, in printed or handwritten form, and the input devices must convert each original token into a corresponding one manipulable by the machine. A similar situation obtains when the original tokens can be used by some machines, but not by the given one.

In their most elementary form, the input operations require direct manual intervention, usually key-punching. The Univac control console is equipped with a typewriter keyboard (Fig. 2), activated by the instruction *10 m*. The execution of this instruction takes place in three steps: first, a visual signal is given to the operator, telling him that the machine is ready to receive a type-in of 12 characters; then, the characters are typed and held in a temporary storage register; finally, actuating a special "word-release" key causes the word to be transferred from temporary storage to register *m*. The conversion of the original tokens into machine tokens is effected during the type-in: when a key associated with an original character is struck, special mechanisms set one of the character positions (a group of bistable devices) in the intermediate storage register into the corresponding configuration (Table 1); simultaneously, a type bar connected to the key prints the original character on a sheet of paper inserted in the typewriter in the normal manner. The operator may therefore check his typing before actuating the word-release key by reading familiar tokens in the customary way.

The importance of this seemingly trivial detail deserves emphasis, because it is difficult to realize, without having had the experience, how tedious and how conducive to error it is to "read" binary patterns of holes in cards, of green spots on oscilloscopes, or of lit and unlit neon lamps, or how time-consuming it can be to transmit signs in an unfamiliar representation to an auxiliary device that will transform them back into their original form for proofreading. This unpleasant feature of many machines brings to mind the following observation about an early form of telegraph output equipment:

The receiving transducer for this system comprised a series of 35 inverted test tubes in which electrolytic decomposition of water was to take place, the symbol transmitted being indicated by the test tube in which a bubble of hydrogen was observed. One may wonder how such a system of signaling could ever have been regarded as practical, but retrospective judgments should probably be rendered with caution and charity; a good many of today's proudest technical achievements may appear just as impractical as soon as someone finds a better way to get the same results. (Hunt, 1954, p. 12.)

Efficient use of a high-speed machine would not be possible if constant manual intervention were necessary. For instance, it would take a good typist 45 minutes to fill the 1000-word memory of the Univac, an enormous span in the millisecond time scale of automatic operation. Moreover, the likelihood of an undetected typing error is quite high. The console typewriter is therefore used chiefly to insert minor corrections when checking the operation of a new program and occasionally for control purposes, but never for bulk input. To enable bulk input at a high rate, the manual token conversion and checking operation is usually separated from that of actual input to the central unit of the system, by means of an intermediate or *buffer* storage medium capable of accepting signs at the manual rate and of delivering them to the central unit at high speed. Punched cards, punched paper tape, and magnetic tape are common buffer media.

It is common to distinguish between *on-line* and *off-line* input and output devices. An *on-line* device, of which the console typewriter is an example, is directly connected to the central unit. Input or output devices not so connected are characterized as *off-line*.

In the Univac system, a special off-line typewriter serves to record on magnetic tape, for each instance of one of the characters of Table 1, a corresponding pattern of magnetized spots. Ordinary typewritten copy is produced simultaneously for proofreading purposes. Any number of typewriters may be used, and the central computer can be working on one problem while the instructions and input data for others are being transcribed onto magnetic tape. Ten tape-reading mechanisms provide the means for transferring signs automatically from buffer storage into the central memory unit. In Fig. 2, the path from the typewriter to the tape units is represented by a dashed line, to signify that information transfer is not automatic, but is accomplished by the manual transportation of reels of tape.

The physical characteristics of the typewriter and tape units are influential in imposing blockette and block grouping on words. Synchronization of the movement of the typewriter carriage and of the recording tape requires that exactly one blockette (ten words) be written on each line. In typing a series of *items*, each composed of several words, it is therefore highly desirable that the length, in words, of each item be an exact submultiple or multiple of ten (Sec. 7.2). Otherwise, the first words of successive items will occupy varying positions on the lines, making proofreading difficult and the use of tabular stops impossible. Details of this kind may seem inconsequential, but they assume prime importance when thousands of items are involved. Faulty manual input procedures can create a most serious bottleneck in any large-scale automatic data-processing operation.

The organization of words into blocks (Sec. 1.3) is a common solution of the problem of transferring information from tape to central memory at a high rate. The operation of "reading" magnetic tape is in effect a token conversion: the patterns of magnetic spots are converted into trains of electrical pulses, a magnetized spot corresponding to a pulse, an unmagnetized spot to no pulse. The conversion process depends on motion of the magnetic material relative to the coils of wire that constitute the reading head. High speed is necessary not only to match the internal operating rate of the machine, but also to induce detectable electric currents in the reading coils. Rapid tape travel is not hard to achieve; in fact, tape speeds of 100 in./sec or more are common. The sources of trouble are the necessity of starting the tape before a group of characters is read and of stopping it afterward, and the speed limits consequently imposed by the inertia of the tape and driving mechanisms. Starting and stopping a tape alone is an order of magnitude slower than an internal word transfer; hence, although the word could be the unit of transfer from tape, the continuous transfer of longer groups is more economical of time. On the Univac, the transfer from tape into memory of a block of 60 words requires about 100 ms, or an average of 1.7 ms per word, which is comparable to the internal transfer time. The necessity of grouping input information into artificial units like blocks is obviated in several newer computers by various artifices.

A further refinement enables the transfer of a block from tape into memory to take place while the operation of the central computer is interrupted for only 3.5 ms. The tape-reading instruction In directs the transfer of a block from the tape on mechanism n, not into the central memory, but into the 60-word buffer register I (Fig. 2). Independent control circuits are provided to supervise the transfer into I once it has been initiated. Therefore, after the 3.5 ms required to initiate the transfer have elapsed, but while register I is being filled, the main control unit may proceed to the next and later instructions, which may be any in the repertoire. The instruction $3 m$ is then used to transfer the contents of register I to 60 consecutive registers beginning with m. If this instruction is reached before the transfer into I is complete, its execution is automatically delayed, otherwise it is executed forthwith. Moreover, a pair of instructions $3 m$, In may be replaced by the single instruction $3n\,m$ which first transfers (I) to registers $m, m+1, \ldots, m+59$, and then initiates the transfer of the next block from tape n to register I, all in 3.5 ms. Clever interposition of useful instructions between one $3n\,m$

instruction and the next can therefore keep the effective block transfer time very close to 3.5 ms, but only at the price of superimposing somewhat artificial constraints on the natural structure of the problem to be solved. These constraints vary in detail among different machines, but they are present in all, and satisfying them can be one of the programmer's most trying tasks.

When it happens that the data for a given problem are recorded on punched cards rather than on magnetic tape, either because the original manual token conversion was performed on a key punch or because the data are the output of punched-card machinery, a card-to-tape converter (Fig. 2) must be used to do an additional token conversion. The operation of this converter is automatic and off-line. Other machines have several types of on-line input units to enable direct reading of information recorded on different media. Each approach has its peculiar advantages and drawbacks.

The problems of output are similar to those of input, and are solved in similar ways. The instruction $50\,m$ provides for automatically typing the word in register m on the console typewriter. For bulk output at high speed, the tape-writing instruction $5n\,m$ is used to record the contents of registers $m, \ldots, m+59$ as a block on the tape mounted on mechanism n. A buffer register intervenes also in this operation, but transfer into and out of this register is automatically controlled by the $5n\,m$ instruction.

The instructions $V\,m$ and $Y\,m$ and their respective companions $W\,m$ and $Z\,m$ can be useful in assembling blocks prior to recording. Execution of the pair of instructions $V\,m$, $W\,n$ transfers the contents of registers m and $m+1$ to n and $n+1$ respectively in roughly half the time required to effect the same transfer word by word with the instruction sequence $B\,m$, $H\,n$, $B\,m+1$, $H\,n+1$. Ten consecutive words may be transferred in less than twice the time necessary for a single word transfer by means of the instruction sequence $Y\,m$, $Z\,n$.

If the output data are needed on punched cards for further processing, a tape-to-card converter (Fig. 2) is available to do the conversion. A tape-actuated high-speed printer, which prints lines of 120 characters at rates up to 600 lines per minute, can be used to produce output information in printed form.

In "automatic" information processing, the manual transcription of printed symbols into a mechanically readable form is still by far the most tedious operation and the most susceptible to errors. Why is it, when one high-speed printer can work as fast as and more accurately than a group of 25 typists, that printed text cannot be read automatically at the same rate? Any printed page gives obvious clues to the answer. Recognizing the many varieties of type fonts, character sizes, punctuation marks, mathematical symbols, and so forth, that occur within one text, let alone several, is something enormously more difficult to accomplish automatically than actuating type bars under the control of carefully chosen configurations of bistable devices. Of the automatic recognition of ordinary handwriting we need not speak.

When the format of the material to be read can be controlled, progress toward fully automatic transcription is possible and has indeed been made in several directions. Anyone who has taken I.Q., aptitude, or other mass-produced tests will recall the injunctions to use only the special pencil, to be careful to make heavy marks between lines next to the answer, and always thoroughly to erase mistakes. The "mark-sensing" machines that read these pencil marks are probably the oldest and most widespread automatic transcription devices now in use. Their operation depends upon the relatively low resistance presented by graphite marks to the passage of electric current.

Photoelectric devices sensitive to the difference in the light absorbed by black and white areas are the basic elements of machines able to read patterns of black dots printed on business documents. Unfortunately, clerks cannot easily read these patterns, and printing the same information independently in ordinary type may lead to inconsistent versions; moreover, smudges and rubber stamps can play havoc with the system, at best leading the machine to reject documents as unreadable.

Designers of check-handling equipment for banks, where worry about minor inconsistencies and errors takes on heroic proportions, have created tokens for numerals readable at once by man and machine. To the eye, these tokens appear as ordinary printed digits, but, because a finely ground magnetic material is one ingredient of the ink, each induces a characteristic electric impulse when it passes under a coil of wire. It is more difficult and more costly to tell apart pulses induced by printed characters than those induced by a simple pattern of magnetized and unmagnetized spots, and the size and shape of the characters must be fairly closely controlled.

Machines capable of reading ordinary typescript or printing are in commercial use. It is claimed that high standards of accuracy and reliability can be met, and that the machines can be adjusted to read a variety of fonts, although apparently only one at a particular setting. Standardized formats remain essential, and interference by smudges or extraneous marks is unavoidable.

A device for automatically converting the sound waves of speech into either some phonematic transcription or the conventional spelling would be a necessary input mechanism for an automatic speech translator. The benefits of such a device to all human communication would be so great that considerable research activity is spreading in all directions likely to lead to its perfection. Unfortunately, the problems met on the way make automatic reading look like child's play.

Underlying these problems is the lack of an adequate system for characterizing the invariant features common to the speech of communities consisting of more than one individual. There exists as yet no solution even to the very basic problem of segmenting the stream of speech into significant units—phonemes, words, or sentences—a problem much less serious for printed texts. Limited success has been achieved, as for instance with a machine able to recognize spoken digits and having obvious applications to telephony. It is typical of the difficulties to be overcome that this machine must be adjusted for each particular voice, and that each digit must be rather carefully enunciated.

We need not dwell here on the highly specialized technical and linguistic problems of automatic reading or speech recognition. The bibliographic notes of Sec. 1.8 will guide the interested reader to more detailed references. For the foreseeable future texts to be translated must be transcribed manually with devices like that described in Sec. 2.8, unless they happen to have been prepared for printing by such devices as the "Teletypesetter", "Monotype", or "Photon" and the intermediate machine-readable media used in these processes are made available to the translating machine.

1.6. The Characteristics of Information-Storage Media

At first glance, the thousand registers constituting the memory of the Univac central computer seem to provide ample information-storage capacity; certainly the programs and the data in the examples of Sec. 1.4 easily fitted in them. Unfortunately, there is a vast quantitative difference between simple program fragments written expressly for illustrative purposes and complete programs for the mass-production operations of commercial data processing or of automatic language translation. It is the rule, rather than the exception, that programs and data together require more storage space than is available in the central memory. This is true even of machines provided with several times the storage capacity of Univac, if only because more complex problems look solvable and longer programs are written as soon as larger memories become available, in a pattern apparently governed by Parkinson's law (1957).

While there is no theoretical limit to the number of storage elements that may be incorporated in a machine, the fact that their cost is high imposes a financial one. In part, the high cost of memory elements stems from the requirement that information transfers must take place in milliseconds or microseconds, a necessity when the completion of many important processes is a matter of hours or days even in this time scale. The discovery of new storage elements and improvements in production techniques have contributed to a slow but steady increase in the storage capacity of high-speed memories. When speed requirements can be relaxed by an order of magnitude or so, economical storage facilities become practically boundless.

The magnetic tapes used as input and output buffer storage media lend themselves readily to broader duties. If the recording techniques are the same, a tape written by the central computer will be as acceptable to an input tape unit as one prepared on the typewriter. Each reel of tape then becomes an auxiliary storage unit of great capacity: on the Univac, a standard reel stores up to 2000 blocks or 120,000 words, at a cost per recorded character of a small fraction of a mil. Most machines are so designed that the input-output buffer medium, magnetic or paper tape, or punched cards, can also be used as an extension of the central memory. The effective limit on storage capacity is then set by the number of reels of tape or decks of cards that can be bought and stored.

Information stored on magnetic tape cannot, in general, be transferred to the operation unit as quickly as that held in the central memory. When a particular segment of tape happens to be near the reading head, the block on that segment may be transferred to central memory in a matter of milliseconds. If, however, the tape is in the rewound state, and the block selected for transfer is at the very end, the tape must travel several *minutes* before this block comes under the reading head. Naturally, this extreme case is usually avoided; on the assumption that any block, of the N on the tape, is as likely to be selected for transfer as any other, a time $t = Nt_0/2$ (t_0 = time required to read one block, including start and stop) will elapse on the average before the selected block is reached from the rewound position.

The mechanisms on the Univac and on some other machines can read a tape moving in either direction. Reading therefore need not always begin in the rewound state, and different average and maximum access times will be obtained depending on where and in which direction scanning is assumed to begin. The fractions of tape length that must be scanned under various assumptions are shown in Table 3 (adapted from Iverson, 1955, p. III-17). Scanning is said to be directed when the tape is moved in the direction in which the desired block is known to lie; when such knowledge is unavailable or cannot be used, scanning is nondirected except in a limited way, as toward the short end, the long end, or a fixed (left or right) end of the tape. Prior to each request, the tape may have one of its ends, its center, or a random position directly under the reading head. The average scanning fractions are calculated on the assumption that any block is as likely to be requested as any other.

The question of scanning time is only one of many factors affecting the

Table 3. Tape scanning fractions.

Type of scanning	Initial position	Average scanning fraction	Maximum scanning fraction
Directed	End	$\frac{1}{2}$	1
Directed	Center	$\frac{1}{4}$	$\frac{1}{2}$
Directed	Random	$\frac{1}{3}$	1
Nondirected	End	$\frac{1}{2}$	1
Nondirected	Center	$\frac{3}{4}$	$1\frac{1}{2}$
Nondirected (toward short end)	Random	$\frac{2}{3}$	$1\frac{1}{2}$
Nondirected (toward long end)	Random	$\frac{2}{3}$	2
Nondirected (toward fixed end)	Random	$\frac{2}{3}$	2

organization of storage facilities for a complex process. For instance, the proportion of information to be held in central memory and in auxiliary storage at any particular time must be carefully determined. Next to input transcription, to which it is intimately related, storage organization is the most serious problem in programming an existing machine or in designing a new one for applications where a large amount of information must be processed; in general, it can be safely assumed that the design of operating and control units will raise relatively fewer and less serious technical and economic problems. The term "peripheral equipment", commonly used to designate input, output, and auxiliary storage devices, is something of a misnomer, for in terms of size, cost, and problems raised this tail strongly wags its dog. A more detailed discussion of storage technology and logical organization is therefore given next, in preparation for the description of automatic dictionary files in Chapters 7 and 9.

The conventional sharp distinction between central memory and auxiliary storage is based chiefly on accidental details of the physical location of storage elements. It seems sounder to think of the storage system as a whole made up of a number of parts, each having different characteristics and therefore best used in different ways.

We have already seen that the time needed to transfer information from a storage medium is one distinguishing characteristic of the medium, called *access time*. Physically, storage units may be *static* or *dynamic, erasable* or *permanent*. In a static storage unit, no energy is required to maintain an established configuration. For example, once spots have been magnetized on a tape, or holes punched in a card, the configuration remains stable indefinitely, unless purposely disturbed. Tapes and cards are therefore static. In a dynamic storage device, energy must be continuously supplied to maintain a particular configuration. Hence a dynamic unit is always erasable, and vulnerable to power failures.

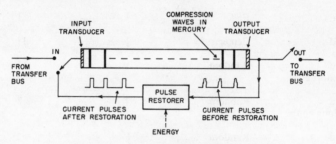

Fig. 7. Mercury-delay-line storage unit.

Magnetic materials, although static, are erasable, since they can readily be returned to a reference state. Individual punched cards, on the other hand, are both static and permanent, if one admits that erasing by plugging holes is not generally practical. Photographic storage techniques are coming into wider use, mainly because the high resolution of modern emulsions enables the concentration of vast quantities of tokens in relatively small areas. Photographic memories are obviously static and quite permanent.

A schematic representation of a Univac I dynamic storage unit is shown in Fig. 7. Characters are represented by sound pulses traveling in a mercury "delay line"; viewed at an instant in time, these pulses form a configuration of compressions and rarefactions. The output transducer converts the sound pulses into electrical pulses, which are amplified, reshaped, and synchronized in a pulse restorer, and then turned back into sound waves by the input transducer. A pulse pattern representing ten words thus circulates periodically around the loop. By closing the out-gate at the moment the first pulse of a word appears at the output transducer, and opening it after the last has passed, this word can be transmitted over the transfer bus. A delay naturally occurs if a word is requested for transfer when it is just entering the mercury. Read-in of new symbols and erasure of the old are simultaneous: if the in-gate is thrown at the proper time, a train of pulses coming from the transfer bus replaces that arriving from the pulse restorer.

Dynamic storage units were used originally because of their high speed of operation; they are being replaced by more recently developed magnetic devices, such as the "core" matrix illustrated in Fig. 8, which also have very high operating speeds and are static besides. Magnetic cores are so called because of their analogy to like-named components of transformers; they are actually small doughnut-shaped pieces of magnetic material. Their mode of operation is complex; it is sufficient to say here that a given core is set in the desired state of magnetization by sending current pulses through the unique pair of horizontal and vertical selection wires which intersect it, and that the state is detected by the presence or absence of a pulse on the output wire when the same selecting wires are once again energized simultaneously.

The notion of *address* may be examined more closely in the light of the physical details presented in the preceding paragraphs. The address of one of *n* registers was defined as an integer between 1 and *n*, and is so written in the machine instructions. Actually, nothing in the structures diagramed in Figs. 7 and 8 directly corresponds to such an integer. Even a register is not palpable in a mercury delay line, but is hypostatized by defining a time interval during which the out-gate is to remain closed. To extract a word from a mercury storage unit, it is necessary to know which unit, and at what time in the circulation cycle the first and last pulses of the word appear at the output transducer. The number of the unit and the location of a time interval with respect to a reference point in the cycle are therefore the "natural coordinates" of the word.

Magnetic-core registers are commonly formed by stacking into a three-dimensional unit as many planar arrays like that of Fig. 8 as the representation of a word requires. The number of registers in the unit is then equal to the number of cores in the basic array, and a register consists of the cores occupying the same position in each planar array, for example, those intersected by vertical selection wire No. 3 and horizontal wire No. 2. The natural coordinates in this case are obvious. It is a very attractive property of magnetic-core registers that reading and writing speeds are very nearly the same for every register in an array at any time.

The magnetic-drum storage unit sketched in Fig. 9 is similar to both magnetic-tape and mercury units. Information is represented, written, and read as with magnetic tape. A circular track on the circumference of the drum is analogous to one mercury channel, except that, whereas the mercury is fixed and pulses circulate, the magnetic spots are fixed on the continuously rotating drum. Each coil of wire serves as both an input and an output transducer; the fact that several transducers may be disposed around each track (two are shown in Fig. 9) means that a given word may be read by the first transducer it passes by, and the delay in access is inversely proportional to the number of transducers provided. The coordinates of a word are three:

the track number, the designation of the nearest transducer, and a time interval.

In most machines, the programmer need not be aware of the fundamental storage coordinates. Integral addresses are the only "visible" coordinates. Permanently wired circuits transform addresses into the fundamental coordinates, and fix the number of characters to be read or written. It is these circuits, rather than the storage units themselves, that establish the visible

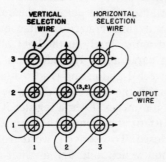

Fig. 8. Magnetic-core matrix
storage unit.

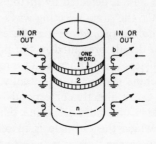

Fig. 9. Magnetic-drum storage
unit.

coordinates, the hierarchy of character strings, and their different transfer properties. Within broad limits, there is nothing intrinsic in the structure of mercury, core, drum, or tape units that dictates what visible coordinates or character groupings should be used, but it is important to realize that in most machines these properties have been fitted to the needs of scientific computation, or, in some cases, to those of commercial information processing, and that a different mode or organization may better suit the peculiar requirements of automatic language translation. For example, in a simple calculation where the programmer can foresee what numbers he will need and when they will be used, he is free to store them in any registers he wishes, and therefore knows exactly what addresses to use in instructions referring to them. The simplicity of the program of Fig. 6 owes much to the fact that consecutive summands are stored in registers with consecutive addresses. The reader need only attempt to write a program for summing numbers stored in randomly chosen registers to convince himself that this is true. Unfortunately, under many circumstances the address of a symbol cannot be defined so readily. Such is the case when the value $f(x)$ of a function is required, but the argument x is not an address; this happens in the Univac, for example, when x is an integer larger than 999. In general any table look-up process is complicated when a table entry (corresponding to $f(x)$) has an address different from the key identifying the entry (the analogue of x). One solution of this problem is suggested by Exercise 1-3, and others are mentioned in Sec. 1.7.

If operations can be performed much faster than information can be brought from memory to the operation unit, this unit will be idle a good part of the time. On the other hand, if transfers from memory take place in a period very short relative to the time in which operations can be effected, costlier rapid-access memory devices are being used where cheaper slow-access ones might do. The latter situation occurs fairly rarely in practice because of the relative ease of building very high-speed operation units. It is useful to describe the relation between a memory unit and the rest of an information-processing system by giving the name *immediate-access memory* to one for which the average access time is negligible with respect to the time in which operations are performed, and *delayed-access memory* to one of which the converse is true. It frequently happens, for example, that transferring information from magnetic tape takes longer than executing desired operations on the information. In such cases, magnetic tape is a delayed-access memory, and the program is said to be *tape-limited*. In some machines, complex circuits are provided for the simultaneous operation of several tapes, thus enabling the clever programmer to use tape as an immediate-access memory.

When several groups of signs are to be transferred in succession out of a storage unit the order in which they are transferred will often affect the access time. For instance, consider two words w_1 and w_2 represented by consecutive pulse trains in a mercury delay line. Instructions for immediately consecutive transfers of w_1 and w_2, in that order, will always be executed faster than if the order were reversed. This is evident from the observation that when w_1 is at the output transducer, w_2 will be there next, whereas when the order of transfer is reversed, nearly a whole cycle must elapse before w_1 becomes available. These statements are not exactly true in practice, because of the time necessary to put the next instruction in the control unit. However, there is *some* w_i, $i = 2, 3, \ldots, 10$, which is available immediately after w_1, and for which the argument applies. A program constructed to take advantage of this kind of idiosyncrasy is called a *minimum-access* program.

Programming for minimum access requires so intense a concentration on machine characteristics that it is easy to lose sight of the original problem. Some efforts have been made to facilitate minimum-access programming by automatic programming devices like those mentioned in Chapter 3, but the use of the process is best reserved for programs likely to be executed so often that every millisecond gained is multiplied into many hours. In the Univac, when it is known in advance that two or ten consecutive words are to travel as a group, this issue may often be sidestepped by using the word-pair and blockette transfer orders (V, W, Y, Z) mentioned in Sec. 1.4. However, we know enough now to see that these instructions cannot be effective unless all words in the group are in the same delay line. Since Univac delay lines store ten words each, there is a restriction, peculiar-looking at first sight, that addresses used in blockette transfer instructions must be multiples of ten.

In some magnetic-drum storage systems the fact that, because of delays in circuit operation, the word succeeding a particular one on a track is usually not the one immediately accessible after the transfer of the first, can be masked by a simple artifice. There is no fundamental reason why the storage register *physically* following that with address n should have address $n+1$. Often, one can simply arrange matters so that the address $n+1$ is assigned to the register whose content is accessible directly after the transfer of (n). Now, the naive but very reasonable expectation that $(n+1)$ should be available immediately after (n) is indeed fulfilled.

It is true of all storage devices that the access time to a particular string of characters depends to some degree on the location of the last string previously transferred, a phenomenon independent of the absolute time scale. For instance, in a desk dictionary open at "idiom", the definition of "idiomatic" is found faster than that of "thermodynamics", although either process is faster than transferring a block from one end of a 2400-ft tape and then one from the other end, and slower than any transfer from mercury delay lines or magnetic cores.

Access time actually depends explicitly or implicitly on several factors, which can be said to determine the *state* of the storage device at the time a transfer is initiated. In Table 3, for instance, the state of a tape mechanism is assumed to be specified by the position of the tape relative to the reading coil. For a given state, the access time naturally depends on what string is to be transferred. When the order in which transfers are requested is either fixed and known or else at the discretion of the programmer, the total access time can be calculated in principle, and conceivably it can be minimized by properly tailoring either the physical memory system or the disposition of information in the system; but in practice the pattern of transfers can be described at best only statistically. A useful general measure of the dependence of access time on the states can be obtained by assuming that every stored string is as likely to be requested as any other, and that the requests are independent, that is, that knowledge of the request for a particular string tells nothing about which string will be requested next. Mathematically, this is expressed by saying that each request is drawn independently from a rectangularly distributed population. The average access time for a string is then calculated under each of the two following standard conditions, theoretically over an infinite number of interrogations:

(*a*) *Random-order interrogation:* as each request is drawn, the corresponding transfer is executed;

(*b*) *Matched-order interrogation:* all requests are drawn first, then ordered in a way calculated to minimize total access time, and only then are the transfers executed.

It is evident that, if the access time is independent, or very nearly so, of the state of the memory, average access time per string will be the same for random-order and matched-order interrogation. When this is not the case, the random-order access time will be longer than the time for matched-order. The ratio of average random-order to average matched-order access time, the *access ratio*, can be taken as the desired measure of dependence on the memory states. A memory device for which this ratio is nearly unity is called a *random-access* device, and one for which it is large, a *serial-access* device. These terms are used very loosely in the literature, and must be interpreted with great caution; a few illustrations should clarify the sense in which they are used here.

Suppose we are using a tape mechanism in directed scanning from a random position, and wish to transfer m distinct blocks from a tape holding a total of N blocks. For random-order interrogation, Table 3 shows that each block will be obtained after scanning over 1/3 of the blocks, on the average. The m blocks will therefore be transferred in an approximate average time $t = (1/3)mNt_0$. The maximum transfer time for matched-order interrogation occurs if the tape is initially at the center, and one of the m blocks sought is at the beginning and another at the end of the tape. In this case $3N/2$ blocks must be scanned, whatever the value of m. The average access ratio is therefore equal to or greater than

$$\frac{mNt_0/3}{3Nt_0/2} = 2m/9.$$

For the large values of m necessary for a meaningful comparison this ratio is much greater than unity, consistent with the view that magnetic tape is a serial-access device.

The access ratio for mercury delay lines is not exactly unity, and the matched-order interrogation achieved by minimum-access programming does yield faster operating programs. However, the ratio is sufficiently close to unity for delay lines to be viewed as random-access devices, especially since access can be regarded as immediate in most applications, when either random-order or matched-order interrogation is used. Magnetic-core arrays are nearly ideal random-access memory units. Magnetic drums occupy an intermediate place, their exact status depending on the number of transducers provided. In the extreme case where there is available only one transducer that must be moved physically from channel to channel, a drum behaves much like tape; at the other extreme, if as many transducers are provided as there are words on the drum, its properties resemble those of magnetic cores. Similar comments apply to a variety of memory devices of the so-called "juke-box" type.

The access ratio of a deck of cards depends very much on how the deck is used. If a person extracts a card by estimating its position in the deck, scanning a few intermediate cards, and allowing the deck to close up once he has found the desired card, the deck will have a fairly low access ratio. Most automatic card-handling machines must scan all cards from the beginning of the deck, which then is analogous in this respect to a magnetic tape, and has a comparably high access ratio.

It should be remembered that the access ratio and the immediate-access or delayed-access designation are distinct characteristics. The first is completely independent of the absolute time scale and refers to a particular device only; the second takes into account the relative speeds of operation of several components of a system. When such different memory devices as magnetic tapes and mercury delay lines are used in one system, they are compatible only under carefully controlled conditions. On the time scale established by the speed of the operation units of the Univac, a machine word whose address is known is normally considered immediately accessible in mercury storage with either random- or matched-order access. From the description of the tape mechanisms, it is clear that the word is immediately accessible on tape only if words are grouped into longer strings for transfer, and if matched-order access is used. The first condition is met by building the block structure into the machine, the second must be satisfied by the programmer, who, if free to do so, must use considerable ingenuity and judgment to dispose information on tape in a satisfactory manner.

1.7. The Organization of Stored Information

The disposition of information in storage is only a mildly complicated business when such "internal" affairs as the location of program segments, constants, and other details of the operation of the program itself are concerned. In this case, the programmer is almost completely free to arrange things as well as he can within reasonable limits of time and effort. It is a much more serious matter to deal with files of information, whose contents are dictated fairly stringently by the nature of a problem, and for which the interrogation pattern can hardly be controlled by the programmer.

Knowledge of the total size of the file, of the distribution of the sizes of individual items in the file, of the rate of insertion and deletion of items, of the relative frequency of interrogation of the items, of the correlation between successive inquiries, and so forth, is of some help, of course, but at present the use of this knowledge to achieve an optimum organization of given or planned machinery is at best only a poorly developed art. Bitter experiences with early attempts at mechanizing commercial accounting processes indicate how easy it can be to underestimate the difficulty of handling huge masses of data; the expenditure of 40 man-years of programming effort to put a payroll system in operating condition was a typical experience.

It is not sufficient, for example, to agree that if the access ratio of a particular storage system is high only matched-order interrogation will be used. Matched-order interrogation is impossible if transfers must be made in the order in which inquiries arrive. In many applications it is possible to accumulate inquiries in batches and to arrange each batch in the order best suited for interrogation. Unfortunately, arranging is itself a complex and time-consuming operation, and a compromise must be worked out between delay in access and the delay and cost of arranging.

It is sometimes startling to the beginner in information processing that so commonplace a process as arranging objects in some order should be regarded as a major problem. It is easy to teach any clerk to arrange a few objects by hand, but when the rather subtle choices and decisions involved in the process must be explicitly formalized and expressed as a program or as a design for a machine capable of processing many thousands of objects, the situation is radically different. A very rough indication of the magnitude of the problem is the fact that the number of possible arrangements of n objects tends to grow factorially or exponentially with n. Fortunately, the actual problem is never quite that bad. The essential objective of research on arranging is to discover methods whereby the number of operations grows as slowly with n as is possible with given equipment or within limits imposed by economic or other design criteria. An account of arranging techniques commensurate with the importance and the complexity of the problem cannot be given within the scope of this book, and the interested reader is urged to consult the works mentioned in the bibliographic notes of Sec. 1.8.

Major difficulties also arise when no direct knowledge is available of the address where a file item may be found. Knowing that "migrate" occurs on page 533 of *Webster's New Collegiate Dictionary*, one could simply turn to that page. In the absence of such knowledge, some fumbling through the dictionary is necessary, more or less tedious according to whether or not the dictionary has a thumb index. This is one example of major differences between procedures feasible when the *address* of an item is known, and those feasible when only a *key* to the item is known. Note that what is sought is a complete dictionary entry. The key used in searching through the dictionary is the English word about which information is wanted. When this key is matched with an identical one at the head of a dictionary entry, the search is at an end.

Analogously, we can speak of *coordinate access* to a memory unit when coordinates or an address are either known or easily obtainable, and of *search access* when they are not. The use of a combination of coordinate and search access is also possible. The thumb index of a dictionary provides coordinate access to sections, and the coordinate of the location of a particular word is given by its first letter. Within sections, search access must be used. It should be noted that direct coordinate access is possible only when the key used to obtain access is itself an address or coordinate. If it is not, but a simple function of it is, it may be more economical to evaluate this function than to resort to search access. It is often difficult to find such a simple function; for example, consider the use of people's names and home addresses as keys. One valuable and well-known hybrid access technique is that of binary partition (Gotlieb and Hume, 1958, p. 218). If this technique is to be used, the items in the file must be in a definite order, such as monotone increasing order of numerical keys, and the search key is matched against that of an item at or nearest the middle of the file. If no match is obtained, the next comparison is made in the middle of the upper or of the lower half of the file, according to whether the search key is greater or less than the key with which it was first matched. For a file of N items, the process terminates after at most $\log_2 N$ comparisons. This technique is most valuable with random-access storage devices. With serial-access devices, matched-order interrogation is usually preferable.

Search access is usually more costly in programming effort and in operating time than coordinate access, when coordinate access is possible. For example, if a word w is known to be in register n of the central memory, a single Univac instruction is sufficient to bring it to the operation unit. If it is known only that w is somewhere in the memory, a search program like that of Exercise 1-3 may be necessary. These observations are valid even when coordinates are not explicitly represented by addresses. As an illustration, the location of a block on tape may be specified by two coordinates, one the number of the tape mechanism on which the tape is mounted, the other, the ordinal number of the block. Only the first coordinate is explicitly mentioned in Univac "read" or "write" instructions, which always move a tape one block away from its initial position. Machines could be designed with instructions mentioning both coordinates explicitly, but since this is not the case with most present machines, a program of several instructions is required to read the kth block of a tape. However, no provision need be made in such a program for ascertaining what symbols are recorded in the blocks scanned to reach block k. Search access is somewhat more complicated. Although the amount of scanning is the same, the search key must be compared for identity (for example, by a Univac Q-order) with the key of each item encountered en route. The time to execute what can be a quite complicated program for key-matching must be added to the scanning time.

Still more problems arise when the content of a file varies with time, either because of changes within individual items or because whole items are added or deleted. Punched-card files are very pleasing in this respect: new cards can be inserted, or old ones taken out, manually and with relatively little effort. There is of course a chance that the wrong card will be removed inadvertently, or a new one inserted in the wrong place. However, if the inevitability of some errors is accepted and if, therefore, the proper precautions are taken, no serious harm need ensue. At the other pole, permanent media such as photographic plates do not lend themselves to change at all. In the short run, a small erasable memory can be used in conjunction with a permanent one to keep track of changes. In the long run, the whole permanent memory must be prepared afresh to incorporate the changes accumulated in the auxiliary store.

Changes are more easily made in erasable memories, especially if the size of items is not affected. In mercury delay line, drum, or magnetic-core memories, the content of any register or group of registers can be changed fairly easily and rapidly. For technical reasons, it is impossible in many machines to change words or blocks in the middle of a magnetic tape, but it takes relatively little labor to transfer the contents of one tape, with the desired changes, onto another. Special programs can be written to control the transcription, which may be effected in a slack period between uses of the file. In general, the prompt removal of items is not as critical as the insertion of new ones, and old items are often allowed to remain until they become numerous enough to overtax storage capacity or to increase access time appreciably.

Providing for the extension of existing items requires very careful planning. For example, adding an extra word to an item without disturbing all the following ones is possible only if there is an empty register between the item and its successor, except with magnetic tapes, where the new word can be inserted during transcription. With all other storage media, whose storage capacity is always limited, there are only two alternatives, both unattractive: either enough empty spaces must be provided to allow for expansion, spaces that are then "wasted" for a long time, or a wholesale transfer of items must be made each time an item is extended. Many compromises between these two extremes are possible but some waste of storage capacity or of motion seems unavoidable.

The insertion of new items is subject to similar considerations. If the items are not ordered in the file, an empty space at the end of the file will suffice, but if order must be preserved, what is true of item extension applies here as well. Determining a good arrangement for a variable file is further complicated by the difficulty of obtaining coordinate access due to the changes in the relative positions of items in a file that occur when extensions or insertions are made. When an item is transferred from the end of one tape to the beginning of another, or from one drum channel to another, because of expansion in the preceding section of a file, the ensuing change in its coordinates must be taken into account. This complication often throws the balance in favor of search access, a process logically unaffected by any change in a file that does not alter the structure of the entry keys.

When items are of different lengths, direct coordinate access is impossible unless by such Procrustean means as fitting each item into a storage space large enough to accommodate the longest item. Again, a properly designed search-access procedure can easily accommodate items of a wide range of length; it may be necessary, however, to include in the file "punctuation" marks to separate successive items.

Most of the problems we have examined are not peculiar to automatic machines, and their solutions may parallel well-known manual methods. In a library card catalog, most items occupy only one card each, some cards obviously being filled more than others. Exceptional items are spread over two or three cards, as a compromise between the impractical extremes of using fractional cards or standardizing on cards much too large for most items. Drawers are never originally filled to capacity, to prevent a surge of items into one drawer from propagating throughout the file. Labels on each file drawer give partial coordinate access, as do the index cards provided within the drawers. The final selection of a card is by some variety of search access. Overflow of one drawer necessitates changing the coordinate labels on subsequent drawers, the number of which is determined by the fraction of the capacity to which they are filled, and by the amount of overflow. All these problems are much more acute in automatic systems, for two reasons: first, automatic storage devices cost so much more than manual access files of equal capacity that a correspondingly higher degree of efficiency in use is essential; second, procedures that could be left to Miss Jones to develop on the spur of the moment must be planned with extraordinary care and attention to detail before an equivalent program can be written. As a science, the efficient automatic processing of large numbers of items is still very much in its infancy.

Exercise 1–4. (*a*) The ordinal number of the last block read from the tape on mechanism No. 5 is stored in register *0*, and that of the next block to be transferred to registers m through $m+59$ is in register 1. Write a program that will effect the indicated transfer. (b) The *word* in register 0 is known to be on the tape on mechanism No. 5. Write a program that will locate this word and transfer it to register m. What precautions are necessary when it is not certain that the word is on the tape?

Exercise 1–5. Each entry in a given file is one block long; the key of each entry is a distinct integer, and is in the first word of the block. It must be assumed that the entries are ordered in increasing numerical order of their keys. Consider a hybrid search procedure consisting of the following steps:

(1) The key of an entry nearest the middle of the file (how is this available?) is compared with the search key; if the search key matches the reference key, the process terminates; if it is greater, the search proceeds in the upper half of the file; if it is smaller, in the lower half.

(2) Same as (1), using the middle entry of the proper half of the file.

(3) The same process is continued with quarters, eighths, and so on. With or without actually writing programs, evaluate the advantages of this method applied (i) to a tape memory, (ii) to a mercury-delay-line memory, relative to the "pure" search methods of Exercises 1–4b and 1–3 respectively.

1.8. Bibliographic Notes

A. *General.* The history of automatic digital computers is surveyed by Alt (1958), and by Wilkes (1956), who gives a scholarly account of the methods of logical design and programming of computers, and also a very thorough survey of the physical details of machine design and construction. A comprehensive bibliography is appended to Wilkes's book. A description of the machine *BESM* which figures in accounts of Russian experiments in automatic translation is given by Lebedev (1956). Weik (1957) has prepared *A Second Survey of Domestic Electronic Digital Computing Systems*, a condensed tabulation of salient characteristics of the machines available at that time. The proceedings of a symposium sponsored by the Association for Computing Machinery (1957) describe machines of somewhat later vintage. The Office of Naval Research *Digital Computer Newsletter*, appended to each issue of the *Communications of the Association for Computing Machinery*, is helpful in keeping abreast of current developments. The October 1953 issue of the *Proceedings of the IRE* (Vol. 41, No. 10) was devoted entirely to tutorial articles on computers, and is still a good source of articles and bibliographies covering several specialities within the computer art. Bowden (1953) is an eminently readable and comprehensive survey work, and Booth and Booth (1956) give an extensive bibliography.

B. *Programming.* Books devoted entirely to the art of programming are still scarce, at least partly because of the difficulty of giving a generally applicable account of the subject. Gotlieb and Hume (1958) and McCracken (1957) have tried to avoid specialized reference to a particular machine by basing their books on hypothetical instruction repertoires, incorporating features common to most machines. McCracken's emphasis on scientific computation precludes the very detailed attention to input-output problems and storage organization necessary in commercial applications and in automatic translation. These problems are treated more thoroughly by Gotlieb and Hume. However, both books should prove extremely useful to the beginner. A more recent book by Jeenel (1959) seems less valuable. The only other book in the field, by Wilkes, Wheeler, and Gill (1957), is based on the late EDSAC I, a machine at Cambridge University. Details regarding the instructions of particular machines may be found in the manuals provided by the manufacturers, but the beginner may have difficulty deciphering many of these unless both a machine and a tutor are at hand. The description of Univac I characteristics given in this chapter is based on a manual by Esch and Calingaert (1957).

C. *Input and Output Technology.* Review articles by Gibbons (1957) and by Carroll (1956) present brief descriptions of a variety of input and output devices. Some of the problems of speech analysis and synthesis are outlined by David (1958), Fry and Denes (1956), and by Dudley (1955); the Bell Telephone Laboratories' digit recognizer is described by Davis, Biddulph, and Balashek (1952) in the *Journal of the Acoustical Society of America*, in which much of the work in this field is published. Devices for reading characters handwritten under carefully controlled conditions are described by Dimond (1957). Greanias and Hill (1957) explore some basic problems in designing recognition devices for printed symbols. Eldredge, Kamphoefner, and Wendt (1956) and Bennett *et al.* (1955) sketch some properties of machines for reading characters printed in magnetic ink. Shepard, Bargh, and Heasly (1957) describe a character-sensing system for business documents, a photo-electric reader for ordinary printed characters is discussed by Shepard and Heasly (1955), and techniques for reading patterns of printed dots are described by Davidson and Fortune (1955). A machine for reading micro-filmed punched cards has recently been developed by the National Bureau of Standards (1957). Several methods of displaying characters on the face of a cathode-ray tube have been invented in recent years; one such method is described by Smith (1955). The production of high-quality printing under automatic control is a possible application of a photographic composing machine (Photon) described in Chapter 19 of Higonnet and Gréa (1955). The Joint AIEE-IRE-ACM Computer Conference held in New York in December 1952 was devoted to a review of input and output equipment, much of which is now standard on many operating machines.

D. *Storage Technology.* A combination of the new technology of magnetic reading and recording and of novel file-handling concepts with the older "unit record" (card) idea is described by Hayes and Wiener (1957). Three approaches to the use of photographic techniques in the design of permanent high-capacity memory systems are presented by Hoover, Staehler, and Ketchledge (1958), Baumann (1958), and King (1955). The search for cheap erasable storage devices of high capacity has led to the development of several magnetic devices representing a wide variety of geometric configurations and patterns of motion. Four such devices are described by MacDonald (1956), Noyes and Dickinson (1956), Welsh and Porter (1956), and Begun (1955). Eckert's (1953) review article summarizes early developments in storage technology. Descriptions of magnetic-tape storage systems are to be found in the references pertaining to input and output technology.

E. *Storage and Arranging Logic.* Peterson (1957) is primarily concerned with minimizing search and maintaining stable coordinates in files whose contents vary. According to Hayes and Wiener (1957), a new type of magnetic card under development can be automatically organized into files where entries are ordered by frequency of interrogation. The problem of economizing storage space in an ordered file is examined by Dean (1957). Oettinger (1957a) describes some criteria for choosing keys suitable for coordinate access as well as for other purposes such as sorting. Giuliano (1957a) considers a variety of file arrangements suitable for an automatic dictionary file. Hollander (1956) discusses the efficient application of a hybrid storage device sharing characteristics of magnetic tapes and drums. Some criteria for a useful file storage device are defined by Eisler (1956), and similar analyses are given by Iverson (1955) and by Oettinger (1955). Postley (1955) considers a novel way of organizing magnetic-tape files. Arranging methods have been analyzed with considerable thoroughness by the following: Hildebrandt and Isbitz (1959), Burge (1958), Demuth (1956), Davies (1956), Friend (1956), Isaac and Singleton (1956), and Ashenhurst (1954). No single systematic exposition of the subject is available, but Hosken (1955) reviews and compares several methods, and gives a comprehensive bibliography.

APPENDIX

Univac I Instructions[1]

Transfer Instructions

Internal ($m = 0, 1, \ldots, 999$)[2]

B	$m \to A, X$
F	$m \to F$
L	$m \to L, X$
H	$A \to m$
C	$A \to m$; clear A
G	$F \to m$
J	$X \to m$
K	$A \to L$; clear A
E	extract $m \to A$ under control of F
V	$m, m+1 \to V$
W	$V \to m, m+1$
Y	$m, \ldots, m+9 \to Y$[3]
Z	$Y \to m, \ldots, m+9$[3]

Input/Output ($n = 1, \ldots, 9, -$)

1n	read tape n forward, transferring one block to register I
2n	read tape n backward, transferring one block to register I
3n	$I \to m$; read tape n forward, transferring one block to register I[3]
4n	$I \to m$; read tape n backward, transferring one block to register I[3]
5n	write on tape n the block stored in registers $m, \ldots, m+59$[3]
6n	rewind tape n
8n	rewind tape n; lock tape unit n
10	stop; call for keyboard input to register m
30	$I \to m$[3]
40	$I \to m$[3]
50	output breakpoint stop; type out one word from register m

Arithmetic Instructions

Algebraic

A	$m \to X$; $X+A \to A$	
X	$X+A \to A$	
S	$-m \to X$; $X+A \to A$	
M	$m \to X$; $L \cdot X \to A$	rounded
N	$-m \to X$; $L \cdot X \to A$	rounded
P	$m \to X$; $L \cdot X \to A, X$	unrounded
D	$m \to A$; $A/L \to A$	rounded
	$\to X$	unrounded

Shifting n Places ($n = 1, \ldots, 9$)

.n	shift A right with sign[4] (all columns)
;n	shift A left with sign[4] (all columns)
$-n$	shift A right without sign[4] (column 1 stays put)
0n	shift A left without sign[4] (column 1 stays put)

[1] Adapted from Esch and Calingaert (1957).

[2] Since there is little danger here of confusing (m) and m, parentheses have not been used in the body of the table.

[3] The least significant digit of m is interpreted as 0, that is, m is treated willy-nilly as a multiple of 10.

[4] Vacated positions are filled with zeros.

Univac I Instructions—*continued*

Sequence Control Instructions

00   no effect (null instruction)

9   stop

U   unconditional transfer of control to *m*

Q   if $A = L$, transfer control to *m*

T   if $A > L$, transfer control to *m*

R   if stored in register *c*, this instruction causes the instruction pair *00 000 U(c+1)* to be placed in memory register *m*

CHAPTER 2

THE STRUCTURE OF SIGNS

In this chapter, we shall examine in detail certain properties of signs that are of importance in the design and application of automatic information-processing machines. In so doing, we shall also be laying the groundwork for an analysis of the problems of language translation. The morphologic and syntactic properties of artificial sign systems are by no means identical to those of natural languages. The simplicity and regularity of some artificial systems are notably absent in natural languages, but the study of simple artificial systems is relevant to linguistic analysis for several reasons: first, because some of the properties of these systems are present in natural languages even if not in pure form; second, because a machine cannot be used to manipulate natural languages without the prior establishment of a satisfactory correspondence between these languages and the artificial sign system of the machine, and when this correspondence is not felicitous serious difficulties arise; finally, a special notation must be introduced to aid in the concise and lucid expression of the problems considered in subsequent chapters.

2.1. Use, Mention, and Representation

In a discussion of signs, a distinction must be made between the *use* of a sign, for instance, as the name of a material object, and the *mention* of a sign, when it itself is the subject of discussion. So-called *autonymous* speech, where a sign may designate itself, can be dangerously confusing. In ordinary discourse, autonymy is avoided informally by circumlocution or other subterfuges. Logicians have been led, by inclination and by necessity, to evolve various systematic methods for distinguishing use from mention (Quine, 1955; Curry, 1953). When automatic machines are applied to information processing, the need for such distinctions is made even more acute by frequent references, not only to the information being processed, but also to each machine's coding language, its token system, and often to one or more intermediate "automatic programming" languages.

Three preliminary examples should clarify the nature of the problem. The first is due to Quine:

'Boston is populous' is about Boston and contains 'Boston'; ' 'Boston' is disyllabic' is about 'Boston' and contains ' 'Boston' ' . ' 'Boston' ' designates 'Boston', which in turn designates Boston. To mention Boston we use 'Boston' or a synonym, and to mention 'Boston' we use ' 'Boston' ' or a synonym. ' 'Boston' ' contains six letters and just one pair of quotation marks; 'Boston' contains six letters and no quotation marks; and Boston contains some 800,000 people.

The second is a syllogistic pun cited by Curry:

'I' is a capital letter.
'me' is the objective case of 'I'.
∴ 'me' is the objective case of a capital letter.

The third is given by the prudent Frenchman who, in exclaiming "Nom de nom de Dieu!" looks to a nest of quotation marks for protection against divine wrath.

The notational conventions developed for use in this book differ from the conventions exemplified by those of Quine in that they are extended to account systematically, not only for use and mention, but also for the distinctions among signs, representations, and tokens that have already been introduced informally in Sec. 1.2 and that will be analyzed in some detail in Secs. 2.6 and 2.7. Because subsequent chapters are concerned primarily with the properties of written languages, there is no need to account for phonetic and phonematic notations, although the conventions could easily be extended further to do so.

A series of examples, rather than formal definitions, will be used to define the notational system; these examples consist of sets of statements each held to be true.

Example 2-1.

(*a*) Boston is a city.
(*b*) /Boston/ is an English word.
(*c*) "Boston" is a representation of /Boston/.
(*d*) [Boston] is a token for "Boston".

Example 2-2.

(*a*) 3 is a number.
(*b*) /3/ is an Arabic numeral.
(*c*) "3" is a representation of /3/.
(*d*) [000110] is a Univac token for "3".

Example 2-3.

(*a*) /B/ is a letter of the English alphabet.
(*b*) "B" is a representation of /B/.
(*c*) [010101] is a Univac token for "B".

Each time a new sign is introduced to mention another, for instance, //3// to mention /3/, or //dog// to mention /dog/, representations and tokens for the new sign must also be introduced. The systematic character of this progression is illustrated in Fig. 10, where successive levels of reference are arranged vertically one underneath the other, and where the hierarchy of representation is displayed horizontally. The following recast version of Quine's example further illustrates the usage of the new notation; the parenthetic statements are not directly expressible in Quine's notation:

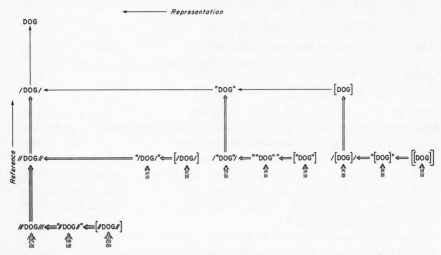

Fig. 10. Use, mention, and representation.

/Boston is populous/ is about Boston and contains /Boston/; /"Boston" has six alphabetic characters/ is about "Boston" (not about /Boston/ nor about "— ... — — — ... — — — — — —.") and contains /"Boston"/. //Boston// designates /Boston/, which in turn designates Boston. (/"Boston"/ designates "Boston", which is the conventional representation of /Boston/, but might be used to represent /Chicago/ for cryptographic purposes.) To mention Boston we use /Boston/ (conventionally represented by "Boston") or a synonym, and to mention /Boston/ we use //Boston// (conventionally represented by "/Boston/") or a synonym. ("/Boston/" contains six alphabetic characters and a pair of slanted lines; "Boston" contains six alphabetic characters and no slanted lines; and Boston contains some 800,000 people.)

The following three true statements, which would be gibberish in ordinary notation, are readily expressible in the new notation:

(1) dog = — ·· — — — — — · = Hund.

(2) /dog/ = /— ·· — — — — — ·/ ≠ /Hund/.

(3) "dog" ≠ "— ·· — — — — — ." ≠ "Hund".

The truth of (1) follows from the reasonable assumption that a single species is being mentioned. If there is but one *English* word /dog/, then /– ·· ––– ––·/ is the same word, although this word itself is mentioned in this sentence by two distinct but synonymous words //dog// and //– ·· ––– ––·//; since /Hund/ is a *German* word, the truth of (2) follows. Since "dog" has three characters, "– ·· ––– ––·" has three dots and six dashes, and "Hund" has four characters, the truth of (3) follows.

In Examples 2-1, 2-2, and 2-3, and in Fig. 10, tokens are associated directly only with the representations of signs, not with signs. It often is simpler, and usually harmless, to consider a token to be directly a token for a sign; in other words, if [Boston] is a token for the representation "Boston" of the word /Boston/, we can say simply that [Boston] is a token for /Boston/.

To avoid cluttering subsequent pages with slanted lines and square brackets, italics and boldface will be used instead. However, to distinguish italics of this kind from the ordinary italics used for emphasis, an asterisk will be added. Because iterated reference and representation cannot be expressed in this modified notation, the techniques illustrated in Fig. 10 will continue to be used in those rare instances, exemplified by several of the preceding paragraphs, where multiple bracketing is necessary. Using the modified notation, Example 2-1 may be rewritten as:

Boston is a city.
*Boston** is an English word.
"Boston" is a representation of *Boston**.
Boston is a token for "Boston".

It is this modified form of the new notation that has already been introduced informally in Sec. 1.2. The usefulness of the notation in expressing certain types of statements that occur frequently throughout the remainder of the book should become apparent from a careful reading of the statements in the following example:

Example 2-4.

(*a*) *dog** has the singular representation "dog" and the plural representation "dogs".

(*b*) "dog" has three characters.

(*c*) That "dogs", in "the dogs bark", represents a plural is indicated by its ending "s".

(*d*) *dog** is ordinarily spelled "dog", printed as **dog**, but stored in the Univac as **010111101001011010**.

(*e*) Although *dog** is ordinarily spelled "dog", when writing about information-processing machines it is sometimes convenient to use the representation "010111101001011010".

(*f*) **dog** is an ink mark on paper, but **010111101001011010** consists of 18 bistable devices, of which the first is in one state, the second in the other, and so on.

(*g*) *B** conventionally has the capital representation "B" and the lower-case representation "b". *B** comes after *A** in the alphabet, but no such order relation necessarily obtains between "B" and "A".

(*h*) *IV** is a Roman numeral which names the number 4. Its usual representation consists of a sequence of the two characters, "I" and "V".

(*i*) "Seven" has five characters, "sieben" has six, and *seven** and *sieben** are the English and German names of seven, respectively.

The new notation does not cure all difficulties. Consider the statements in the following example:

Example 2-5.
(*a*) *I** is a personal pronoun.
(*b*) *I** is a Roman numeral.
(*c*) *I** is the ninth letter of the English alphabet.

These three statements are consistent with the defining Examples 2-1, 2-2, and 2-3, respectively, yet their conjunction is manifestly absurd. The trouble is that there is nothing in our notation that guarantees the absence of homography. We remain free to associate signs with what they designate in any way we please. It happens to be quite convenient to use //I// homographically to designate three distinct kinds of elements, for the same reason that *Bill** is used to designate untold numbers of men. Usually the context adequately completes the definition of the association. It is only when two Bills are in a room that we find it necessary to call them by the names *Bill Smith** and *Bill Doe**, or *Susan's Bill** and *Bill with the red hair**. We are also free to modify our nomenclature so that homography will be eliminated, as in the following example:

Example 2-6.
(*a*) I_w* is a personal pronoun.
(*b*) I_n* is a Roman numeral.
(*c*) I_i* is the ninth letter of the English alphabet.

With such a change, Curry's syllogism clearly turns out to be invalid:

I_i* is a capital letter.
me* is the objective case of I_w*.
∴ *me** is the objective case of a capital letter.

The notation defined in this section will be useful in later sections of this chapter, but particularly in Chapter 5 and subsequent chapters, whenever signs, representations, and tokens must be mentioned simultaneously. Practicing mathematicians and linguists can usually avoid painful confusion due to carelessness about distinguishing between use and mention or between signs and their representations without feeling a need for special notation and without excessively cumbersome circumlocutions. In discourse about the application of information-processing machines to languages, where contiguous references to several varieties of signs, representations, and tokens are unavoidable, a convenient, unambiguous notation is essential. The reader who remains doubtful is invited to recast the statements in Example 2-4, most of which are quite typical, into "plain" English, or to translate them into equivalent statements in another language!

Still, ordinary discourse has a familiarity and simplicity which should not be lightly abandoned. Therefore, in situations where clarity and unambiguity cannot suffer, familiarity and simplicity will be given the upper hand.

2.2. Sets and Isomorphisms

Modern mathematics, algebra in particular, has developed several concepts that can be usefully adapted in the study of sign systems, at least for heuristic purposes. There is a growing body of literature in mathematical linguistics dealing with the application of several branches of mathematics, for example, set theory and statistics, to the formulation of grammars. No attempt will be made in this book to survey this literature. We shall restrict ourselves to an examination of a few elementary concepts, essential for a concise and lucid exposition of certain properties of sign systems. The interested reader will find some key references to the literature of mathematical linguistics in the bibliographic notes of Sec. 2.9.

A *set* is a collection of elements, called *members* of the set. Capital letters, and strings such as "$\{a, b, c, \ldots, x, y, z\}$", will be used as names of sets. For example, the English alphabet $\{a^*, b^*, c^*, \ldots, x^*, y^*, z^*\}$ is a set of 26 elements, while the integers from 0 to 99, $\{0, 1, 2, \ldots, 97, 98, 99\}$, and the corresponding set of Arabic numerals $\{0^*, 1^*, 2^*, \ldots, 97^*, 98^*, 99^*\}$ form sets of 100 elements each.

Table 4. Correspondences between integers and
letters of the English alphabet;
$I = \{1, 2, 3, 4, 5\}; \quad A = \{a^*, b^*, c^*, d^*, e^*\}.$

(a)	(b)	(c)
$1 \leftrightarrow a^*$	$5 \leftrightarrow a^*$	$2 \leftrightarrow a^*$
$2 \leftrightarrow b^*$	$4 \leftrightarrow b^*$	$1 \leftrightarrow b^*$
$3 \leftrightarrow c^*$	$3 \leftrightarrow c^*$	$4 \leftrightarrow c^*$
$4 \leftrightarrow d^*$	$2 \leftrightarrow d^*$	$3 \leftrightarrow d^*$
$5 \leftrightarrow e^*$	$1 \leftrightarrow e^*$	$5 \leftrightarrow e^*$

Two sets are said to be in *one-to-one correspondence* if and only if every element of one set is associated with one and only one element of the other. The associated elements are called *images* of one another. For example, the three correspondences, defined in Table 4, between the set $I = \{1, 2, 3, 4, 5\}$

of the first five positive integers and the set $A = \{a^*, b^*, c^*, d^*, e^*\}$ of the first five letters of the English alphabet are all one-to-one in the sense of the definition. Expressions such as "$1 \leftrightarrow a^*$", introduced in Table 4, are shorthand for "1 is in one-to-one correspondence with a^*", and so on.

Under certain circumstances, the relation between two sets in one-to-one correspondence is called an *isomorphism*. One-to-one correspondence is a necessary condition for an isomorphism, but the additional requirement is frequently imposed that something of interest be *preserved* by the correspondence. The number of elements in two sets that are in one-to-one correspondence is necessarily always preserved; hence any one-to-one correspondence might be called a *membership-preserving* isomorphism; but since the preservation of number is an obvious consequence of one-to-one correspondence, sets related by such a correspondence usually are simply said to be isomorphic. The following is an example of an isomorphism which preserves not only the *number* of elements but also, in a sense which will be made clear presently, the *order* of the elements. Let α and β be any two elements of the set A defined in the preceding paragraph, and let "$\alpha \mathscr{F} \beta$" stand for "α follows β in alphabetic order". Given that $\alpha \leftrightarrow x$, and $\beta \leftrightarrow y$, where x and y are members of the set I also defined in the preceding paragraph, we can ask whether or not $x > y$ implies that $\alpha \mathscr{F} \beta$ and vice versa. This is obviously not the case under the correspondence of Table 4b, where $b^* \mathscr{F} a^*$ but $4 \not> 5$, nor is it true under the correspondence of Table 4c since $b^* \mathscr{F} a^*$ but $1 \not> 2$. However, we can easily verify that under the correspondence of Table 4a, whenever one of a pair of integers is greater than the other, its alphabetic image follows the alphabetic image of the other. The relation between I and A defined by the correspondence of Table 4a is therefore said to be an *order-preserving isomorphism*.

The practical significance of this relation is that the two sets can be regarded as equivalent—or like in form, as the etymology of *isomorphism** suggests—in many significant respects save the names of their elements. For instance, suppose that each of a deck of five cards were labeled in the left-hand corner with a numeral from the set $\{1^*, 2^*, 3^*, 4^*, 5^*\}$, and in the right-hand corner with a letter from the set $\{a^*, b^*, c^*, d^*, e^*\}$, each chosen so that the integer named by the numeral and the letter are images under the correspondence of Table 4a. If the cards are shuffled, a person reading only the left-hand corner and arranging the cards in increasing numerical order will return them to precisely the same order as another person reading the right-hand corner and arranging the cards in increasing alphabetic order. So far as distinguishing and ordering five objects is concerned, the set I, with the relation $>$, and the set A, with the relation $\mathscr{F}$, are equivalent.

The notion of isomorphism is conveniently extended as follows. Consider two sets S and S' that can be put in one-to-one correspondence, one having a relation $\mathscr{R}$ and the other a relation $\mathscr{R}'$ defined between pairs of their elements. A one-to-one correspondence between these sets is said to define a *relation-preserving isomorphism* if $x \mathscr{R} y$ implies that $x' \mathscr{R}' y'$, and vice versa, whenever $x \leftrightarrow x'$ and $y \leftrightarrow y'$.

Exercise 2-1. Let α and β be elements of A, and "$\alpha \mathscr{P} \beta$" stand for "α precedes β in alphabetic order". Show that the correspondence of Table 4a defines a relation-preserving isomorphism between A with the relation $\mathscr{P}$ and I with the relation $>$. (Hint: identify $\mathscr{P}$ with $\mathscr{R}$ and $>$ with $\mathscr{R}'$ in the definition given in the text.)

Common logarithms provide us with an example of an isomorphism of great practical importance. To every positive number x, there corresponds a unique number log x, the logarithm of x, and every number is the logarithm of a unique positive number, that is to say, the correspondence between positive numbers and their logarithms is one-to-one. The usefulness of logarithms derives from the fact that the logarithm of the *product* of two numbers is the *sum* of the logarithms of each, in other words, log $(x \cdot y) =$ log $x +$ log y. Rather than multiplying two numbers, a tedious operation for large numbers, we can find their logarithms in a table, add these, and again consult the table to find the number of which the sum is the logarithm. Thus, if $\mathscr{R}$ is the relation between pairs of numbers and their products, and $\mathscr{R}'$ the relation between pairs of numbers and their sums, the correspondence between numbers and their logarithms is a relation-preserving isomorphism. By passing from the domain of numbers to the domain of logarithms, substituting addition for multiplication, and then returning again to numbers, much tedious labor is saved, and yet the isomorphism guarantees that the end result will be the same as if the multiplication had been performed in the usual way.

Two sets will simply be called isomorphic when it is obvious which of their characteristics are preserved by the isomorphism between them. Occasionally, it will be of interest to preserve only certain selected characteristics. In these cases, the scope of the isomorphism will be indicated by an explicit statement of the characteristics that are preserved, regardless of whether or not other characteristics chance to be preserved also. For example, to emphasize the preservation of the order relation by the isomorphism between the set I with the relation $>$ and the set A with the relation $\mathscr{F}$, we can describe it as an isomorphism between the sets $\{>; I\}$ and $\{\mathscr{F}; A\}$.

When several elements of a set A may correspond to a single element of a set B, but each element of A is associated with only one element of B, the correspondence between A and B is called *many-to-one*. Many-to-one correspondence is illustrated in Fig. 11a. The mirror image of Fig. 11a gives Fig. 11b, which illustrates a *one-to-many* correspondence between the sets A'

and B'. The sets A'' and B'' of Fig. 11c are in *many-to-many* correspondence. The correspondence between the letters of the alphabet and the tokens for these letters appearing on this page is an illustration of a one-to-many correspondence.

The relation between two sets in *many-to-one* correspondence is called a *homomorphism from* the *many* set *to* the *one* set. For example, the mapping of Fig. 11a is a homomorphism from A to B. Isomorphism may be regarded as a special case of homomorphism.

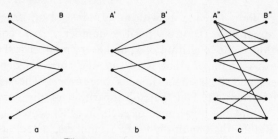

Fig. 11. Multiple correspondences.

As a relation, an isomorphism has three important properties: it is *reflexive, symmetric*, and *transitive*. A set S is isomorphic to itself (reflexivity); if S is isomorphic to S', then S' is isomorphic to S (symmetry); and if S is isomorphic to S', and S' is isomorphic to S'', then S is isomorphic to S'' (transitivity). A *proper* homomorphism, that is, one which is not an isomorphism, is only transitive, for a set cannot be in one-to-many (many-to-one) correspondence with its whole self, and a set in many-to-one (one-to-many) correspondence with another cannot, according to our definition, also be in one-to-many (many-to-one) correspondence with it. However, we can speak of the preservation of operations and relations under a homomorphism in precisely the same sense as under an isomorphism.

2.3. Transformations

Substituting isomorphic or homomorphic image elements, operations, and relations for given elements, operations, and relations to simplify a problem is a very common technique. A rule for replacing an element of one set by its image in a corresponding set is called a *transformation*. The set of original elements is called the *domain* of the transformation, and the set of image elements, its *range*. A transformation is sometimes called a *mapping*, which maps the domain into the range. With every isomorphism, we can associate the transformation that maps each element of one set into the image defined by the isomorphism. Such a transformation has the valuable property of having a unique *inverse* transformation, a rule for mapping the image or *transform* of an element back into that element. The properties preserved under the isomorphism are said to be *invariant* under the corresponding transformation.

For example, denoting the transformation corresponding to the order-preserving isomorphism of Table 4a by T, and its inverse by T^{-1}, we can express the result of applying T to an element of $\{>; I\}$ as $Tx = \alpha$, and the result of applying T^{-1} to an element α of $\{\mathscr{F}; A\}$ as $T^{-1}\alpha = x$, where $\alpha \leftrightarrow x$. The transformation T is itself the inverse of T^{-1}, since the result of applying T to the result of applying T^{-1} to an element α is again α, that is to say, $T(T^{-1}\alpha) = Tx = \alpha$.

Applying T to each member of the ordered set $\{1, 2, 5, 4, 3, 3, 2\}$ transforms this set into a set $\{a^*, b^*, e^*, d^*, c^*, c^*, b^*\}$. In other words, this transformation replaces *numbers* by the *letters* which are their images under the isomorphism of Table 4a.

We can also define a one-to-one correspondence between the set of *characters* $\{$"1", "2", "3", "4", "5"$\}$ and the set of *characters* $\{$"a", "b", "c", "d", "e"$\}$, such that the string of characters "1254332" is mapped onto the string of characters "abedccb" by the associated transformation. Such transformations are the basis of what cryptographers call *simple substitution ciphers*. For example, by simple substitution, the representations "bad" and "ace" of the words *bad** and *ace** can be replaced by the representations "214" and "135" respectively. The correspondence being known, it is easy to return from these representations to the original ones. If the correspondence is not known, returning is not so easy, which is the whole point of cryptography. The basis for returning, when the correspondence is not known, is the knowledge or the assumption that strings like "214" and "135" represent *words*, not *numbers*. Simple substitution is a mere change in the *representation* of fixed elements, not an exchange of *elements* like that in the preceding paragraph. The failure to distinguish between a transformation from one representation of a *fixed element* to another, and a transformation of one element into another in a *fixed representation* system is the source of much confusion (Sec. 4.2). Students of physics are familiar with the *alias/ alibi* problem with vectors: is a discussion about one vector in two different coordinate systems, or two different vectors in one coordinate system?

Using "214" in one instance as a representation of the word *bad** and in another as a representation of the number 214 is mere homographic usage. We are often driven to homographic usage by the lack of a large enough repertoire of basic signs or characters. This is harmless enough, unless it leads to the mistaken notion that, because some sets of elements happen to have the same representations, these sets are therefore identical or isomorphic. We are at liberty to use "1", "2", "3", "4", and "5" not only to represent 1, 2, 3, 4, and 5 as usual, but also to represent, in an appropriate context, *horse**, *dog**, *pear**, *table**, and *rock**, respectively. We should not therefore conclude that *horse** plus *dog** make *pear**, or worse, that a horse plus a dog make a pear, nor should we claim that a rock is greater than a table.

Mistaken identification is the basis of the notion that machines can be used to translate languages only if "words are somehow turned into numbers". When using certain machines for linguistic work, it is often convenient to use a numerical *representation* of words, namely, a simple substitution cipher. To the extent that words have properties, like ordering, that are also properties of numbers, the change of representation may harmlessly be viewed as a transformation replacing words by elements of $\{>; 1, 2, 3, \ldots\}$. However, a significant isomorphism between words and $\{+, >; 1, 2, 3, \ldots\}$ does not exist, since there is no significant operation on words that corresponds to the addition of numbers.

> *Exercise 2-2.* Replace the elements of the standard representation "abedccb" of the ordered set of letters $\{a^*, b^*, e^*, d^*, c^*, c^*, b^*\}$ by their images under the correspondence "a"$\leftrightarrow$"2", "b"$\leftrightarrow$"1", "c"$\leftrightarrow$"4", "d"$\leftrightarrow$"3", "e"$\leftrightarrow$"5". Imagine that the resulting transform of "abedccb" is a representation of an ordered set of integers. Rearrange this set into increasing numerical order, writing the representation of the rearranged set. Now imagine that the representation of the rearranged set of numbers is that of a string of letters, and find the standard representation of this string by applying the inverse of your original transformation. How does this representation compare with a string representing the ordered set $\{a^*, b^*, b^*, c^*, c^*, d^*, e^*\}$? Repeat these processes with the correspondence "a"$\leftrightarrow$"1", "b"$\leftrightarrow$"2", "c"$\leftrightarrow$"3", "d"$\leftrightarrow$"4", "e"$\leftrightarrow$"5". Can you explain the results?

2.4. The Identification of Indiscernibles

In the earlier parts of this chapter, and in Sec. 1.2, we have busily introduced distinctions, for instance, among signs, representations, and tokens. Much useful discourse can be carried on without distinctions such as these: indeed, they may be felt to obscure rather than to clarify. In addition, we have not made these distinctions everywhere, and openly declared at the end of Sec. 2.1 the intention to make them only when convenient. Such shifty behavior requires some justification.

The root of the trouble is the old problem of *identity*. It can be argued that no two things in the universe have exactly the same specific properties. It can even be argued that something at time t is not the same thing it was before nor the same thing it will be later. Such an extreme position is untenable in practice, because any discourse whatsoever becomes impossible.

One way out is to ignore irrelevant properties. The child may distinguish a bright penny from a dull one. The coin collector looks at dates. For a parking meter, any penny's a penny and it may even be a slug. In some instances, deliberate ignoring may not be necessary: no difference may be discernible. Trouble then arises when better instruments or prolonged thought reveal differences.

Philosophers, Leibniz among them, who have pondered this problem often resort to a Principle of Identification of Indiscernibles. Quine (1953, p. 71) has propounded such a principle in the following form:

> Objects indistinguishable from one another within the terms of a given discourse should be construed as identical for that discourse. More accurately: the references to the original objects should be reconstrued for purposes of the discourse as referring to other and fewer objects, in such a way that indistinguishable originals give way each to the same new object.

Quine clearly recognizes that what is or can be distinguished varies with circumstances:

> Our maxim of identification of indiscernibles is relative to a discourse, and hence vague in so far as the cleavage between discourses is vague. It applies best when the discourse is neatly closed . . . but discourse generally departmentalizes itself to some degree, and this degree will tend to determine where and to what degree it may prove convenient to invoke the maxim of identification of indiscernibles.

Perhaps a Principle of Identification of Indiscernibles should be complemented by a Principle of Distinction of Discernibles. The notational problems connected with either principle are quite similar. We may speak of pennies, dimes, and quarters, but when merely describing the source of the jingling noise in our pockets, the word *coin** is useful, and would, if it did not exist, have to be invented. Conversely, while most of us can name only a few colors of the rainbow and still live happily, a recent dictionary of color names (U.S. Department of Commerce, 1955) has some 7500 entries ranging from *absinthe green** through *French nude** and *languid lavender** to *Zuni brown**.

Both identification and distinction tend to increase the population of the universe. In addition to the ontological problems raised by this tendency, which we shall try to ignore here, serious notational problems are raised, which we cannot ignore.

If we multiply the number of distinct signs we use, the notation becomes difficult to learn or to print. If we limit ourselves to a few signs serving many purposes, homographic usage is inevitable, together with ensuing confusion. There is particularly great danger in assuming that identity of name or representation implies identity of what is named or represented. When two sets are indeed isomorphic, we can replace the name of an element of one set by the name of its image, and imagine not only with impunity but sometimes to advantage that we are then naming the image; but the identity of names implies the existence of an isomorphism only when no more is to be preserved by the isomorphism than the number of members in the corresponding sets: any set of n signs will do merely to name n elements.

Moreover, isomorphism is not identity. Speaking of isomorphism enables us to regard certain sets, for example, $\{>; I\}$ and $\{\mathscr{F}; A\}$ of Secs. 2.2 and 2.3, as interchangeable under certain circumstances, without thereby implying the identity of the set of all integers with that of all letters. The relation between a set of signs and a set of representations of these signs on the one hand, and the relation between the set of representations and a set of tokens on the other, are normally *defined* as isomorphisms which preserves naming. For example, if *dog** names a species, "dog" and **dog** may be construed when desirable as naming the same species, thereby greatly simplifying certain areas of discourse. Such definitions are arbitrary, and there is nothing inconsistent about using "102", for example, to designate in different contexts an integer, a prisoner, or a locomotive.

2.5. Models

When applied to the study of phenomena of nature, rather than developed merely for its own sake, a system of signs serves as a *model*. That which is being modeled is sometimes called an *interpretation* of the system of signs. The interpretation of a system of signs may well be another system of signs. In ordinary English, the word *model** is used in a number of different senses. The specialized sense in which it is used here is closest to that defined by Webster as "a miniature representation of a thing; sometimes, a facsimile". The key idea is that of "representation", of "abstraction", but a good model may also play more active roles.

An architect's model of a house is obviously not the same thing as the house itself. It is useful, however, because it faithfully represents certain selected characteristics of the house. Although its dimensions are not those of the house, the proportions of all dimensions are preserved; the texture of various building materials and landscape detail may be faithfully imitated to give an idea of how the house itself will look. The relation between such a model and its interpretation is an isomorphism preserving selected characteristics of the interpretation, for instance, proportions but not absolute size. The fact that not all characteristics of the interpretation are represented in the model is fundamental. The only thing possibly identical with a given house is the house itself; and likewise with the idea of a house. This selectivity is, of course, the source of the usefulness of models; but, at the same time, there is a danger of misinterpretation when reasoning is based on properties unique to the model or to the interpretation, that is to say, properties not preserved by the isomorphism. This may be illustrated by the following example.

Say we have a box of matchsticks. Each matchstick is made of wood; some may have splinters; all have a light strawy color, but closer examination shows that each is spotted in a different motley way; all feel "woody"; and if they smell at all it is not a particularly interesting smell. If we take them in a bunch, stand them on a table, and put a flat piece of cardboard over them, we see that the tops of all of them touch or come very close to touching the cardboard. This seems interesting, and we decide to describe the situation by saying that the matchsticks have a property of "length", and that this length is very nearly the same for all. Let us now glue two matchsticks together end to end. If we do this with several pairs of matchsticks, we find that if these are stood on end and covered with a cardboard, again, all of the tops touch or come very close to touching the cardboard. We may now try to glue a third matchstick on each of the pairs. We have a choice of gluing it either on the matchstick near the table, or on that near the cardboard. We try it both ways, and find that it makes no difference in the length of the combination.

Now someone clever comes along and tells us how silly we are to play with matchsticks. He can play the same game on a piece of paper, and not mess with glue. Here is how he does it. He takes one matchstick and says that henceforth he will associate this matchstick with the number 1. Since all matchsticks behave the same way under the cardboard, one might as well associate each of them with the number 1 as shown in Fig. 12.

Now, the act of gluing a pair of matchsticks together will be associated with the addition of the numbers corresponding to each, thus yielding a new number $1 + 1 = 2$, and then the glued pair will be associated with the number 2. If we do this for all pairs of matchsticks, we get the set shown in Fig. 13. The number corresponding to three matchsticks glued together may be obtained either as the sum $1 + 2 = 3$ or as the sum $2 + 1 = 3$. Thus, the commutative law of addition adequately models the irrelevance of the order in which we glue matchsticks. Furthermore, our friend claims, if 2353 matchsticks are glued to 786 others, it makes no difference whether the 2353 or the 786 are on the table. We are overwhelmed and agree, but not without a mental reservation: after all, to test this, we need two separate sets of matchsticks, and if those of one set are all chosen from among those which exactly touched the cardboard, and those of the other sets from among the ones which fell a bit short of touching the cardboard, the agreement may not be perfect. But then, this is only a game, a difference of a matchstick or so in 3139 will not disturb us, and we would much rather add 2353 to 786, knowing we may be off somewhat, than glue all those matchsticks together.

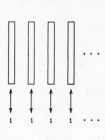

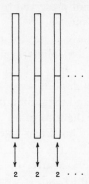

Fig. 12. A correspondence between matchsticks and numbers.

Fig. 13. A set of glued matchsticks.

What we have done, essentially, is summarized in Fig. 14. We have established an isomorphism between matchsticks and integers that preserves certain important relations. Among all qualities of matchsticks, we have abstracted one. We have ignored the material of the matchsticks, their color, the temperature of the room they are in. We have assumed that the length of the matchsticks does not change with time, that they are not

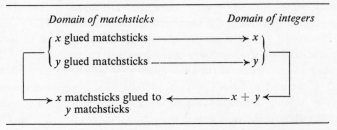

Fig. 14. An isomorphism between matchsticks and integers.

compressed when standing on end. By paying this price in simplification, we have been able to transfer our operations from the domain of matchsticks, where they are difficult, to the domain of integers, where we can perform equivalent but easier operations, and predict results we can or could observe in the original domain. In this sense, the domain of integers is a *mathematical model* of that of matchsticks.

The model actually has some properties in addition to those claimed. For example, it models weight as well as length. But since nothing in the model represents color or texture, no valid inferences regarding the color or texture of matches may be drawn from it.

The reader may wonder what happened to the "system of signs", since the matchstick model has numbers as elements. In practice, our model is a composite one. The matchsticks are modeled by numbers, but these numbers in turn are associated with numerals, themselves associated with representations and tokens. What we actually manipulate, when using our models, is tokens for numbers. Because a transitive chain of homomorphisms links tokens with numbers, the gist of the preceding exposition is unaffected by the choice of level of discourse. For purposes of that exposition, as for most ordinary discourse, numbers, numerals, character representations of numerals (or of numbers), and tokens of character representations (or of numerals or numbers) are indiscernible.

2.6. The Structure of Representations

The writing system of many languages is based on a small alphabet. Representations of words are formed by concatenating characters. We shall find it useful to consider the general class of sign systems in which signs are represented by strings of characters chosen from a small set of primitive characters, called the *alphabet* of the system.

A representation consisting of a single character will be called *irreducible over* the given alphabet, and any longer string will be called *reducible* or *factorable over* the given alphabet. A sign will be considered irreducible when it has no more than one property of interest, and reducible otherwise. Tokens may also be regarded as irreducible or reducible in a very literal sense, namely, according to whether or not they are physically analyzable into component tokens, whenever such an analysis is relevant.

The relative character of the concept of reducibility should be noted. As might be suspected, this concept is intimately connected with the identification of indiscernibles and its inverse. The representation "dog" of *dog**, for example, is reducible over the alphabet $L = \{$ "a", "b", "c", ..., "x", "y", "z"$\}$ but irreducible over the alphabet $W = \{$ "aardvark", "aardwolf", ..., "dog", ..., "zymotic", "zymurgy"$\}$. The representation "John came here" of the sentence *John came here** is reducible over both L and W. Over L, it is a string of twelve elements of L; over W, it is a string of three elements. The token for a text recorded on Univac magnetic tape may, under different circumstances, be considered to be factorable into block, blockette, word, or character tokens (Sec. 1.3).

The distinction between irreducible and reducible representations is useful in analyzing certain important properties of strings serving as names. For example, the strings "Plato", "the founder of the Academy", and "the author of the *Republic*" essentially all name the same person when used in a sufficiently unambiguous context; however, over the alphabet W of words, the first is irreducible, while the last two are reducible. The irreducible string simply refers; unless we know precisely what it refers to, and are therefore enabled to recall or otherwise to determine relevant characteristics of the referent, nothing more can be deduced about it by merely examining the string. The reducible strings refer to the same person, but their individual components also tell us about what he did and what he wrote.

As a second example, consider the number twelve. This number has many characteristics of interest to the mathematician: it is the successor of eleven, it is even, its prime factors include two instances of two and one of three, it is divisible by four, and so on. None of these properties become evident by a mere examination of the string "twelve". This string happens to be reducible over L, but, since none of its components are names of numbers, it is irreducible over the alphabet {"one", "two", "three", . . .}. On the other hand, the notation "111111111111" clearly suggests how twelve is related to eleven and to thirteen, evenness is reflected in the notation "2 × 6", the prime factors are denoted as components of "2^2 × 3", divisibility by four is obvious from the representation "4 × 3", and the string "12" is designed to convey the fact that $12 = (1 \times 10^1) + (2 \times 10^0)$. The structure of each representation models selected properties of the number 12; all name this number.

The differences between the properties of irreducible and reducible strings have important practical consequences. Suppose that we are required to determine the parity of any integer, between 1 and 10,000, represented in one of the notation systems shown in Table 5. The first notation is based on an alphabet of 10,000 characters, and the representation of every integer in the set is irreducible by definition. (Only the first 26 of these characters are in common use. The selection of the remaining 9974 is left as an exercise for the reader.) Given such a character, the only way to determine the parity of the corresponding integer is to consult a table containing 10,000 entries.

The alphabet for the second notation includes only ten characters, the familiar decimal digits. The rules for forming strings of characters and associating them with integers are such that if the character "d" occurs at the rightmost position of the string for an integer i it occurs also as the rightmost character of the string for the integer $i + 10$. Furthermore, we observe that whenever the string for an integer ends with an instance of "d" the parity of that integer is the same as that of the integer between 0 and 9 represented by a single instance of "d". We can therefore make a table of ten entries, indicating that the digits "0", "2", "4", "6", and "8" represent even integers, and "1", "3", "5", "7", and "9" represent odd integers, and then determine

Table 5. Irreducible and composite representations of the integers.

Notation 1	Notation 2	Notation 3	Notation 4	Parity[1]
A	0	2 × 0.0	0	E
B	1	2 × 0.5	1	O
C	2	2 × 1.0	10	E
D	3	2 × 1.5	11	O
E	4	2 × 2.0	100	E
F	5	2 × 2.5	101	O
G	6	2 × 3.0	110	E
H	7	2 × 3.5	111	O
I	8	2 × 4.0	1000	E
J	9	2 × 4.5	1001	O
K	10	2 × 5.0	1010	E
L	11	2 × 5.5	1011	O
M	12	2 × 6.0	1100	E
N	13	2 × 6.5	1101	O
.	.	.	.	.
.	.	.	.	.
.	.	.	.	.

Alphabets

{"A","B","C"...}   {"0","1",...,"9"}   {"0","1",...,"9"," × ","."}   {"0","1"}

[1] $E^* = even^*$, $O^* = odd^*$.

the parity of any integer by the following *algorithm* (*algorithm** is the name nowadays commonly given to an explicit recipe for a symbolic process): (1) isolate the last digit of the representation of the integer; (2) use the isolated digit as a key for look-up in the parity table; (3) associate with the given integer the parity determined from the table.

There are twelve characters in the alphabet of Notation 3, and only two in that of Notation 4. The parity of an integer expressed in either of these notations is again specified by the rightmost digit of its representation. In both notations, the choice of this digit is restricted to two alphabetic elements, and a table with only two entries is sufficient. The complete algorithm is similar to that outlined for the second notation.

The pronounced influence of notation on the process for determining the parity of integers is a remarkable but by no means isolated phenomenon. Instances where a happy choice of notation proved to be a key to progress are numerous in all disciplines. (For an illuminating and amusing analysis of this question, see Henle (1949); also Ashenhurst (1956).) The factors determining the choice of notation are many, and they are interrelated with some complexity. *The preceding examples indicate how table length can be traded for algorithms more sophisticated than simple table look-up, merely by substituting suitably reducible representations for irreducible ones.* Fundamentally, the known properties of any *finite* set of objects can be listed in a table, which may then be used to determine whether or not a particular object has a certain property. This approach has the merit of conceptual simplicity and wide applicability, but there obviously is a limit to the size of tables that can be used effectively in practice. It is our good fortune, therefore, when interesting properties of an element can be modeled by the structure of its representation, and then determined, when necessary, by simple operations on the representation.

The reduction of table size made possible by the application of algorithms to factorable representations is used to advantage in the construction of dictionaries, especially for inflected languages. In principle, dictionaries could be built by using every inflected form of every word as a key to a distinct entry, but the size of English dictionaries would be more than doubled thereby, and even larger factors would apply for highly inflected languages like Latin, Russian, or German. In conventional dictionaries only one standard form is usually listed to represent an inflectional paradigm, and special algorithms must be used to reduce the other forms to the standard form for look-up.

Example 2-7. The inflected forms of every noun in a hypothetical language *L* are constructed by adding to an invariant stem the suffixes listed in Table 6. The stems of nouns in *L* are known to have no more than ten letters. A noun is stored in register *0*, the suffix characters occupying positions 11 and 12. We are required to write a program that will (*a*) place the initial character of the name of the *case* of the given noun in position 12 of register *1*, and space-fill in other positions; (*b*) place the initial character of the name of the *number* of the given noun in position 12 of register *2* and space-fill in other positions.

Table 6. Suffixes for inflection of a
hypothetical language.

Case	Number	
	Singular	Plural
Nominative	"SN"	"PN"
Genitive	"SG"	"PG"
Dative	"SD"	"PD"
Accusative	"SA"	"PA"
Instrumental	"SI"	"PI"
Prepositional	"SP"	"PP"

Auxiliary Storage Registers

0	[ ]	Noun (initially and after execution of the program)
1	[ ]	Case initial (after execution of the program)
2	[ ]	Number initial (after execution of the program)
3	0 0 0 0 0 0 0 0 0 0 1	
4	[ ]	Register for temporary storage
5	△△△△△△△△△△△	

Program

100	B	0			Noun to register *A*
			.1		Shift noun right one position
101	C	4			Shifted noun to register 4
			F	3	Extractor to register *F*
102	B	5			Spaces to register *A*
			E	4	Extract number initial from register 4 into register *A*
103	C	2			Number initial to register 2
			B	5	Spaces to register *A*
104	E	0			Extract case initial from register 0 into register *A*
			C	1	Case initial to register 1

Exercise 2-3. (*a*) Write a program meeting the requirements of Example 2-7, but based on the suffixes given in Table 7. Compare your program with that of the example.

(*b*) Write a program that will generate all inflected forms of a noun, given the nominative singular, according to (i) Table 6; (ii) Table 7.

In Sec. 2.4, we noted that the choice of alphabet had important consequences relative to the question of identification of indiscernibles. We now see that the size of the alphabet of a notation system also affects the length, in characters, of composite representations constructed with the alphabet. These factors, in turn, affect the size of the tables or the structure of the

Table 7. Suffixes for inflection of a
hypothetical language.

Case	Number	
	Singular	Plural
Nominative	"AB"	"EN"
Genitive	"OS"	"MU"
Dative	"AR"	"OK"
Accusative	"TI"	"SA"
Instrumental	"LU"	"IF"
Prepositional	"FY"	"UD"

algorithms necessary in certain processes. The degree of reducibility afforded by different notation systems is an instance of a general problem expressed by Quine (1953, p. 26) in the following words:

In logical and mathematical systems either of two mutually antagonistic types of economy may be striven for, and each has its peculiar practical utility. On the one hand we may seek economy of practical expression—ease and brevity in the statement of multifarious relations. This sort of economy calls usually for distinctive concise notations for a wealth of concepts. Second, however, and oppositely, we may seek economy in grammar and vocabulary; we may try to find a minimum of basic concepts such that, once a distinctive notation has been appropriated to each of them, it becomes possible to express any desired further concept by mere combination and iteration of our basic notations.

Some examples drawn from arithmetic may help further to clarify the general problem.

Consider the problem of obtaining the sum $N_1 + N_2$ of a pair of integers, where $0 \leq N_1, N_2 \leq N_{max}$. The largest possible value of $N_1 + N_2$ is $2N_{max}$. If the integers are represented according to the first notation of Table 5, the required number of alphabetic characters is $2N_{max} + 1$, and the representation of every integer between 0 and $2N_{max}$ is irreducible. To obtain the representation of $N_1 + N_2$, given that of N_1 and of N_2, it is only necessary to consult once an addition table with $N_{max} + 1$ rows and an equal number of columns. This pure table look-up algorithm is simple but grows impractical for large N_{max}, and *no number pair not explicitly mentioned in the table can be summed.*

Ordinary decimal arithmetic is based on the fact that any integer N can be expressed by

$$N^* = a_n 10^n + a_{n-1} 10^{n-1} + \cdots + a_2 10^2 + a_1 10^1 + a_0 10^0 *,$$

where the "a_i" stand for integers between 0 and 9. The further observation that the "10^i" may be suppressed without confusion if the familiar conventions about the position of decimal digits are adopted leads to a notation of the well-known type: $N^* = a_n a_{n-1} \cdots a_2 a_1 a_0 *$ (for instance, Notation 2, Table 5). The number of characters in the alphabet is now reduced to ten, but the representation of every $N \geq 10$ is composite, comprising $n + 1$ instances of primitive characters. The number n obviously grows with N and can be shown to be the integral part of $\log_{10} N$. When N_1 and N_2 are expressed in this notation, finding the representation of $N_1 + N_2$ requires only a 10×10 table *regardless of the value of* $N_{\max}$, and there is nowhere explicit mention of the numbers that can be summed. One can simply use this table repeatedly, combining the characters obtained from the table according to an *iterative* column-by-column algorithm, which is learned by every schoolboy along with the table, but is more complex than a simple table look-up.

It can be shown that any integer N can also be expressed by

$$N^* = b_n 2^n + b_{n-1} 2^{n-1} + \cdots + b_2 2^2 + b_1 2^1 + b_0 2^0 *,$$

where each "b_i" stands for either "0" or "1", leading to the binary notation $N^* = b_n b_{n-1} \cdots b_2 b_1 b_0 *$ (Notation 4, Table 5). The 2×2 addition and multiplication tables necessary to do arithmetic in this notation are shown in Fig. 15a. The algorithm for obtaining the numeral for the sum of two numbers is illustrated by the example in Fig. 15b. The binary algorithm is quite similar to that for decimal numerals: the sum digit in the rightmost column is obtained by table look-up, and a carry, if necessary, is made to the next column; successive columns are then added in turn, two table look-ups being necessary in a column having a carry from its predecessor. Carries are shown as primed numerals. An example of multiplication is given in Fig. 15c.

As was already pointed out in Sec. 1.2, it is natural to associate binary representations of numbers with the tokens given by the configurations of sets of bistable devices. The relation between the arithmetic algorithms defined on binary representations and physical operations easy to carry out on bistable devices can readily be shown to be a homomorphism. It is therefore often convenient, in discourse about automatic information-processing machines, to substitute binary representations of signs for their conventional representations.

Exercise 2-4. How many instances of primitive characters are there in the binary numeral for N? Compare this number with that for the decimal notation. Devise an algorithm for transforming a binary representation into the corresponding decimal representation, and apply this algorithm to check the examples of Fig. 15.

The preceding observations also shed some light on the question of what the usefulness of automatic machines really is. This question can be raised quite seriously in view of the fact that much human time and effort are necessary to program the solution of each problem. This time and effort, it can be argued, might well be applied to solving the problem directly.

---

a. Operation tables

+	0 1		×	0 1
0	0 1		0	0 0
1	1 0¹		1	0 1

¹ Carry 1 to next column.

b. Addition

$$1'\,1'\,1'\,1'$$
$$101101$$
$$+\quad 110110$$
$$\overline{\;1100011}$$

c. Multiplication

$$101101$$
$$\times\quad 110110$$
$$\overline{}$$
$$101101\,\cdot$$
$$101101$$
$$101101\,\cdot$$
$$101101$$
$$\overline{}$$
$$100101111110$$

Fig. 15. Binary arithmetic.

Imagine the following automatic procedure for translating *War and Peace* into English: a machine is so arranged that Constance Garnett's translation of *War and Peace*, previously recorded in the machine's memory, is printed by an output mechanism each time a certain button is pushed. It is so much simpler and cheaper to get a copy of this translation from a library or to order it from a bookseller that the suggested "automatic" procedure is obviously preposterous. Similar "automatic" procedures have been used on a smaller scale to give impressive demonstrations of automatic translations of excellent quality (Dostert, 1955), but that is another matter.

In the realm of scientific computation, the usefulness of automatic machines has now been established beyond doubt. Yet it might be objected that building an automatic calculator is the hard way to do sums, because the addition tables and algorithms have to be prepared and built into the machine. Moreover, not only do elementary texts on arithmetic contain addition tables for anyone to consult, but each of us carries at least a small one in his head, along with the rules for using it. For these reasons it is clear that no one in his right mind would produce a program of 201 instructions (Sec. 1.4), let alone build a million-dollar machine, only to obtain once the sum of a set of 200 *specific* numbers. The iterative program of Example 1-3 was worth writing only because, once written, it can control the summation *of practically any 200 numbers* and, with minor modifications, the summation of *any number of numbers* within the limits of machine storage capacity. More fundamentally, the addition unit of a computing machine is worth building because, as was demonstrated in this section, it is based on an algorithm for summing *any* pair of numbers that does not make *explicit* special provision for the summation of every possible pair. This kind of generality is the key to the usefulness of automatic information-processing machines.

2.7. The Structure of Tokens

Tokens are configurations of physical objects and, as such, are subject to the laws of nature. It follows that the choice of tokens and of the way in which they are used is governed to a considerable degree by technological factors. A stylus on clay, a paint brush and India ink on rice paper, a goose-quill pen on parchment, movable type slugs on modern paper, paper punches on pieces of cardboard, and electrical switches with vacuum tubes are used to best advantage in quite different ways.

For example, in *The Printing of Mathematics* (Chaundy, Barrett, and Batey, 1954, p. 22), the authors point out:

In 'Monotype', with its precise units of type-width as well as the point units of body-depth, we can regard composition as if it were set out on graph paper. On the other hand the mathematician's pen is free to roam, enlarging or compressing his script, filling blank space with symbols, constructing marvels of formulae—even, perhaps, ringing a character or inserting one character within another in a way that cannot easily be directly imitated with metal type.

One kind of physical factor considered in the design of automatic calculators is mentioned in the following quotation (Staff of the Computation Laboratory, 1951, p. 153):

In this system, n decimal digits can require not less than n nor more than $2n$ unit binary digits ("1"s) for their representation. This minimal fluctuation has been shown to be an advantage in the operation of magnetic delay line storage devices.

In a similar vein (Weik, 1958, p. 7):

It is interesting to determine the minimum ratio of holes to total bit spaces in punched paper tape and cards that would be required to store and transmit English language text. Minimization of holes can reduce punch wear, reduce power requirements, and ease trouble shooting.

As may well be expected, tokens designed to satisfy one requirement may do violence to another, and the actual choice of tokens is often based on compromise. When an existing machine is to be adapted to a new application, the existing compromise may be found to differ considerably from what seems best under the new circumstances.

In the Univac the 63 characters of the machine's alphabet are represented by irreducible groups of six bistable devices (Table 2). In some other machines groups are reducible and the single bistable device is the irreducible element; in other words, there exist instructions causing operations to be performed on selected individual bistable devices. Machines of this kind are commonly called *pure binary* machines. Machines in which tokens are arranged in irreducible groups each representing a single decimal digit are often called *binary-coded decimal* machines. Those which, like the Univac, have irreducible tokens not only for decimal digits but also for characters of the English alphabet are termed *alphanumeric* machines. Some recent machines have been designed for use in any of the alphanumeric, binary-coded decimal, or pure binary modes, at the option of the programmer. In the pure binary mode, great flexibility is available, but detailed instructions are necessary. In the alphanumeric mode, some flexibility is sacrificed to achieve economy of instructions. Each mode has advantages and disadvantages of the type described in Sec. 2.6.

Even in machines where character tokens are irreducible relative to programming, they may be treated as reducible for such engineering purposes as the detection of machine malfunction. The failure of a machine component may change a token for one character to a token for another character. For instance, a token for "B" will change into a token for "A" (Table 2) if the rightmost bistable unit changes from **1** to **0**. In binary-coded decimal machines (Table 1), where only ten of the sixteen possible configurations of four bistable units are used as tokens, a change leading to one of the six unused configurations can be detected by checking circuits that test for the presence of one of these configurations and stop the machine if it occurs.

Extra bistable devices are often added to character storage units to provide systematically for the detection of malfunctions. In the Univac seven bistable devices are actually used per character, not six. The configuration of the seventh device is chosen so that a configuration of seven units corresponds to a character if and only if an odd number of units are **1**'s. A few of the seven-unit configurations are shown in Table 8, where the seventh or *checking*

Table 8. An error-detecting token system.

Tokens	Characters
1 000000	"i"
0 000001	"△"
0 000010	"–"
1 000011	"0"
0 000100	"1"
1 000101	"2"
.	.
.	.
.	.

unit is set apart. The remaining units are in the configurations given in Table 2. In this system, a change of the configuration of any one of the seven units from **0** to **1** or vice versa leads to a token not associated with any character, and the error can be detected.

A simultaneous change of the configuration of two units (in fact, of any even number of units) leads to an acceptable token and cannot be detected. Systems capable of correcting as well as detecting a single error, or of detecting and even correcting two or more simultaneous errors, have been devised (Hamming, 1950); the cost of the additional bistable units required by these systems is so high, and the probability of several simultaneous errors so low, that in practice only single-error detection has been widely used.

The association between tokens and characters is determined with some rigidity by the physical structure of a machine. For example, actuating the *A*-key of the Univac input typewriter puts a section of the temporary input storage register in the configuration **1 010100**; conversely, when a section of the appropriate output register is in the configuration **1 010100**, the console typewriter prints a token for "A". The correspondence between a given token and a character also depends on what output unit is interpreting a particular configuration, on the setting of certain switches on the output unit, and even on the preceding token! The details of the correspondence between central computer tokens of the Univac and characters whose tokens can be recognized or produced by the several input and output units are given in Table 9. For example, actuating the quotation-mark key on the input typewriter produces a token **1 100001** on tape. Such a token causes the high-speed printer to leave a space when a certain switch is in the "normal" position; when this switch is in the "computer digit" position, the high-speed printer prints a "0". As for the console typewriter, when a certain switch is in the "normal" position, it may do nothing, as explained under "option" in Table 9, or it may print a token for ";" or one for ":" depending on which of the tokens **1 101101**, **0 101111**, or **0 111101** immediately preceded it. The actions of the console typewriter with the switch in the "computer digit" position and of the two converters are also described in Table 9. There is no point in dwelling on the details of Table 9; it is sufficient to indicate that the correspondence among various sets of signs, their representations, and their tokens can obviously be a matter of active concern to the machine user. In most machines, establishing the correspondences best suited for a particular process can require special programming, setting of switches, wiring of printer plugboards, and so on, singly and in a wide variety of combinations.

The circuits of the Univac operation units impose a definite order relation on the tokens given in Table 9. Let us imagine, for the sake of simplicity, that a Univac register stores only a single character rather than twelve. The circuits for executing the Q instruction are so arranged that control is transferred if and only if register A is in the same configuration as register L. This instruction thus enables the machine to determine when instances of the same character are stored in registers A and L. When register A is in a configuration occurring in Table 9 below that of register L, the execution of a T instruction will cause a transfer of control. When the two registers A and L are in the same configuration, or when register A is in a configuration listed in Table 9 above that of register L, control is not transferred.

If signs having an order relation of their own are associated with tokens in such a way that the mapping from tokens to signs is an order-preserving homomorphism, the machine is enabled to choose alternative courses of action according to whether a particular pair of sign instances are in order or not. It is evident from inspection of Table 9 that the mapping from tokens to numbers preserves numerical order, and that from tokens to English letters preserves alphabetic order. As a by-product, the letters are ordered relative to the numbers and an order is *imposed* even on the punctuation marks and other special signs.

Table 9. Interpretation of token configurations by Univac I input and output units (adapted from Esch and Calingaert, 1957). *Single symbol in center of box:* print that symbol. *Breakpoint:* if typewriter breakpoint switch is on, stop; otherwise, print indicated symbol, if any. *Option:* if typewriter stall switch is on, typewriter stalls without printing, and next character is admitted by hitting space bar, if stall switch is off, print indicated symbol. *sp:* space, no printing. *es:* error stop.

TOKEN	INPUT CHARACTER (INPUT TYPEWRITER KEY LABEL)	EFFECT ON HIGH-SPEED PRINTER NORMAL	EFFECT ON HIGH-SPEED PRINTER COMPUTER DIGIT	CONSOLE NORMAL ACTION	CONSOLE NORMAL LOWER-CASE	CONSOLE NORMAL UPPER-CASE	CONSOLE COMPUTER DIGIT ACTION	CONSOLE COMPUTER DIGIT LOWER-CASE	CONSOLE COMPUTER DIGIT UPPER-CASE	TAPE-TO-CARD CONVERTER	CARD-TO-TAPE CONVERTER
000000	i	sp	5	NONE				X	X	es	
000001	Δ	sp	6	sp			sp			es	Blank, Or No Column
000010	-	-	-		-	_		-	_	11	11
000011	0	0	0		0	)		0	)	0	0, Blank, Or No Column
000100	1	1	1		1	‡		1	‡	1	1
000101	2	2	2		2	"		2	"	2	2
000110	3	3	3		3	#		3	#	3	3
000111	4	4	4		4	$		4	$	4	4
001000	5	5	5		5	%		5	%	5	5
001001	6	6	6		6	*		6	*	6	6
001010	7	7	7		7	&		7	&	7	7
001011	8	8	8		8	'		8	'	8	8
001100	9	9	9		9	(		9	(	9	9
001101	'	'	'	OPTION	;	:	OPTION	;	:	es	
001110	&	&	&	OPTION	;	:	OPTION	;	:	12	12
001111	(	(	(	OPTION	;	:	OPTION	;	:	es	
010000	r	sp MULTILINE	E	CARRIAGE RETURN				/	?	es	
010001	,	,	,		,	'		,	'	es	
010010	.	.	.		.	.		.	.	es	
010011	;	;	;		;	:		;	:	0 and 12	
010100	A	A	A		A	A		A	A	1 and 12	1 and 12
010101	B	B	B		B	B		B	B	2 and 12	2 and 12
010110	C	C	C		C	C		C	C	3 and 12	3 and 12
010111	D	D	D		D	D		D	D	4 and 12	4 and 12
011000	E	E	E		E	E		E	E	5 and 12	5 and 12
011001	F	F	F		F	F		F	F	6 and 12	6 and 12
011010	G	G	G		G	G		G	G	7 and 12	7 and 12
011011	H	H	H		H	H		H	H	8 and 12	8 and 12
011100	I	I	I		I	I		I	I	9 and 12	9 and 12
011101	#	#	#	OPTION	;	:	OPTION	;	:	es	
011110	¢	sp	C	OPTION	;	:	OPTION	;	:	es	
011111	@	sp FAST FEED I	D	OPTION	;	:	OPTION	;	:	es	
100000	↑	sp	N	TAB				V	V	es	
100001	"	sp	O				OPTION	;	:	es	
100010	|	sp FAST FEED II	P	OPTION	;	:	OPTION	;	:	es	
100011	)	)	)	OPTION	;	:	OPTION	;	:	0 and 11	
100100	J	J	J		J	J		J	J	1 and 11	1 and 11
100101	K	K	K		K	K		K	K	2 and 11	2 and 11
100110	L	L	L		L	L		L	L	3 and 11	3 and 11
100111	M	M	M		M	M		M	M	4 and 11	4 and 11
101000	N	N	N		N	N		N	N	5 and 11	5 and 11
101001	O	O	O		O	O		O	O	6 and 11	6 and 11
101010	P	P	P		P	P		P	P	7 and 11	7 and 11
101011	Q	Q	Q		Q	Q		Q	Q	8 and 11	8 and 11
101100	R	R	R		R	R		R	R	9 and 11	9 and 11
101101	$	$	$	SHIFT LOCK				Z	Z	es	
101110	*	*	*	OPTION	;	:	OPTION	;	:	es	
101111	?	sp FAST FEED III	M	UNSHIFT				8	8	es	
110000	Σ	sp STOP	V	STOP						es	
110001	β	sp BREAKPOINT	W	BREAKPOINT			BREAKPOINT	Y	Y	es	
110010	:	:	:	OPTION	;	:	OPTION	;	:	es	
110011	+	+	+		+	@		+	@	es	
110100	/	/	/		/	?		/	?	es	
110101	S	S	S		S	S		S	S	1 and 0	1 and 0
110110	T	T	T		T	T		T	T	2 and 0	2 and 0
110111	U	U	U		U	U		U	U	3 and 0	3 and 0
111000	V	V	V		V	V		V	V	4 and 0	4 and 0
111001	W	W	W		W	W		W	W	5 and 0	5 and 0
111010	X	X	X		X	X		X	X	6 and 0	6 and 0
111011	Y	Y	Y		Y	Y		Y	Y	7 and 0	7 and 0
111100	Z	Z	Z		Z	Z		Z	Z	8 and 0	8 and 0
111101	%	%	%	SINGLE SHIFT				-	_	es	
111110	=	sp FAST FEED IV	T	OPTION	;	:	OPTION	;	:	es	
111111	(NONE)	es	es	OPTION	;	:	OPTION	;	:	es	

Exercise 2-5. Three arbitrary characters are stored in position 12 of registers *1*, *2*, and *3*, respectively. The other positions of these registers are filled with spaces. Write a program that will place the three characters in registers *4*, *5*, and *6*, in the order defined in Table 9.

The behavior of the arithmetic units confirms the relation between certain tokens and integers partially established by the behavior of the input and output units and by that of the *T* instruction. For example, if the same positions in registers *A* and *X* are tokens for 1 and 2, in either order, that position of register *A* will become a token for 3 following the execution of an *X* instruction. In this fashion, physical operations on tokens are associated with abstract arithmetic operations so as to establish an operation-preserving homomorphism from the set of tokens to that of numbers. Although any pair of machine words may be "added" by the execution of a *B* instruction followed by an *A* instruction, the resulting content of register *A* may not always have a meaningful interpretation as the representation of something.

In the Univac, an attempt to "add" two words *both* having literals in the same columnar position is automatically detected and the machine is stopped; but, when a particular position in one word is occupied by a literal and the same position in the other word is occupied by a numeral, the numeral is ignored in "addition" and that position in the "sum" word is filled by the unaltered literal. This facility is provided especially to permit the modification of instruction addresses as illustrated in Example 1-3. That the same physical addition unit and the same *B* and *A* instructions are used, for economic and technical reasons, for both *addition* and *instruction modification* should not obscure the fundamental differences between these two operations.

The possibility of using the addition unit to modify instructions is based on the use as addresses of consecutive integers, rather than of letters, number-pairs, or "natural" storage coordinates. Indeed, upon detecting a literal, the addition unit changes itself from an adder to a device that behaves like an adder only for the word positions occupied by addresses, but like a mere copying instrument for positions occupied by instruction names. Careless use of the addition unit for address modification may produce "sum" words that are wrong instructions or not instructions at all. In many machines, especially the newer ones, it has been found advantageous to recognize formally the differences between addition and instruction modification by providing separate facilities for the convenient execution of the latter operation.

Occasionally a physical operation on tokens which normally leads to uninterpreted resultant tokens may be given significance by an alert programmer. Turning an accident of engineering into an advantage is often no more than a clever trick. Sometimes the trick is discovered to be a fundamental operation with useful interpretations in many realms and then is built in as a legitimate instruction in later models of machines.

As an example, consider the "multiplication" of letters. At face value, such an "operation" is obviously meaningless. However, there are useful applications of the result of multiplying G^* by 1. Let $(L) = 0000000000G0$ and $(5) = 010000000000$. The instruction P 5, when executed, causes the "product" $(L) \times (5)$ to appear in register A. With the content of registers 5 and L as indicated, this "product" will be $(A) = 000000000007$. In fact, the same result will be obtained if G^* is replaced by P^* or X^*. This phenomenon can be expressed by the following rule, with reference to Table 2: the product of any character by 1 is the character in the same row as the given character and in the first column of Table 2. For example, multiplying D^*, M^*, or U^* by 1 gives 4. One application of this odd operation is illustrated in Sec. 2.8. This application is of sufficient importance for an equivalent operation to have been included explicitly in the instruction repertoire of several machines.

Since the structural and reducibility properties of tokens are governed in some measure by physical constraints, they are not always fully adaptable to the properties of the signs that the tokens represent. For instance, although the printed page is normally an irreducible physical entity, its configuration and the physical processes of normal reading are such that individual letters, words, or sentences are easily read by the trained person. On the other hand, arranging the words occurring on a page into alphabetic order is not so easily accomplished. If the page is expendable, it may be cut into pieces each carrying one word; otherwise each word must be copied onto a single slip of paper, and these word-tokens can then be arranged in the desired order. Writing words on individual slips suffers from the disadvantage that the sequence of words cannot be read as quickly as when printed on a page, and, while the slips are easily arranged in order, they may be all too easily disarranged by being dropped on the floor.

A birthday greeting or business communication is most conveniently written and read in the ordinary alphabet. The cost of transmission, however, is roughly proportional to the number of characters in the representation of the message, or better, to the number of tokens that must be transmitted. Standard representations are therefore converted to shorter strings by reference to a code book, and tokens for these shorter strings are transmitted. The inverse of the coding transformation must, of course, be performed at the receiving end to yield a message intelligible to its ultimate addressee.

Much of the physical equipment in an automatic information-processing system and much programming effort are devoted to the business of transforming representations or tokens convenient for one purpose to others more convenient for some other purpose. Some factors affecting the structure of tokens have been described in the present section and in Secs. 1.5 to 1.7. Illustrations of the relative advantages and disadvantages of various systems of notation have already been given in Sec. 2.6. Here we shall illustrate the problem of matching a set of representations with a set of tokens. For instance, this sentence could be stored in the Univac in either of the ways shown in Fig. 16. Because of the physical structure of the Univac, machine words are irreducible relative to transfer operations. If English words are viewed as reducible over the ordinary alphabet, there is nothing incongruous about storing them as shown in Fig. 16a. In fact many manuscripts and early printed texts followed just such a convention. However, just as the artificial constraints introduced by page width make printers struggle with justification and secretaries with hyphenation, so do the machine word boundaries introduce breaks in the sequence of characters that seem artificial to us.

0	FOR△INSTANCE
1	,△THIS△SENTE
2	NCE△COULD△BE
3	△REPRESENTED
4	△IN△THE△UNIV
5	AC△IN△EITHER
	⋮

a

0	FOR△△△△△△△△△
1	INSTANCE△△△△
2	,△△△△△△△△△△△
3	THIS△△△△△△△△
4	SENTENCE△△△△
5	COULD△△△△△△△
	⋮

b

Fig. 16. Packed and unpacked English words.

If, on the other hand, the English words are themselves regarded as alphabetic elements of a higher order and hence as irreducible, the arrangement of Fig. 16a presents serious difficulties, and that of Fig. 16b is preferable. Corresponding to the structure of the machine tokens, we have in Fig. 16a such artificial irreducible elements as ",△THIS△SENTE", which cannot be identified with an element of the alphabet of English words; to reconstitute irreducible tokens for alphabetic elements such as "sentence", it is necessary to appeal to extract and shift orders. In Fig. 16b, a token irreducible relative to transfer is used for each alphabetic element, but at the expense of wasting storage with space-fill.

The use of a whole register as a token for a single comma illustrates one of the dilemmas facing the programmer. Treating the comma as in Fig. 16b is wasteful, but including it in a register with an English word precludes the straightforward use of that register as an irreducible token for either the word or the comma. Frequently both the methods illustrated in Fig. 16 may be used in the same process. To save storage space and tape-reading time texts might be stored in the first way on tape, but rearranged when brought into the central memory to facilitate the subsequent transfer of individual words. This rearrangement is necessary often enough on most machines to have been called *unpacking*. The inverse process of rearranging individual

English words into the more compressed configuration is known as *packing*.

Exercise 2-6. Write a program for unpacking English words stored as in Fig. 16*a*, and storing them as in Fig. 16*b*. Write a program for the inverse packing process. It will simplify matters to assume that there is no punctuation in the text and that no English word is represented by more than 12 characters.

The problem of packing and unpacking arises also when character strings identifiable with alphabetic elements of a higher order cross over the boundaries of such token structures as blockettes and blocks; for instance, a sentence which may be regarded as an element of an alphabet of sentences and occasionally treated as a unit may, when represented by a string based on the alphabet of letters, occupy more than a blockette or a block. When a representation that should be irreducible relative to certain operations exceeds the capacity of a single register or block, special precautions must be taken to insure that it will indeed be handled as a unit. For example, *polymerization**, stored as "POLYMERIZATI ON△△△△△△△△△△" in two registers, could be transferred as a unit by using V and W orders.

In some machines special facilities are provided to adjust token structure to sign structure. Individual machine characters are addressable rather than words and a string to be transferred is specified by giving the address of its first character. This character and the ones following are transferred in turn until a special partition character is encountered denoting the end of the string. In theory, therefore, the length of the "register" defined by two successive partition symbols can always be set precisely to the length of the string it is to store.

In practice this mode of operation is convenient only for search access, since the location of the beginning of a string cannot ordinarily be specified in advance, and coordinate access to strings is therefore impossible. Consequently some compromise is usually necessary; nevertheless, the flexibility gained by not fixing the token structure mechanically but instead allowing it to be induced by that of the represented signs can be advantageous. Word length and block length are fixed in some machines, but instructions are available to transfer the contents of an arbitrary number of registers as a unit from buffer storage to internal memory and within the internal memory. This arrangement is obviously the analogue, at a higher level of organization, of the "register" of adjustable length.

2.8. Code Conversion: Adapting the Univac for Processing Russian

The input and output devices of an automatic information-processing machine establish a relation between machine tokens and the set of characters which the tokens are intended to represent. The alphabet participating in this relation may be called the *natural* alphabet for the machine. The natural alphabet of the Univac has been described in Secs. 1.3 and 2.7.

Any machine can be adapted to process information coded in an alphabet other than its natural one, provided only that a suitable, practically sound homomorphism can be established between tokens and the alternative alphabet. Of course, a machine could be constructed for which the alternative alphabet would be the natural one, but a sound design for such a machine is best evolved through preliminary experiments on some existing, suitably adapted machine. Since the Univac at the Harvard Computation Laboratory has been adapted to process Russian, it will be used to illustrate the problems and methods of adaptation. Although details will vary for other machines, the problem and its solution are of sufficient importance and generality to merit discussion here.

Two representations of the Cyrillic alphabet used in experimental work are defined in Table 10. The alphabet to be represented is given in column 1. The representation used for automatic processing is shown in column 2, and a transliterated representation is shown in column 3.

Russian material is entered into the machine via magnetic tape prepared on a specially adapted version of the Unityper (Oettinger, 1958), which is a typewriter-like input device associated with the Univac (Fig. 2). A photograph of this machine is shown in Fig. 17, and its keyboard layout is displayed in Fig. 18.

In lower-case, the machine serves as a Cyrillic typewriter, and in upper-case as a Roman typewriter. From the point of view of immediate typing ease and accuracy, the most desirable keyboard layout is one that is in standard use, and therefore familiar to typists used to working with ordinary typewriters. The Roman layout of Fig. 18 is the standard one, while the Cyrillic layout is very close to one of the two commonly used. Three of the least frequently used Cyrillic characters are not in their usual positions as a result of a necessary technical compromise. A sample of the copy produced by this typewriter for proofreading purposes is shown in Fig. 19. Having both Roman and Cyrillic characters available on a keyboard facilitates the transcription of texts to be translated, a matter described more fully in Chapter 8. Type slugs capable of this dual function are not generally available and had to be cast from dies made specially for the purpose.

When the key for a Cyrillic character is struck, a conventional token for the character is printed on paper in the normal way and, simultaneously, a piece of magnetic tape is placed in a corresponding configuration of magnetized and unmagnetized spots. The correspondence between keys, representations, and tokens is given in Fig. 20. For example, when key No. 1 is struck in lower-case, the character "й" is printed, and the token **0 010011**

Fig. 17. Cyrillic Unityper.

ъ	"	#	$	%	*	.	—	⊙	(	)
I	2	3	4	5	6	7	8	9	⊙	Ц

0   3   7   11   15   19   23   27   31   35   39

Q	W	E	R	T	Y	U	I	O	P	+
Й	У	К	Е	Н	Г	Ш	Щ	З	Х	Э

1   5   9   13   17   21   25   29   33   37   41

A	S	D	F	G	H	J	K	L	'	Δ
Ф	Ы	В	А	П	Р	О	Л	Д	Ж	Δ

2   6   10   14   18   22   26   30   34   38   42

Z	X	C	V	B	N	M	:	;	.
Я	Ч	С	М	И	Т	Ь	Б	Ю	,

4   8   12   16   20   24   28   32   36   40

Fig. 18. Keyboard layout of the Cyrillic Unityper.

THE TEXT ON THIS PAGE HAS BEEN RECORDED ON MAGNETIC TAPE BY MEANS OF A MODIFIED UNITYPER.

TO DISTINGUISH SPACES FROM BLANK PAPER, THE SYMBOL "." IS PRINTED WHENEVER THE STANDARD SPACE BAR OR A SPECIAL SPACE KEY IS STRUCK.

WHEN THIS UNITYPER IS SET FOR LOWER-CASE TYPING, CYRILLIC CHARACTERS ARE PRINTED AND RECORDED IN A SPECIAL CODE... ROMAN CHARACTERS AND ARABIC NUMERALS ARE AVAILABLE IN THE UPPER-CASE SETTING.

THE AVAILABILITY OF ROMAN AND CYRILLIC CHARACTERS ON THE SAME KEYBOARD FACILITATES THE TREATMENT OF EQUATIONS AND OTHER SPECIAL SYMBOLS OCCURRING IN TEXTS, E.G.:

(1) РАССМАТРИВАЮТСЯ КОНЕЧНЫЕ МНОЖЕСТВА ОБЪЕКТОВ $A SUB 1,..., A SUB N$, КОТОРЫЕ...
(2) ...МОЖНО ЗАПИСАТЬ В ВИДЕ МАТРИЦЫ $EQUATION 3$ ГДЕ...

THE DOLLAR SIGNS IN THE EXAMPLES ARE USED AS BRACKETS WHICH SIGNAL THE TRANSLATING MACHINE TO TREAT THE SYMBOLS WITHIN BRACKETS IN A SPECIAL WAY.

В ГЛ. 1 ЭТОЙ КНИГИ ДАНЫ НЕКОТОРЫЕ ВЕКТОРНЫЕ СООТНОШЕНИЯ И ВЫРАЖЕНИЯ * , КОТОРЫЕ ПОЯСНЯЮТСЯ НИЖЕ . * . СКАЛЯРНОЕ ПРОИЗВЕДЕНИЕ ДВУХ ВЕКТОРОВ $CAP A BAR TIMES CAP B BAR$. * . ЭТО . * . В ПРЯМОУГОЛЬНЫХ КООРДИНАТАХ ВЕЛИЧИНА . * . СКАЛЯРНАЯ ВЕЛИЧИНА . * . В ПРЯМОУГОЛЬНЫХ КООРДИНАТАХ $EQUATION 1$ ГДЕ $A SUB X$ ... КОМПОНЕНТА ВЕКТОРА $A BAR$ НА ОСЬ X , $B SUB X$ ... КОМПОНЕНТА ВЕКТОРА $B BAR$ НА ОСЬ X И Т Д ... ВЕКТОРНОЕ ПРОИЗВЕДЕНИЕ ДВУХ ВЕКТОРОВ ЕСТЬ ВЕКТОР . * . ВЕКТОРНОЕ ПРОИЗВЕДЕНИЕ ДВУХ ВЕКТОРОВ $A BAR$ И $B BAR$. * . ПРИ ИЗМЕНЕНИИ ПОРЯДКА ПЕРЕМНОЖЕНИЯ МЕНЯЕТСЯ ЗНАК ПРОИЗВЕДЕНИЯ . * . ТАКИМ ОБРАЗОМ . * , $EQUATION 2$. * . В ПРЯМОУГОЛЬНЫХ КООРДИНАТАХ КОМПОНЕНТЫ ВЕКТОРНОГО ПРОИЗВЕДЕНИЯ РАВНЫ $EQUATION 3$. * . . .

$END OF TEXT$

Fig. 19. Hard copy produced by the Cyrillic Unityper.

is recorded on magnetic tape. Inasmuch as **0 010011** is a token for ";" in the natural alphabet of the Univac, it is convenient to say that "й" is represented by ";". This is the convention adopted in Table 10, where "й" is found in column 1, and ";" in column 2. A similar convention is followed in Fig. 20,

Fig. 20 — Tokens corresponding to Cyrillic and Roman characters

For each key the upper line of the TOKEN columns is the lower-case token and the lower line is the upper-case token.

KEY No.	LOWER-CASE	UPPER-CASE	1	2	3	4	5	6	7
0	①	Ⓢ b	0	0	0	0	1	0	0
			1	1	1	0	1	0	1
1	й Ⓙ	Ⓠ	0	0	1	0	0	1	1
			1	1	0	1	0	1	1
2	ю Ⓜ	Ⓐ	1	1	0	0	1	1	1
			1	0	1	0	1	0	0
3	②	Ⓝ "	1	0	0	0	1	0	1
			1	1	0	0	0	0	1
4	я Ⓨ	Ⓩ	0	1	1	1	0	1	1
			1	1	1	1	1	0	0
5	у Ⓛ	Ⓦ	0	1	0	0	1	1	0
			1	1	1	1	0	0	1
6	ы Ⓣ	Ⓢ	1	1	1	0	1	1	0
			1	1	1	0	1	0	1
7	③	Ⓟ #	1	0	0	0	1	1	0
			1	1	1	1	1	0	1
8	ч Ⓠ	Ⓧ	1	1	0	1	0	1	1
			1	1	1	1	0	1	0
9	к Ⓑ	Ⓔ	0	0	1	0	1	0	1
			1	0	1	1	0	0	0
10	в ③	Ⓓ	1	0	0	1	0	1	0
			1	0	1	0	1	1	1
11	④	Ⓢ $	0	0	0	0	1	1	0
			1	1	0	1	1	0	1
12	c Ⓙ	Ⓒ	0	1	0	0	0	1	1
			0	0	0	1	1	1	0
13	E ⑥	Ⓡ	0	1	0	1	0	0	0
			1	1	0	1	1	0	0
14	A ①	Ⓕ	0	0	0	0	1	0	0
			0	0	1	1	0	0	1

KEY No.	LOWER-CASE	UPPER-CASE	1	2	3	4	5	6	7
15	⑤	Ⓟ %	0	0	0	1	0	0	0
			0	1	1	1	1	0	1
16	M Ⓓ	Ⓥ	1	0	1	0	1	1	1
			0	1	1	1	0	0	0
17	H Ⓔ	Ⓣ	1	0	1	1	0	1	1
			1	1	0	1	0	1	0
18	п Ⓗ	Ⓖ	1	0	1	1	0	1	1
			0	0	1	1	0	1	0
19	⑥	⊛	1	0	0	1	0	0	1
			1	1	0	1	1	1	0
20	и ⑨	Ⓑ	0	0	1	0	1	0	1
			1	1	1	0	1	0	1
21	г ④	Ⓨ	0	0	0	0	1	1	1
			0	1	1	1	0	1	1
22	p ①	Ⓗ	0	0	1	1	1	0	0
			1	0	1	1	0	1	0
23	⑦	⊙ ·	1	0	0	0	1	1	0
			1	0	1	0	0	1	0
24	T Ⓚ	Ⓝ	0	1	0	0	1	0	1
			1	1	0	1	0	0	0
25	Ш Ⓡ	Ⓤ	0	1	0	1	0	0	1
			0	1	1	0	1	0	0
26	0 Ⓖ	Ⓙ	0	0	1	1	0	1	0
			1	1	0	0	1	0	0
27	⑧	⊖ −	0	0	0	0	1	0	1
			0	0	0	0	0	1	1
28	b Ⓤ	Ⓜ	0	1	1	0	0	1	1
			1	1	0	0	1	1	1
29	щ ⊕	Ⓙ	1	1	1	0	0	1	1
			0	0	1	1	1	0	0

KEY No.	LOWER-CASE	UPPER-CASE	1	2	3	4	5	6	7
30	л Ⓒ	Ⓚ	0	0	1	0	1	1	0
			0	1	0	0	1	0	1
31	⑨	Ⓞ	1	0	0	1	1	0	0
			1	0	0	0	0	1	1
32	б ②	⊙ :	1	0	0	0	1	0	1
			0	1	1	0	0	1	0
33	3 ⑧	Ⓞ	0	0	0	1	0	1	1
			0	1	0	1	0	0	1
34	д ⑤	Ⓛ	0	0	0	1	0	0	0
			0	1	0	0	1	1	0
35	⑥	⊙ (	1	0	0	0	0	1	1
			1	0	0	1	1	1	1
36	ю Ⓧ	⊙ ;	1	1	1	1	0	1	0
			0	0	1	0	0	1	1
37	x Ⓝ	Ⓟ	1	1	0	1	0	0	0
			0	1	0	1	0	1	0
38	ж ⑦	⊙ '	1	0	0	1	0	1	0
			0	0	0	1	1	0	1
39	ц Ⓟ	⊙)	0	1	0	1	0	1	0
			0	1	0	0	0	1	1
40	⊙	⊙ .	1	0	1	0	0	1	0
			0	1	1	0	0	1	0
41	3 ⊙	⊕ +	0	1	1	1	0	0	0
			1	1	1	0	0	1	1
42	⊘	⊘	0	0	0	0	0	0	0
			0	0	0	0	0	0	1

NOTES:

1) 0 = NOTCH CUT, 1 = NO CUT

2) WHEN THE CHARACTER PRINTED ON THE HARD COPY IS DIFFERENT FROM THAT RECORDED ON MAGNETIC TAPE, THE FORMER IS SHOWN TO THE SIDE OF THE APPROPRIATE CIRCLE.

Fig. 20. Tokens corresponding to Cyrillic and Roman characters.

where the natural alphabetic correspondent of each token is encircled, and the Cyrillic representation is given to the left, whenever applicable. These conventions are useful since all units of the Univac system, other than the Unityper, were left unmodified for economic reasons, and hence interpret the tokens in terms of their natural alphabet. For example, the word *ядро**, when typed on the modified Unityper, would appear on the printed copy as indicated in row 1 of Fig. 21, and on magnetic tape as indicated in row 2.

Table 10. Computer and transliterated representations of Cyrillic characters; column 1, Cyrillic; column 2, ranked computer representation; column 3, transliterated representation.

1	2	3	1	2	3	1	2	3
А	1	A	М	D	M	Ш	R	SH
Б	2	B	Н	E	N	Щ	+	SHCH
В	3	V	О	G	O	Ъ	S	#
Г	4	G	П	H	P	Ы	T	Y
Д	5	D	Р	I	R	Ь	U	'
Е	6	E	С	)	S	Э	V	EH
Ж	7	ZH	Т	K	T	Ю	X	JU
З	8	Z	У	L	U	Я	Y	JA
И	9	I	Ф	M	F	△	△, 0[1]	△
Й	;	J	Х	N	X	.	.	.
К	B	K	Ц	P	TS	%	%	—
Л	C	L	Ч	Q	CH			

[1] See Sec. 7.6

If the recorded tape is subsequently "played back" on the console typewriter or on the high-speed printer (Fig. 2), "Y5IG" is printed, as indicated in row 3 of Fig. 21. This somewhat annoying substitution cipher effect could be eliminated simply by replacing the natural type slugs on the printers by Cyrillic ones, but for experimental purposes another alternative—transliteration—proved more practical.

1	Cyrillic	"Я"	"Д"	"Р"	"О"
2.	Tokens	0 111011	0 001000	0 011100	0 011010
3	Natural alphabet (ranked representation)	"Y"	"5"	"I"	"G"
4	Transliteration	"JA"	"D"	"R"	"O"

Fig. 21. Representations of ЯДРО*.

The transliteration system adopted in our experiments is described by column 3 of Table 10. The transformation from the ranked computer representation to the transliterated representation is readily effected by a machine program. Whenever a tape is to be printed, a short computer run suffices to convert Cyrillic material on the tape from one representation to the other. Following such a run, the tokens for *ядро** given in row 2 of Fig. 21 would be replaced by a sequence of tokens for the elements "j", "a", "d", "r", and "o" of the natural alphabet. The latter tokens cause "JADRO" to be printed, that is, they correspond to the transliterated representation given in row 4 of Fig. 21. In normal operation the user of the machine sees only the representations given in rows 1 and 4. The rest is of no direct concern to him once the equipment has been modified, and operating programs have been written and checked out.

Example 2-8. One of the characters of "москва" (stored as "DG)B31" in representation 2, Table 10) is in position 12 of register *1*. Place the corresponding character of representation 3, Table 10, in position 12 of register *2*.

Auxiliary Storage.

1	△ △ △ △ △ △ △ △ △ △ X	Character in representation 2
2	[ ]	Character in representation 3
3	△ △ △ △ △ △ △ △ △ △ D	
4	△ △ △ △ △ △ △ △ △ △ G	
5	△ △ △ △ △ △ △ △ △ △)	
6	△ △ △ △ △ △ △ △ △ △ B	
7	△ △ △ △ △ △ △ △ △ △ 3	
8	△ △ △ △ △ △ △ △ △ △ 1	
9	△ △ △ △ △ △ △ △ △ △ M	
10	△ △ △ △ △ △ △ △ △ △ O	
11	△ △ △ △ △ △ △ △ △ △ S	
12	△ △ △ △ △ △ △ △ △ △ K	
13	△ △ △ △ △ △ △ △ △ △ V	
14	△ △ △ △ △ △ △ △ △ △ A	
15	E R R O R △ S T O P △ △	

Program.

100	B	1			108	B	9		
			0					U	114
101	L	3			109	B	10		
			Q	108				U	114
102	L	4			110	B	11		
			Q	109				U	114
103	L	5			111	B	12		
			Q	110				U	114
104	L	6			112	B	13		
			Q	111				U	114
105	L	7			113	B	14		
			Q	112				0	
106	L	8			114	C	2		
			Q	113				9	
107	50	15							
			9						

Exercise 2-7. The words of a hypothetical Slavic language are all represented by strings of six characters chosen from an alphabet comprising the first four Cyrillic characters of Table 10. Write a program that will store in register *2* the image in representation 3 (Table 10) of a word stored in register *1* in representation 2.

The correspondence between Cyrillic characters and their ranked computer representation as given in Table 10 was not selected at random. A comparison of column 2 of Table 10 with the second column of Table 9 indicates that the mapping is an order-preserving one, hence the name *ranked computer representation**. The choice of correspondence was further influenced by a desire to make the recognition of the terminal letters of words an easily programmable, rapidly executed process (Giuliano, 1957*a*). The identification of terminal letters is important in identifying desinences of Russian words, a process which will be discussed in Chapters 5 and 10. All that need be said here is that the terminal character of a word may belong to one of two subsets of the alphabet: characters in the first subset may belong to desinences, those in the second subset do not. The identity of a character belonging to the first subset must be determined precisely, to determine what desinence, if any, it belongs to. No identification is necessary for a character found to belong to the second subset, namely, a *nonfactorable* character.

An obvious way of identifying characters and choosing alternative program branches accordingly is to use a series of *L* and *Q* instructions in the manner illustrated by Example 2-8. The number of *L* and *Q* pairs required is equal to the size of the first subset, which has 14 members. If no transfer of control has occurred when the last *Q* order in the series has been executed, the character being identified may be presumed to belong to the second subset, and a transfer can be made by a *U* order to a program designed to take the appropriate action. A *maximum* of 14 comparisons is therefore necessary to identify a member of the first subset, and *exactly* 14 comparisons must always be executed before a character is found to belong to the second subset. The *average* number of comparisons necessary to identify a member of the first subset may be reduced to 6 by using knowledge of the expected frequencies of characters in terminal positions, such as is given in Fig. 22. It is sufficient to arrange the *L Q* pairs so that the most frequently selected branch will be selected by the first pair, the next most frequent by the second pair, and so on. All 14 comparisons must still be made before a character is found to belong to the second subset, but, since the frequency of any member of the second subset is comparatively low, the number of comparisons per character, averaged over all characters, remains close to 6.

By properly choosing the representation of the Cyrillic alphabet, and taking advantage of the "multiplication" operation on letters described in Sec. 2.7, some reduction in the average number of comparisons executed to

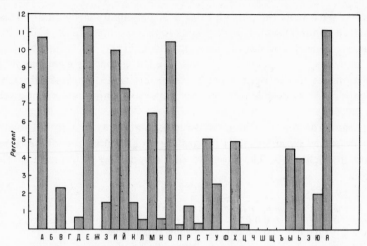

Fig. 22. Final letter distribution based on a 6000-word sample of texts.

identify a character or to assign it to the second subset is possible. The mapping between the Cyrillic alphabet and the natural Univac alphabet, already given in Table 10, is displayed in Table 11 in a layout patterned after that of Table 2. As in the latter, tokens are represented in two parts, of which one, consisting of two characters, is called the *zone* part, and the other, the

Table 11. Definition of the ranked computer representation.

Decimal digit	Excess-three part	Zone part				Number of secondary comparisons
		0 0	0 1	1 0	1 1	
0	0 0 1 1	0 0	Й′ ;	С)	Щ +	1
1	0 1 0 0	А′ 1	— A	— J	— /	0
2	0 1 0 1	Б 2	К B	Т′ K	Ъ S	1
3	0 1 1 0	В″₂ 3	Л C	У″₁ L	Ы′ T	3
4	0 1 1 1	Г 4	М′ D	Ф M	Ь″ U	2
5	1 0 0 0	Д 5	Н E	Х′ N	Э V	1
6	1 0 0 1	Е′ 6	— F	— O	— W	0
7	1 0 1 0	Ж 7	О′ G	Ц P	Ю″ X	2
8	1 0 1 1	З 8	П H	Ч Q	Я′ Y	1
9	1 1 0 0	И′ 9	Р I	Ш R	— Z	1

excess-three part. Only the part of Table 2 displaying tokens with an excess-three part in common with that of tokens for the decimal digits is displayed in Table 11. In each column of the zone part of Table 11, elements of the Cyrillic alphabet are displayed to the left, the corresponding elements of the natural alphabet to the right. Characters belonging to the first subset are shown with single or double primes. The remaining characters constitute the second subset.

The reasons for defining the ranked representation as shown in Table 11 are best described by an example illustrating an algorithm for identification based on the definition. The program of Example 2-9 is the core of this algorithm.

Example 2-9.

Partial Program for Ending Identification

1	0 0 0 0 0 0 0 0 0 0 X 0				Character determining branch selection
2	0 1 0 0 0 0 0 0 0 0 0 0				Multiplication factor
3	0		U	1 0 0	Basic branch instruction for planting
4	0 0 0 0 0 0 0 0 0 0 0 0				
5	0 0 0 0 0 0 0 0 0 0 ; 0				
6	0 0 0 0 0 0 0 0 0 0) 0				
7	0 0 0 0 0 0 0 0 0 0 + 0				

50	*L*	*1*			
			P	*2*	$(1) \times (2) \rightarrow (A)$
51	*A*	*3*			
			C	*52*	$(A) + (3) \rightarrow (52)$
52	[				
			U	]	Planted branch instruction
100	*B*	*1*			
			U	*150*	to secondary selection
101	*0*				
			U	a_1	to action ("a")
102	*B*	*1*			
			U	s_2	to secondary selection
103	*B*	*1*			
			U	s_3	to secondary selection
104	*B*	*1*			
			U	s_4	to secondary selection
105	*B*	*1*			
			U	s_5	to secondary selection
106	*0*				
			U	a_6	to action ("e")
107	*B*	*1*			
			U	s_7	to secondary selection
108	*B*	*1*			
			U	s_8	to secondary selection
109	*B*	*1*			
			U	s_9	to secondary selection
150	*L*	*5*			
			Q	a_{150}	to action ("й")
151	*0*				
			U	a_{151}	to action (member of second subset)

The character to be identified is stored in position 11 of register *1*. Assume that it is "A". When the instruction-pair in register *50* has been executed, (*A*) will be 000000000001, which is the product of 1 and "A" according to the rule given in Sec. 2.7. Executing the pair of instructions in register *51* therefore plants the pair *0, U 101* in register *52*. The planted instruction is executed next, and in turn transfers control to the instruction in register *101*. The latter instruction can be reached only if the character to be identified is "A", since "*A*" *is the only Cyrillic character mapped onto 1 by the* "*multiplication by one*" *transformation*. Register *101* therefore contains an instruction $U a_1$, where (a_1) is the first instruction of the program which should be followed if the terminal character is "A".

If the character to be identified is "й", the instruction *U 100* will be planted in register *52*, since "й" is mapped onto 0 by the "multiplication by 1" transformation. However, "o", "c", and "щ" are also mapped onto 0 by the same transformation. A secondary selection is therefore necessary. Accordingly, the instruction-pair in register *100* transfers the character to be identified, (*1*), to register *A* where it is available for comparison, and transfers control to (*150*). The *Q* order in register *150* will effect a transfer of control, since the character to be identified is assumed to be "й". If either "o", "c", or "щ" had been in register *1* originally, all the foregoing would apply, except that the *Q* order in register *150* would *not* transfer control, and the instruction $U a_{151}$ would transfer control to $(a_{1\overline{5}1})$, the first instruction of the program to be followed if the terminal letter belongs to the second subset. Since "й" is the only character mapping onto 0 which is also primed or double-primed in Table 11, only a single *L Q* pair was necessary in the secondary selection.

The number of secondary comparisons required for each of the ten possible images of characters under the "multiplication" transformations is given in the last column of Table 11. The largest number is 3, when the image is 3. At most three comparisons are therefore sufficient to recognize a character as belonging to the second subclass.

Where more than one character in a row of Table 11 belongs to the first subset, the most frequent is primed, tested by the first *L Q* pair in the secondary selection program for the row and therefore recognized after only one comparison. The double-primed characters, of which there are only four, are tested in order of decreasing frequency by succeeding *L Q* pairs. All double-primed characters are less frequent than primed characters, and two of the most frequent primed characters are identified without any secondary selection at all. The average number of comparisons per character is therefore close to one.

The average of about one comparison per character in the system of

Example 2-9 compares favorably with the average of six required under the "obvious" method. However, in addition to the one comparison (two instructions), the instructions in registers *50–52* and one instruction-pair among those in registers *100–109* must inevitably be executed for every character. The average number of *instructions* required in the system of Example 2-9 is therefore ten, while that under the other system is twelve. This comparison is somewhat less favorable. By adroit programming the average of ten can be reduced to six, but not further.

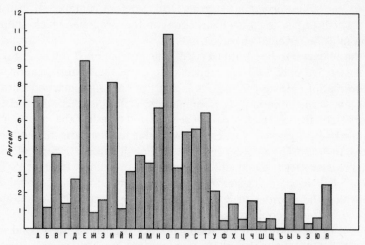

Fig. 23. Total letter distribution based on an 18,000-word sample of texts.

Relative to the Univac, therefore, the procedure of Example 2-9, or subtler variations of it, are tricks of some value but they do not produce spectacular results. On machines where, as indicated in Sec. 2.7, operations analogous to the "multiplication by 1" have been provided as explicit instructions to be used in character identification, the trick emerges as a true method of considerable general value.

Exercise 2-8. Consider the problem of fully identifying each element of the Cyrillic alphabet. Estimates of the over-all frequencies of members of this alphabet are given in Fig. 23.

(*a*) What is the average number of comparisons (and of instructions) necessary to identify a character using the straightforward *L Q* method?

(*b*) What is the smallest average number of comparisons (and of instructions) necessary to identify a character by a program analogous to that of Example 2-9, assuming that elements of the alphabet must be represented in the ranked code of Tables 10 and 11? (Hint: Ignore the primes and double primes in Table 11, and arrange the order of testing for elements in a row in accordance with the data of Fig. 23.) Is the method of Example 2-9 more advantageous relative to the *L Q* method in this case than it was for the identification of terminal letters?

(c) Assuming you are free to redefine the correspondence between the Cyrillic alphabet and those elements of the natural alphabet available in Table 11, by how much can you reduce the number of comparisons over that necessary in (b)?

2.9. Bibliographic Notes

The statement of the problem of use versus mention given by Quine in his *Mathematical Logic* (1955, pp. 23–26) is one of the best. He points out that carelessness in this matter is common in mathematical writings, chiefly because the resultant confusion has not been felt as a practical obstacle in most directions of mathematical inquiry. Problems of notation are also carefully considered by Church (1956) and by Kleene (1952). Menger (1957), who, like Quine, recognizes that the difficulties due to confusion of use and mention and to allied problems are negligible to active mathematicians, makes a strong case for their crucial importance in the teaching of mathematics. Beberman (1958) and his associates at the University of Illinois (University of Illinois Committee on School Mathematics, 1959) are attempting to introduce precise language into secondary school mathematics, in very much the same spirit in which this chapter was written.

No discussion of the theory of signs is complete without mention of Charles Sanders Peirce. The interested reader will find an account of Peirce's views readily accessible in the collection *Philosophical Writings of Peirce* edited by Buchler (1955). Susanne Langer's *Philosophy in a New Key* (1948) provides an excellent introduction to general problems in the theory of signs. Chapter 6 of Cherry's *On Human Communication* (1957) surveys contributions to the theory of signs made by several disciplines. This book is also a good introduction to a wide range of topics in communication theory and linguistics, and contains an extensive bibliography.

More about some of the mathematical techniques discussed in this chapter may be found—in ascending order of precision, thoroughness, and difficulty—in Kemeny, Snell, and Thompson (1957), Andree (1958), Kemeny, Mirkil, Snell, and Thompson (1959), and Birkhoff and MacLane (1950).

The application of mathematical and other models to languages is described by Kulagina (1958), Chomsky (1957, 1956), Oettinger (1957*b*), Belevitch (1956), Herdan (1956), Fairthorne (1958), Apostel, Mandelbrot, and Morf (1957), and by Yule (1944). The critical bibliography of Guiraud (1954) is a good source of references to works on statistical linguistics.

CHAPTER 3

FLOW CHARTS AND AUTOMATIC CODING

3.1. Flow Charts

For all but the very simplest problems it can be quite difficult to jump directly from the statement of a problem in English or in mathematical notation to an algorithm for its solution in the form of a complete program. A program written directly is at best likely to be wasteful of storage space and of operating time, but there are much greater dangers. Only rarely is a problem stated initially with sufficient precision to enable the programmer to foresee all contingencies for which his program must provide. It is therefore not unusual for the programmer to discover that he has, so to speak, painted himself into a corner, and that his only way out is to scrap most of what he has written and to start afresh. When writing a very long program, it is not uncommon to arrive near the end of the program having forgotten all about the detailed properties of the beginning, thereby risking the introduction of fatal inconsistencies. Similarly, it is extremely difficult to master the details of a program written by someone else, or even of a program one has written oneself but laid aside for a few weeks. The reader who has not looked at the short program of Example 1-3 for some time may find it revealing to test his comprehension of it now.

The obscurity of programs is due in some measure to their expression in notations, often called *machine languages*, designed by engineers primarily for the convenience of machines. Also, much of the detail of a program is devoted to bookkeeping operations related to packing and unpacking, order modification, and the like. These operations are rarely mentioned explicitly in the original statement of a problem in a *problem language*, which, be it English or mathematical notation, is organized primarily for the convenience of persons.

Fundamentally, however, the problem arises from a basic, pervasive limitation of speech or writing, expressed in another realm by Klee (1948) in the following words:

It is difficult enough, oneself, to survey this whole, whether nature or art, but still more difficult to help another to such a comprehensive view. This is due to the consecutive nature of the only methods available to us for conveying a clear three-dimensional concept of an image in space, and results from deficiences of a temporal nature in the spoken word.

A variety of techniques based on the use of *flow charts* has been developed

84

to facilitate the transition from problem language to machine language, and to enable the concise and lucid expression of the salient features of program organization. The *two-dimensional* character of flow charts enables the easy description of complicated processes that are very difficult to describe with essentially *linear* strings of words or of machine instructions.

The very properties of flow charts that make them such useful tools are unfortunately not easily described verbally, because a formal verbal description requires precisely the kind of intricate prose that flow charts are intended to replace. However, a working knowledge of the art of interpreting and drawing charts can be acquired quite readily by studying a few examples and through some practice.

Figure 24a shows a very simple flow chart. After entering the chart at the circle labeled "start", and following the arrow, our eyes are led into a box containing the statement of a problem. Since the only arrow leading out of the box points to a circle labeled "stop", we may infer that the statement in the box purports to express the whole problem. This problem has already been stated and solved in Example 1-3, but we shall solve it again to illustrate how flow charts might have been used to derive the program of Fig. 6. Comparing the statement of the problem in Example 1-3 with that in Fig. 24a, we observe that the former is buried in a text, the latter stands out, illustrating one important but not unexpected property of flow charts.

The charting of more complex problems might begin with a single box stating no more than "Solve this problem!" The next step would be to write a more detailed chart with several boxes each containing a prose statement of an intermediate problem to be solved or of some operation that must be performed on the way to the ultimate goal. Directed lines connect these boxes to indicate the sequence to be followed. The eye can then encompass at a glance a long and tortuous path that would be difficult to describe in words. A chart of this kind has considerable heuristic value, but is highly informal, in the nature of scratch work. The ultimate goal is a chart which is not a statement of a problem, but a detailed description of the steps necessary to solve it, and of the precise order in which these steps are to be taken. Such a chart is clearly akin to a program.

It is convenient to regard the process leading from the initial proposal of a problem to the completion of a program for solving it as consisting of three stages, *analysis*, *programming*, and *coding*. Often, when a problem is first stated, its statement may not be as explicit as in Fig. 24a, especially if it is only a part of some larger problem. For instance, the exact number of quantities to be summed may not yet be specified, and it may be difficult to allocate these quantities to definite storage registers until over-all storage requirements have been estimated. Furthermore, the choice of the machine on which the problem is to be solved may itself depend on the expected complexity and length of the program, which can often be estimated only after a sufficiently detailed analysis of the problem has been made. It is therefore desirable, in the initial stages of analysis, to cast the statement of a problem

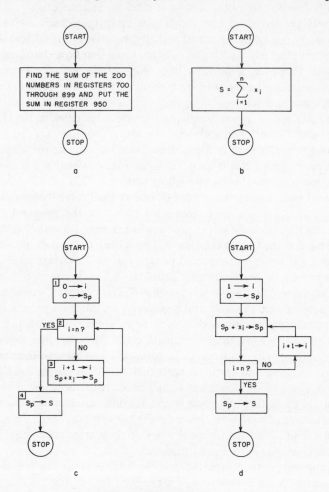

Fig. 24. Flow charts.

into a precise, explicit form as independent of specific machine characteristics and as free from unnecessary restrictions as possible.

When the analysis is complete, the outline of an algorithm may be sketched in gradually greater and greater detail. This programming phase is conveniently done with the aid of flow charts. While greater attention is paid to the properties of machines at this stage than in analysis, specific machine instructions are rarely introduced. Finally, in the coding phase, the algorithm represented by sufficiently detailed flow charts is translated into a machine program. Roughly speaking, analysis can be performed by persons thoroughly familiar with the problem but less familiar with machines, programming can be done by persons with a thorough knowledge of the properties of machines and a capacity to understand the problem, while coding requires only a knowledge of the instruction repertoire of a specific machine, and the ability to read flow charts.

In Fig. 24b, the problem is stated in somewhat greater generality as that of summing n numbers $x_1, x_2, \ldots, x_n$, and is expressed in the more compact mathematical notation. The expression "$\sum_{i=1}^{n} x_i$" is merely a convenient abbreviation for "$x_1 + x_2 + \cdots + x_{n-1} + x_n$" and "$S$" is defined as a still more compact synonym of "$\sum_{i=1}^{n} x_i$". To a trained mathematician, the chart of Fig. 24b is explicit because its interpretation is familiar. He knows that he is to find the sum of several numbers, but that in a specific situation he must first find out exactly how many, and also what they are. Once he has this information, he recalls that one algorithm he knows is an algorithm for finding the sum of two numbers, and that to use it he must first sum two of the given numbers, add a third number to the partial sum, then a fourth to the partial sum of the first three, and so on, until he has calculated the full sum required.

An automatic machine obviously lacks similar interpretive powers; it can follow practically the same steps, but these must be explicitly spelled out, as in Fig. 24c. Consider the loop comprising boxes 2 and 3. The expression "$S_p + x_i \rightarrow S_p$" in box 3 may be paraphrased as "Evaluate the sum of the ith number and the partial sum S_p, and let this sum become the new partial sum S_p"; it prescribes the formation of a new partial sum. Following the directed line leaving box 3, we are led to box 2. There the expression "$i = n$?", paraphrased as "Is i equal to n? If so, follow the path labeled 'yes', if not, that labeled 'no' ", prescribes how the decision to terminate the process is to be made.

Initially, the *index i*, which specifies which one of a series of n numbers is to be added to the partial sum at a given stage, is set equal to 0 as indicated by the expression "$0 \rightarrow i$" in box 1. The expression "$0 \rightarrow S_p$" indicates that the partial sum S_p is also to be 0 before summation begins.

The nature of the summation algorithm represented by the flow chart of Fig. 24c may be understood by following a few steps in the summation process. When we leave box 1, $i = S_p = 0$. Assuming n to be greater than 0, $i \neq n$

when we first enter box 2. The exit path labeled "no" is therefore chosen. Because the expression "$i + 1 \rightarrow i$" precedes "$S_p + x_i \rightarrow S_p$" in box 3, the latter must be interpreted as "$S_p + x_1 \rightarrow S_p$". The resulting new partial sum is therefore $0 + x_1 = x_1$. After returning to box 3 via box 2, "$S_p + x_i \rightarrow S_p$" must be interpreted as "$S_p + x_2 \rightarrow S_p$". The resulting new partial sum is now $x_1 + x_2$. Eventually, "$S_p + x_i \rightarrow S_p$" will be interpreted as "$S_p + x_n \rightarrow S_p$". Upon return to box 2, i will be found to equal n, and the path to box 4 will be chosen for the first time. The expression "$S_p \rightarrow S$" indicates that the partial sum S_p is now $\sum\limits_{i=1}^{n} x_i = S$.

Figure 24d is an equivalent flow chart for the summation process; the reader may find it instructive to compare this chart with that of Fig. 24c.

The parallel between the notation used in the flow charts of Fig. 24 and that introduced in Sec. 1.4 facilitates the transition from flow chart to program. The flow chart of Fig. 25a is an intermediary between that of Fig. 24c and the program of Fig. 6. (How would a program derived from Fig. 24d differ from that of Fig. 6?) It is indicated that the numbers $x_1, \ldots, x_{200}$ are to be stored in registers 700 to 899, that the partial sum S_p is in register 103, and that the final sum S is to be put in register 950. The significance of the notation "$\textcircled{T}$" in Fig. 25 is similar to that of "(m)" in Sec. 1.4. The reader should find it instructive to follow a few steps in the algorithm described by Fig. 25a as was done for that of Fig. 24c in the preceding paragraph.

Figure 25b shows how the summation algorithm may be embedded as a component in a more complex algorithm, where it is used to perform several different summations. New elements, called "connectors", appear in Fig. 25b. The *fixed* connector is simply a device to simplify drawing. When boxes are widely scattered on a page, or appear on different pages, drawing connecting lines between them becomes messy or impossible. Connectors, represented by circles enclosing characters, can be used instead of lines. For example, the exit connector α_1 of box 1 simply indicates that the path leads to the entry connector α_1 of box 4.

The *variable* connector is a much more significant device. Its function as an exit connector is to indicate that the path to be followed is not fixed, but depends in a definite way on preceding *connector setting* operations. The use of variable connectors may be illustrated by following through the chart of Fig. 25b, beginning with box 1. The expression "$200 \rightarrow n$" defines the number n of summands to be 200. The address r_0 of an initial register is defined by the expression "$699 \rightarrow r_0$" to be 699. The expression "$\alpha_2 \rightarrow \Phi$" *sets* the variable connector Φ (exit of box 7) to α_2. Following the path defined by the fixed connectors α_1, we next enter box 4. The reader may verify that under the conditions just outlined the flow chart comprising boxes 4 through 7 defines an algorithm equivalent to that defined in Fig. 25a. Since Φ has been set to α_2, the path out of box 7 leads to box 2.

The function of box 2 is analogous to that of box 1. The parameters n and r_0 are so defined in box 2 as to adapt the summation algorithm to the

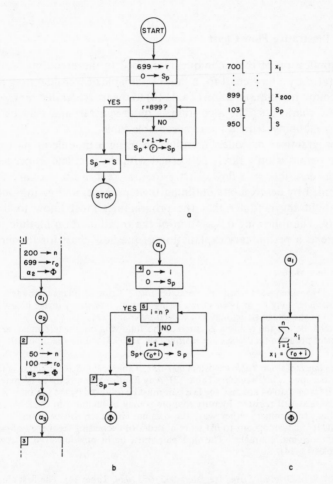

Fig. 25. Flow charts.

summation of 50 numbers stored in registers *101* through *150*. Setting Φ to α_3 establishes a path from box 7 to box 3, following the summation. Boxes 1, 2, and 3 may be regarded as portions of a master program, in which a single summation algorithm serves as a "subroutine" or "module" to be used repeatedly to perform different summations. Modules of this kind are commonly coded once and for all, and kept in a library for use in a variety of programs; when this is the case, it is sufficient to indicate the use of the module in a master flow chart by some abbreviated chart like that of Fig. 25c.

3.2. An Illustrative Flow Chart

The application of the techniques described in the preceding section will be illustrated by a flow chart for a module capable of transliterating from the ranked computer representation (Table 10) to the transliterated representation of Cyrillic characters. The algorithm is quite general, and may be used to effect any mapping defined by an acceptable table.

The programmer or coder interested in using a module as part of some master program is only rarely concerned with the detailed properties of the module as described in a flow chart. Modules kept in libraries are therefore accompanied by descriptions outlining their purpose, and giving only those details about the modules that the programmer must know to use them effectively. The following description of the transliteration module will also serve here as a preliminary explanation of the flow chart for the module.

Transliteration Module

Purpose. Transliterates a string of characters into a string of image characters. A single original character may be mapped into a string of $n \leq 4$ arbitrary image characters. The total length of the image string may not exceed 60 characters.

Parameters. Before the module is entered, the following parameters must be specified by the master program (addresses are *relative* (see Sec. 3.3) within the module):

(1) *Mapping Definition Table* (registers *000–031*, Tables 13 and 14). Each original/image pair occupies a half register (Table 12). Any number of pairs up to 64 may be used, but if fewer than 64 are used the last pair must be followed by the sentinel "ŻZZZZZ" in the next half register. Identity mappings may be included in the table or ignored by the programmer; either way, the original character will be transliterated into itself. It is advantageous to list pairs in order of decreasing frequency of occurrence of the original character. The first pair must be in positions 1–6 of register *000* (Tables 13, 14).

(2) *Location of Original String* (registers *032, 033, 034*, Table 14). The first character of the original string must be in column 1 of the register whose *absolute* (see Sec. 3.3) address must be put in register *032* (Table 14) by the master program. The number of the last of the set of consecutive registers storing the original string (relative to the first which is designated "0") must be put in register *033* (Table 14), and the number of the position of the last character of the string in the last register must be put in register *034* (Table 14). The number in register *033* is called the *original major coordinate*, that in register *034* the *original minor coordinate* of the last character of the original string.

Table 12. Structure of mapping definition table; "O" stands for an original character; "x x x x" stands for images, for example, "S H C H", "△ △ J A", "△ △ △ E"; "n" stands for the number of characters ($1 \leq n \leq 4$) in the image.

Register	Position											
	1	2	3	4	5	6	7	8	9	10	11	12

000 | n x x x x O | n x x x x O |

$k \leq 31$ | n x x x x O | Z Z Z Z Z Z |

or

| Z Z Z Z Z Z | Arbitrary |

Table 13. Mapping definition table, ranked computer representation to transliterated representation.

000	1 △ △ △ △ △	1 △ △ △ △ 0
001	1 △ △ △ O G	1 △ △ △ E 6
002	1 △ △ △ I 9	1 △ △ △ N E
003	1 △ △ △ A 1	1 △ △ △ T K
004	1 △ △ △ R I	1 △ △ △ S)
005	1 △ △ △ L C	1 △ △ △ V 3
006	1 △ △ △ M D	1 △ △ △ P H
007	1 △ △ △ D 5	1 △ △ △ K B
008	1 △ △ △ U L	2 △ △ J A Y
009	1 △ △ △ Y T	2 △ △ C H Q
010	1 △ △ △ Z 8	1 △ △ △ ' U
011	1 △ △ △ J ;	1 △ △ △ G 4
012	1 △ △ △ X N	1 △ △ △ B 2
013	2 △ △ J U X	2 △ △ Z H 7
014	4 S H C H +	2 △ △ S H R
015	2 △ △ E H V	2 △ △ T S P
016	1 △ △ △ F M	1 △ △ △ # S
017	1 △ △ △ : $	1 △ △ △ - %
018	Z Z Z Z Z Z	Z Z Z Z Z Z

Table 14. Register allocation in the transliteration module.

f_0	000	[ ]	$a_0 - a_{31}$ Mapping Definition Table
	.		
	.		
	.		
	031	[ ]	
	032	[ ]	b_0 address of the first word of the original string
	033	[ ]	b_1 original major coordinate
	034	[ ]	b_2 original minor coordinate
f_1	035	[△△△△△△ △△△△△△]	c_0
	036	[.]	c_1
	037	[.]	c_2 output string
	038	[.]	c_3
	039	[△△△△△△ △△△△△△]	c_4
	040	[0 0 0 0 0 0 0 0 0 0 0 0]	$i = 0(1)-$ running original major coordinate
	041	[0 0 0 0 0 0 0 0 0 1 1]	$j = 11(1)0 -$ running original minor coordinate
	042	[0 0 0 0 0 0 0 0 0 0 0 0]	$k = 0(1)4 -$ running image major coordinate
	043	[0 0 0 0 0 0 0 0 0 1 1]	$l = 11(1)0 -$ running image minor coordinate (adjusted to 1(1)12 before exit)
	044	[0 0 0 0 0 0 0 0 0 0 0 0]	$m = 0(1)31 -$ running major transliteration table coordinate
	045	[0 0 0 0 0 0 0 0 0 0 1]	$n = 1(1)0 -$ running minor transliteration table coordinate
	046	[ ]	x_1 ith word of original string
	047	[ ]	x_2 current original character, current image
	048	[ ]	x_3 mth word of transliteration table
	049	[ ]	x_4 current transliteration table entry
	050	[ ]	x_5 number of characters left in image string
	051	[ ]	x_6 kth word of image string
	052	[△△△△△△ Z Z Z Z Z]	d_0 end of table sentinel
	053	[0 0 0 0 0 0 0 1 0 0 0 0]	d_1
	203	△△△△△△ △△△△△△	e_1

Example.

Original string in registers:	530	x x x x x x x x x x x x	To *032: 530*
			To *033:* 1
	531	x x x x	To *034:* 4

Original string in register:	986	x x x x x x x x x x x x	To *032: 986*
			To *033:* 0
			To *034:* 12

(3) *Exit*. A control transfer (U) order for returning to the master program should be planted by the master program in register *202* of the module prior to entry (setting of the variable exit connector Φ, Fig. 26).

(4) *Entrance*. Register *070* of the module contains the first instruction of the module (entry connector α_1, Fig. 26).

Output. The transliterated string will be found in registers *035–039*, with the first character in position 1 of register *035*. The image major coordinate will be in register *042*, the image minor coordinate in register *043*.

A flow chart for a module answering to the preceding description is shown in Fig. 26. It may be regarded as the final chart of a series, the one in closest correspondence with machine language.

The variables, parameters, and constants used in the flow chart are defined in Table 14. Since many of the operations to be performed by the program are operations on individual characters, it is useful to define variables standing for single characters. Thus, if "x_1" stands for the whole 12-character string stored in a register, "x_1^{12}" is defined to stand for the character in position 12 only, and "x_1^j" ranges over the set of variables $\{$"x_1^1", ..., "x_1^{12}"$\}$. It is evident therefore that the program corresponding to an expression such as "$x_1^j \rightarrow x_2^{12}$" must include shift and extract instructions in a machine where individual characters are not addressable directly. Similarly, the variable "$x_4^{7,\ldots,12}$" stands for a string occupying positions 7 through 12 of a register, and if the scope of "n" is the set $\{$"0", "1"$\}$, the scope of "$x_3^{(7,\ldots,12)-6n}$" is defined as the set $\{$"$x_3^{1,\ldots,6}$", "$x_3^{7,\ldots,12}$"$\}$. For convenience in programming, the indices "l" and "j" stand for numerals associated with character positions in the manner specified in Fig. 26.

The variables "A", "F", and "L" are homographic with the names of the Univac registers A, F, and L. They are used in expressions such as "$x_2 \rightarrow A$" when it is known that A will in fact be stored in register A by the actual program. The flow chart of Fig. 26 is thereby somewhat specialized with regard to the properties of the Univac. In more general terms, the question in box 8 could obviously have been expressed by "$x_4^{12} = x_2$?", but the actual form of the expression suggests the necessary program instructions. To be quite rigorous, an expression such as "$x_1^j \rightarrow x_2^{12}$" (box 3) should be accompanied by the expressions "$\triangle \rightarrow x_2^1$", ..., "$\triangle \rightarrow x_2^{11}$" to specify the characters of x_2 completely, but the latter need not be given explicitly if "$x_1^j \rightarrow x_2^{12}$" is interpreted as giving them implicitly. Likewise, the superscript is omitted from "L" in "$x_4^{12} \rightarrow L$" (box 8), subject to the interpretation that it is the set "$x_4^{12} \rightarrow L^{12}$", "$\triangle \rightarrow L^1$", ..., "$\triangle \rightarrow L^{11}$" which is referred to. It is according to these conventions that the sequence of expressions "$x_1^j \rightarrow x_2^{12}$" (box 3), "$x_4^{12} \rightarrow L$", "$x_2 \rightarrow A$", "$A = L$?" (box 8) must be interpreted. It is true of practical flow charting, as it is of applied mathematics, that, while rigorous understanding of principles is essential, informality must be tolerated in practice if concrete results are to be achieved in a reasonable time span.

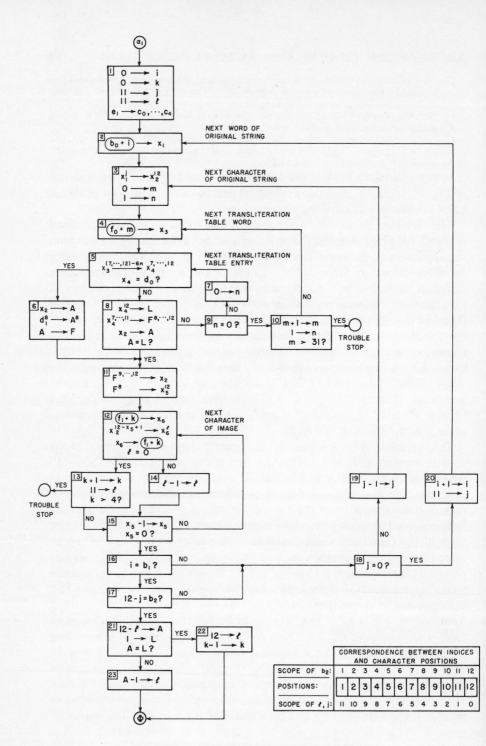

Fig. 26. Flow chart for the transliteration module.

With these remarks in mind, and with reference to the verbal description of the function of the module, the reader should have no excessive difficulty in analyzing the process described by the flow chart of Fig. 26. In so doing, it is best to work inside out from the innermost loop, and backward, by sketching a less detailed chart from which that of Fig. 26 can be derived. For instance, the loop between the entrance to box 4 and the exits of boxes 6 and 8 represents the process of transliterating a single character. From the entrance to box 11 to the exit of box 15 the image is packed in the output registers. Boxes 16 and 17 test for the end of the original string. If the string is not finished, a machine word boundary may have been reached, and is tested for in box 18, from which two alternative paths lead to the extraction of the next original character. The initial values of certain variables are set in box 1, and boxes 21 through 23 restore the natural correspondence between the scope of "*l*" and character position labels.

Experienced programmers may note that a program coded from the flow chart of Fig. 26 is likely to be quite inefficient in view of the highly frequent repetition of the shift and extract instructions required in the innermost loops. For a fixed mapping definition table, more efficient "tree" techniques may be used. An example of the application of such techniques is given in Sec. 5.5.

3.3. Program Modules and Relative Addressing

The relation between program modules and primitive instructions is similar to that between words and letters. Modules, which may be viewed as reducible over an alphabet of instructions as words are reducible over an alphabet of letters, also may fruitfully be considered as irreducible over an alphabet of modules. The hierarchy of machine control actually is based on the elementary gate opening and closing operations described in Fig. 1. The process governed by a single machine instruction is compounded of these elementary operations, and appears irreducible to a programmer only because the sequence of elementary operations is implicit in the physical structure of the machine, and can usually be altered only with pliers and soldering iron. Instructions may, however, be freely combined into programs. Under appropriate circumstances the programmer can choose to regard a group of instructions as a module, which in turn may be freely combined with other modules. The welding of modules into higher-order units, which combine freely among themselves, is obviously a further possibility.

At what level in the hierarchy the irreducibility of an alphabet is to be consummated by an inalterable physical linkage of tokens for its elements is a matter of choice. The range of applications for which a machine is intended is influential in determining this choice, as are a variety of technological and economic factors. The level of complexity of Univac instructions, and their number, are fairly typical of the repertoires of most of the contemporary general-purpose information-processing machines.

Machines are being designed in which more elementary instructions can be combined into instructions of the typical degree of complexity by a process called "microprogramming", but in most machines the instruction repertoire is fixed. Too few instructions of too simple a kind in the repertoire necessarily lead to programs longer in terms of number of instructions and more tedious to write than those based on a larger repertoire of more complex instructions. However, the larger the repertoire, the more difficult it becomes to remember and to use efficiently, and the greater the investment in physical apparatus. Machines with very few but very complex instructions have been built economically, but in the limit one such machine can be used only for a single process. This dilemma is yet another instance of the mutual antagonism between different criteria of symbolic economy described in Sec. 2.6.

The compromise which seems to be evolving, so far as general-purpose machines are concerned, is based on the use of repertoires of 20 to 100 or more instructions of moderate complexity, coupled with facilities for using modules, once they have been written, in a variety of applications.

The flow chart of Fig. 25b describes a module that may be freely combined with the other modules in the manner indicated. The program of Fig. 6 is not, however, a proper module program, being specialized to the summation of precisely 200 numbers stored in definite registers. It is not difficult to modify this program into one capable of serving as a module. The master program need merely plant different instructions in registers 100 and 101 to achieve the equivalent of specifying the parameters r_0 and n as shown in Fig. 25b. The equivalent of setting Φ is also easily accomplished by having the master program plant a control transfer (U) instruction in register 007 to replace the stop instruction. The use of two nonadjacent sets of registers $(000-007, 100-103)$ is unfortunate, and a program intended to be a module is best organized so as to fit into a set of consecutive registers.

More serious is the reference within the program to definite addresses, which precludes storing the program anywhere but in the registers indicated in Fig. 6. For this reason, it is common to write such programs using *relative* addresses, in other words, addresses referred to the address of the first register occupied by the routine. Instead of using a numerical address for the first register, a parameter R_0 ranging over permissible numerical addresses is defined. With respect to R_0 taken as the *origin*, the first register is assigned the relative address 0, the next the relative address 1, and so on. To distinguish relative addresses from ordinary *absolute* addresses, we shall write them as "*R0*", "*R1*", and so on, although this notation is superfluous in contexts where it is clear what kind of address is being used, as, for example, in Table 14.

The program of Fig. 6 is reproduced in Fig. 27 in proper module form. We note that the square-bracket notation, introduced informally in Chapter 1, serves to distinguish registers used to store variables or parameters from those used to store constants. Within the brackets, the first significant

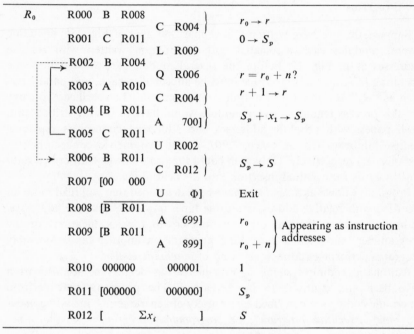

R_0	R000	B	R008			$r_0 \to r$
				C	R004	
	R001	C	R011			$0 \to S_p$
				L	R009	
	R002	B	R004			
				Q	R006	$r = r_0 + n$?
	R003	A	R010			$r + 1 \to r$
				C	R004	
	R004	[B	R011			$S_p + x_1 \to S_p$
				A	700]}	
	R005	C	R011			
				U	R002	
	R006	B	R011			$S_p \to S$
				C	R012	
	R007	[00	000			
				U	$\overline{\Phi}$]	Exit
	R008	[B	R011			r_0
				A	699]	} Appearing as instruction
	R009	[B	R011			$r_0 + n$ } addresses
				A	899]	
	R010	000000		000001		1
	R011	[000000		000000]		S_p
	R012	[	Σx_i	]		S

Fig. 27. Summation module.

specification of each variable is displayed. The parameters are shown specified as necessary to make the module of Fig. 27 equivalent to the program of Fig. 6. The interpretation of "$S_p \to S$" as "*B R011 C R012*" is somewhat absurd in this context, since executing the pair of instructions will merely copy (*R011*) into register *R012* before returning control to the master program. Actually, the operation $S_p \to S$ more properly belongs in the master routine, since the only point in performing it would be to make the sum available in a register where the master routine requires it. This was the assumption behind the original requirement that the sum be transferred to register *950*. The expression "$S_p \to S$" could reasonably be left in the module if S were associated with a register address to be specified by the master program prior to entry into the module. In this case, the instruction in register *R006* would have to be planted by the master program. Methods for replacing relative addresses by absolute addresses and for combining modules into complete programs are discussed in the next section.

3.4. Automatic Coding; The Description of Problems, Algorithms, and Programs

Suppose that we have written a flow chart like that of Fig. 25*b*, including several modules each associated with a program written with relative addresses as in Fig. 27. Before the several modules are welded into one operating program, storage space must be allocated to each, and parameters such as "R_0" and input and output connectors must be specified. To carry out this process manually, every module would have to be transcribed onto fresh paper, with absolute addresses and addressed connectors replacing relative addresses (for instance, "*R012*") and parametric connectors (for instance, α_1) respectively. Only then could the programs be transcribed onto tape or cards for eventual insertion into the central memory.

Programs known as *assemblers* have been developed that will read modules recorded with relative addresses on one tape, introduce absolute addresses, addressed connectors, and other parameters to the specifications of the programmer, and record a working program on another tape. Assembly programs perform a rudimentary kind of *automatic coding*.

Automatic coding systems of ever-increasing degrees of sophistication have been and continue to be developed. The programs that perform automatic coding are described by a variety of generic terms, including *meta-program**, *executive program**, or *programming program** (Russian *программирующая программа**).

As the stock of modules grows, they may be accumulated on a library tape. Programmers may find that new algorithms can be partially or wholly synthesized in terms of complex operations that modules available in the library can perform. For example, a programmer needing to sum *n* numbers in the course of some process may recall that the module of Fig. 25*b* is in his library, and that he can adapt it to his purpose by merely specifying the parameters *n* and r_0. An expression such as "$\sum_{i=1}^{n} x_i$; $n = 30$, $r_0 = 100$" may then be used in a "program" as a *pseudo-instruction* or *pseudo-code* equivalent to the set of "real" machine instructions of Fig. 27. By associating an appropriate pseudo-instruction with every library module, "programs" consisting entirely of pseudo-instructions may be written, and it remains only to replace each pseudo-instruction by a real program before executing it. Programs consisting of pseudo-instructions are sometimes called *pseudo-programs**, or *metaprograms**, while those consisting of conventional instructions are called *object programs**.

Programming programs that convert metaprograms into object programs are of two principal types, *compilers* and *interpretive programs*. A compiler reads pseudo-instructions from one tape, extracts the appropriate modules from the library tape, or manufactures them in some simple instances, specifies parameters, and compiles on an output tape an object program ready for operation. An interpretive program mimics the operation of the machine control unit by "executing" each pseudo-instruction in turn, just as the control unit does with ordinary instructions. When the interpretive program recognizes a particular pseudo-instruction, it transfers control to an appropriate module stored in memory. The ordinary instructions in the module are executed as usual, and the last one restores control to the interpretive program, which then turns to the next pseudo-instruction. Compilers normally are also designed to accept new modules introduced *ad hoc* by the programmer, when no adequate ones are already in the library.

While compilers or interpretive programs do not eliminate the need for flow charting altogether, they do reduce the detail in which flow charts must be written. For example, given a compiler, the chart of Fig. 25c would be sufficient to describe the summation problem. Detailed manual program writing and assembly are eliminated to a large extent, and with them a fertile source of mistakes. When modules stored in a library may be deemed free of mistakes, the tedious business of tracing errors in a program, or "debugging", which otherwise occupies a considerable portion of the time between the formulation of a problem and its solution by a machine, is reduced to a minimum. If library modules are not free of mistakes, or not well described, the library is worse than useless, for the time required to find a mistake in someone else's module may well exceed the time required to write and to debug one's own module.

The prevalent automatic coding techniques do not eliminate the necessity of analysis and of programming, although much tedious coding can be avoided. Ideally, a system of automatic programming should perform the whole process of converting the statement of a problem in problem language to an operating program without human intervention. Present techniques fall far short of this ideal, but nevertheless perform very valuable time- and labor-saving functions. It is only a small step for the imagination to conceive of more powerful programming programs capable of accepting statements in problem language as pseudo-instructions, of analyzing them to determine the combination of modules necessary to perform a process implied by the statement, and then assembling the modules into a working program. Given such programming programs, machine users could be far less concerned about the myriad details which have preoccupied us throughout the first two chapters.

Systems with pseudo-instructions expressed as English phrases have been devised, but require close adherence to rigid rules of morphology and syntax: a misspelled word or irregular word order is sufficient to cause the pseudo-instruction to be rejected by the compiler as unrecognizable or else to be accepted and to introduce a mistake. Moreover, the repertoire of pseudo-instructions is limited. Programming programs for mathematics have been designed that accept statements of problems in a notation much closer to the conventional mathematical notation than any machine instructions. More powerful programming programs will continue to be developed, and increasingly satisfactory approximations to a variety of problem languages can be expected to emerge.

The identification in a flow chart of process segments that are likely candidates for distinction as modules is an important step in experimental work in new areas of machine applications. The saving of labor that ensues from the use of modules has already been mentioned, but equally important to research is the fact that fundamental operations may emerge as recurrent patterns in many flow charts, and suggest themselves not only as modules, but eventually as ordinary instructions to be wired into a machine especially designed for a given application. As experience accumulates, modules are readily modified until they assume satisfactory configurations, an experimental process much more difficult and costly to realize by altering hardware.

The widespread adoption of *pseudo-code** or *pseudo-instruction** to designate components of metaprograms in opposition to the "real" instructions of object programs is unfortunate in that it suggests the same kind of absurd distinction that *real number** and *imaginary number** suggest to beginners in mathematics. There is nothing false, seeming, or professed but not real about a pseudo-instruction. Rather, as suggested by the terms *metaprogram** and *object program**, there is a hierarchical relation between metaprograms and object programs somewhat like that between flow charts showing increasing amounts of detail, or between any reducible element and its components. Although *metaprogram** and *object program** parallel the logicians' *metalanguage** and *object language**, the relation between metaprogram and object program does not parallel that between metalanguage and object language; hence this terminology also is not as felicitous as one might wish.

It is useful also to distinguish among descriptions of problems, algorithms, and programs. The expression "find the value of the polynomial $3x^3 + 5x^2 + x + 7$ if $x = 2$" states a problem. It says nothing about how the problem should be solved, as, for example, by adding 2 to 7, squaring 2 and multiplying the result by 5, and adding the product to the sum of 2 and 7, and so on. The expression "cube 2, multiply the result by 3; add to this the product of 5 and 2 squared, and so on" also describes an algorithm for solving the problem, different, incidentally, from that mentioned in the preceding sentence. Either algorithm may be realized by a variety of programs, depending on the machine to be used and the taste of the coder.

The distinction is useful, because ignoring it is common and leads to trouble. We are used to thinking of expressions like "$3x^3 + 5x^2 + x + 7$" as representations of algorithms: "Why, just cube x and multiply it by 3, then add . . . ". This is dangerous. The equivalent expression "$(((3x + 5)x + 1)x + 7)$" also suggests an algorithm: "multiply x by 3, add 5, multiply the sum by x, add 1, . . . ". The latter algorithm gives the value of the polynomial for three multiplications and three additions, while the former requires five multiplications ($3 \cdot x \cdot x \cdot x + 2 \cdot x \cdot x + x + 7$) and three additions. Such reductions in the number of operations can be very significant in computing practice, especially for higher-order polynomials. Incidentally, a proof that the nesting technique requiring only n multiplications for a polynomial of degree n is "best" in the sense that no other technique uses fewer multiplications and that any technique using n multiplications reduces to the nesting technique has not been found for $n > 4$ (Ostrowski, 1954).

It is a pity that the distinction need be made. For if one expression could be used homographically to describe problem, algorithm, and program, then analysis, programming, and coding could be reduced to a single operation. This may be regarded as an ideal toward which research in automatic programming is striving. This ideal is difficult to attain in general. A simple illustration will show that it is not altogether impossible to attain, at least in favorable circumstances.

Let us define the expressions "Axy" and "Mxy" as synonymous with "$x+y$" and "$x \cdot y$", respectively. Then, "$x+(y+z)$" is synonymous with "$AxAyz$" and "$(x+y)+z$" is synonymous with "$AzAxy$". The expression "$x+y+z$", which cannot properly describe an algorithm since it leaves in doubt which pair should be added first, has no unique synonym in the new notation. This notation, known variously as the parenthesis-free, Polish, or Łukasiewicz (1951) notation, has a number of remarkable properties, one of which is that it can be made to describe both certain problems and the algorithms for solving them. The nested polynomial described earlier is equivalent to "$A7MxA1MxA5M3x$" in this notation. Moreover, the design of machines for which such expressions can serve as programs is surprisingly simple (Burks, Warren, and Wright, 1954; Miehle, 1957; Oettinger, 1956b).

While no such remarkable tool is yet at hand for describing the problems of automatic language translation, or the algorithms and the programs for solving them, automatic programming systems are being developed (Yngve, 1958; Berson, 1958; Razumovskij, 1957; Giuliano, 1958, 1959b), which, when available, will reduce drudgery not only in work on automatic translation, but also in general linguistic studies and in allied areas where automatic information-processing techniques are finding growing applications. The choice of an apt notation has often made the difference between progress and stagnation in research, as illustrated, for example, by the Leibniz and Newton notations for the calculus. By stimulating the invention and facilitating the use of well-designed, felicitous notations, the development of automatic programming may make contributions to the growth of knowledge far transcending in importance its by no means negligible labor-saving effects.

3.5. Bibliographic Notes

The theory and practice of automatic coding are developing at a rapidly increasing rate, and it is too early for any definitive works on these subjects to have appeared. The journal *Communications of the Association for Computing Machinery* is a good source of up-to-date references. Only a few typical illustrations of the avenues being explored can be given here.

A joint committee of the Association for Computing Machinery and of the Gesellschaft für Angewandte Mathematik und Mechanik has set itself the task of developing a language for expressing computational algorithms which would: (1) be as close as possible to standard mathematical notation and be readable with little further explanation, (2) be usable for the description of computing processes in publications, and (3) be translatable into machine programs by automatic means. A first report on the synthesis of such a language has been given by Perlis and Samelson (1959).

Automatic coding techniques especially designed to aid with problems of linguistic analysis and of automatic language translation are described by Yngve (1958), Giuliano (1958, 1959b), Berson (1958), and Razumovskij (1957).

Research on programs for discovering proofs of theorems in mathematics or logic, for game playing, and so forth, is beginning to yield valuable results in the field of automatic coding, which may displace the original objectives as a primary goal of research. The work of Shaw, Newell, Simon, and Ellis (1958) may be cited in this connection. Karp (1959) applies the methods of logical syntax to the analysis of programming languages, and Janov (1958) gives a rigorous analysis of the properties of certain classes of algorithms expressible in a concise notation developed by a group of Russian mathematicians to represent programs.

The book *Problemy Kibernetiki* (Problems of Cybernetics) edited by Ljapunov (1958), in which Janov's paper appears, contains several other interesting papers on automatic programming, and also a section on mathematical linguistics. According to the editor, this volume is the first of a series to be published at irregular intervals. The collection *Voprosy Teorii Matematicheskix Mashin* (On the Theory of Mathematical Machines) edited by Bazilevskij (1958) also contains several papers on the theory of programming.

Gorn (1957), Young and Kent (1958), and Oettinger (1955*a*, 1956*a*) consider a variety of methods for expressing information-processing algorithms in a form independent of the machine by which the algorithms are to be executed. The problem of controlling two or more operations to be executed simultaneously, an approach to machine design that appears to be growing in popularity, is examined by Gill (1958). The proceedings of a symposium held at the Franklin Institute (1957) provide an accessible source of information about several typical automatic coding systems for specific machines.

CHAPTER 4

THE PROBLEM OF TRANSLATION

4.1. Translation

Translating may be defined as the process of transforming signs or representations into other signs or representations. If the originals have some significance, we generally require that their images also have the same significance, or, more realistically, as nearly the same significance as we can get. Keeping significance invariant is the central problem in translating between natural languages.

According to our definition, transforming a printed message into Morse code, transliterating from the Cyrillic to the Roman alphabet, enciphering for cryptographic purposes, and replacing decimal numerals by binary numerals belong to one family with translating *Macbeth* into German and with interpreting into English a speech by a Russian delegate to the United Nations. In the terminology introduced in Sec. 2.3, every member of this family is a transformation whose domain is a set of original elements, with whatever operations, relations, or significance are defined on the set, and whose range is a set of likewise accompanied image elements. The differences distinguishing members of this family one from another may be characterized in terms of the respective domains and ranges, of the knowledge we have about these, of the nature of the correspondence between them, of the invariants to be preserved under translation, and of effective rules for translating.

An analysis of the distinctive characteristics of several of the transformations mentioned in the preceding paragraph should clarify why translating Homer into English seems so much harder (some say impossible) than translating a description of a scientific experiment from Russian into English, a process less tractable in turn than converting some numeral received at a telegraph office into the frozen birthday greeting for which it stands. We shall focus also on common traits of these transformations in order to make plausible the main articles of faith underlying research on automatic translation: first, that complex enough sequences of simple transformations can eventually provide algorithms for translating from one natural language into another automatically and, in certain realms of discourse, as well as if no better than conventional translators; second, that the search for these

104

algorithms can, if conducted in a scholarly fashion, yield valuable new insights into the structure and function of languages, as well as new techniques for research in linguistics.

4.2. Types of Translations

Translating a printed message into Morse code for telegraphic transmission is an example of one of the simplest kinds of translations, whose domain may be characterized simply as a set of strings of elements of the Roman alphabet. An image alphabet is *defined* so as to contain elements (combinations of dots and dashes) in one-to-one correspondence with elements of the original alphabet; the image of a string is taken, again *by definition*, to be the string of the images of the symbols in the original, ordered in the same way. The rule for translating strings is, in theory, a trivial consequence of the definition of the correspondence between strings, and it is simple even in practice because the alphabets are small enough for the table of correspondence between their elements to be memorized. The only relation between a pair of original elements pertinent to the process of transmission is their order of occurrence in the string, and this is preserved by definition.

Something is always lost in translation, namely the original elements. When translating into Morse code, it is preferable perhaps to speak of exchange and not of loss, since the whole point of the process is to substitute tokens whose geometric structure matches the configuration of electric impulses in time for the more familiar ink-on-paper tokens. However, because of this limited objective, we are left complete freedom to ascribe to the image of a string any and all properties of the original that do not depend on the fine structure of its alphabetic elements, in other words, properties relative to which these elements are irreducible. Whatever the significance of the string "red red rose" may be, that of the string

$$\text{``}\, .\, \_\, .\, \quad .\, \_\, .\, .\, \quad .\, \_\, .\, \quad \_\, .\, .\, \quad .\, \_\, .\, \quad \_\, \_\, \_\, \quad .\, .\, .\, \quad .\,\text{''}$$

can be *defined* to be identical, and the general acceptance of this definition is the basis of the successful use of the Morse code as a tool of communication. In other words, *red red rose** and all its attributes are invariant, the transformation being defined on representations, not on words.

The simplicity of translation *into* Morse may be contrasted with the complexity of a superficially similar situation. Imagine that we are listening to dots and dashes on a radio receiver. Far from being able to define a translation as we wish, we must question whether there is anything significant to translate. We may be listening to an aircraft guidance beacon or to the telemetering signal from an earth satellite. On the *hypothesis* that the transmission is in ordinary Morse code and a representation of English, we can *experiment* and attempt to recover a conventional English representation by applying the inverse of our simple transformation. If this hypothesis fails, it may be necessary to draw upon more sophisticated cryptographic techniques to *discover* the correspondence, if any, between the supposed representation and a hypothetical original. The situation is characterized by an almost total absence of a priori knowledge of domain, range, correspondence, invariants, and rules for translation. This problem and that of translating into Morse code, although compounded of the same ingredients, occupy opposite ends of any reasonable scale of difficulty, chiefly because of differences in kind and accessibility of knowledge regarding these ingredients.

Transliteration, as from Cyrillic to Roman, is a relatively simple kind of translation which nevertheless exhibits to a marked degree some of the characteristics which make translation between natural languages such a difficult and controversial art; a brief analysis of the problems of transliteration can therefore illuminate the problems to be encountered in translating from one natural language to another. Transliteration is like translation into Morse code in that the original and image strings are simply different representations for the same words. There is, however, less freedom in choosing the image alphabet since the basic purpose of transliterating is in this case to obtain representations of Russian words in the Roman alphabet. The correspondence is thereby necessarily complicated, since the Roman alphabet has fewer members than the Cyrillic. As a dodge, some transliteration schemes employ diacritic marks like those used in Czech, thereby extending the alphabet, but precluding the use of conventional English typewriters. If, however, the Roman alphabet is assumed, some Cyrillic characters, for instance, "ж", "ц", "ч", "ш", "щ", "ю", and "я", must be mapped into strings of Roman characters, "zh", "ts", "ch", "sh", "shch", "ju", and "ja" (Table 10). As a consequence, the rule for translating must make provision for handling images of varying lengths; the loop including boxes 12 and 15 in Fig. 26 illustrates one method of mechanizing such a rule. It is obvious that the length of strings is not an invariant of such a transformation.

Returning from a transliterated representation to the original Cyrillic requires more than the simple inversion by which rules for translating from Morse code back into the Roman alphabet can be obtained, trivially in theory and not so trivially under certain practical circumstances (Gold, 1959), from those for translating into the Morse code. For instance, while "т", "с", and "ц" map unequivocally into "t", "s", and "ts", respectively, "ts" may be the image either of "ц" alone or of the string "тс". Returning to

Cyrillic from such a transliterated representation raises scansion problems as serious as many of the syntactic scansion problems to be faced when translating from Russian into English.

On the basis of the empirical knowledge that the digram "тс" is extremely rare as an initial digram in the representation of Russian words (Ushakov (1935) lists only *тсс** meaning *hush!**) a partial but already complex algorithm for inverse transliteration may be sketched (Fig. 28). We enter the chart of Fig. 28 through the connector α_1 to the test $x_1 = $ "t"?. The selection process represented by the flow chart directs a choice among four alternative exit paths. The path through α_2 is taken if $x_1 \neq$ "t", and leads to a test of the identity of x_1 with some character other than "t". If the exit through α_3 is chosen, the next step in general would be to let the Cyrillic image y_i of x_i be "т". The path through α_4 should lead to a process for deciding what to do about the string "tss", which could be the exceptional case of an image of Cyrillic "тсс". However, if $x_i = $ "t" and $x_{i+1} = $ "s", it is most likely that $x_{i+2} \neq$ "s", hence the string "ts" is usually mapped into Cyrillic "ц" and α_5 is used as the exit connector. The rule $i+2 \to i$ ensures that the next character to be tested will be that following "ts".

The completion of the algorithm of Fig. 28 to account for such pairs as "присутствие"–"prisutstvie", "продаются"–"prodajutsja", and "граница"–"granitsa" is a matter of some difficulty, not worth pursuing further merely for the sake of illustration. In contrast, returning from the string "zh" to an original is somewhat simpler, since Roman "h" is the image of no Cyrillic character. By referring to Table 10, the reader should find it easy to sketch a flow chart for the inverse transliteration of strings including "z" or "zh".

Inverse transliteration, as illustrated in Fig. 28, provides a relatively simple and limited example of one role of *context* in translation. Context must be called into play because of the absence of a one-to-one correspondence between the alphabets of the domain and the range. A given character in a transliterated string, for instance, "h", may not correspond to any single literal of the original, and the interpretation of others, such as "s", may not be fixed. It is likely that a one-to-one correspondence could be established between complete representations of Russian words and their

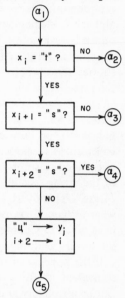

Fig. 28. Inverse transliteration.

Roman transliterations, but the use of word representations as irreducible alphabetic elements would entail the use of excessively large translation tables. The algorithm of Fig. 28 represents a compromise between rules for simple one-for-one substitution of characters, which would not be adequate, and rules based on an alphabet of words, which would be uneconomical in practice. The compromise is based on the observation that, if a transliterated representation is regarded as a reducible string rather than as an irreducible alphabetic element, the elements of the Cyrillic representation can be determined by examining groups of transliterated elements in the context of their neighbors. Because the groups that must be examined usually turn out to be shorter than whole-word representations, the process is more economical than one based on whole words.

The aim of transliteration tends to overlap that of phonetic or phonematic transcription, which often attempts to produce images providing explicit guides to pronunciation. The choice of correspondents is accordingly constrained to associate, insofar as possible, Roman elements with Cyrillic ones allied to similar sounds, as for example, Roman "n" with Cyrillic "н", or the string "shch" with Cyrillic "щ", pronounced somewhat as in "fresh cheese". This overlap leads to a conflict of interests, and to the question of what a "good" transliteration should be. Numerous examples have already been given in Chapter 2, illustrating that a single system of notation cannot serve every legitimate purpose at once, but this point deserves further emphasis.

If the purpose of transliteration is simply to give a character-for-character equivalent of Russian spelling in the Roman alphabet, no transliteration can be absolutely good, since there are not enough characters in the Roman alphabet. "Good enough" might be defined as "not using more long strings than absolutely necessary", but then the system given in Table 10 is not good enough, for why use "shch" when two-character strings have not been exhausted? Obviously phonetic considerations played a role in this choice. In fact, why use pairs of characters and why extend the alphabet by adding " ' " and "#" when "c", "h", "q", and "w" do not appear as correspondents? The likelihood of confusion between Roman "c" and Cyrillic "с" (transliterated "s"), and Roman "h" and Cyrillic "н" (transliterated "n") may be part of the answer; moreover, the problem of reading "zh", "ch", "sh", and "shch" correctly would be greatly complicated if "c" and "h" were used individually as correspondents. The sound associated with "w" is associated with no single Cyrillic character, and "q" would suggest the sound associated with "к". The use of " ' " is traditional, and explained perhaps by the desire to use as unobtrusive as possible an image for an original which is not associated so much with a sound as with a modification of the sound associated with the preceding character. The character "#" is used rather than the traditional " " " simply because the latter cannot be printed on the Univac high-speed printer (Table 9).

There are, of course, several systems of transliteration in common use that differ in varying degrees from the one used in this book, and that meet, presumably to someone's satisfaction, some criterion of good transliteration. The name *Xrushchev** (*Хрушев**) would be represented by "Khrushchev" or "Xruščëv" in some of these systems, and mathematicians may have wondered whether Tschebycheff and Chebyshev (Чебышев) were the same person. A leading character (Берсенев) in Turgenev's *On the Eve* is labeled "Bersenyev", "Bersieneff", "Bersenev", and "Berseneff" by four different translators.

Transliterated strings are also in constant danger of confusion with translated strings for the same reason that the interpretation of the string "1001" is ambiguous in discourse where both the binary and the decimal notations are used. Nothing differentiates an isolated instance of the string "algebra" obtained as the image of Cyrillic "алгебра", from an instance of the same string representing the English word *algebra**. However, under the ground rules of transliteration, a transliterated string represents precisely the same word as the untransliterated one, namely a Russian word. That the same string should have two possible interpretations is merely an inter-lingual instance of homography of the kind exemplified within a single language by strings such as "stand", "fast", and so forth. It happens that the English translation of *алгебра** is indeed *algebra**, and such coincidences can, if proper precautions are taken, be useful in the development of translation rules. The need for precautions to guard against nonsafe failures in interpreting transliterated strings is illustrated by the strings "most", "sad", "do", "glad", "net", which represent the Russian words *мост**, *сад**, *до**, *глад**, *нет**, translated into English as *bridge**, *garden**, *until**, *hunger**, *no**!

French provides many good examples of the fundamental difference between transliteration and language translation. Transliterating French into characters available on an English typewriter is very nearly an identity transformation, and it is therefore easy to read French words as if they were English. Savory (1957, p. 95) points out the danger:

> French, in fact, is full of examples of what has been called illusory correspond-ence, and the would-be translator is obliged to learn very soon that *brave* does not mean 'brave', *honnête* does not mean 'honest', *joli* does not mean 'jolly', and so on. Before long it becomes an article of faith that if a French word looks like an English word it is certain to have a different meaning.

The results of repeated transliterating can be surprising. For example, Majakovskij's line "Кофе Максвел, гуд ту ди ласт дроп" turns out as "Kofe Maksvel, gud tu di last drop".

4.3. The Translation of Natural Languages

Interlingual translation can be defined as the replacement of elements of one language, the domain of translation, by *equivalent* elements of another language, the range. Research on automatic translation is concerned with the explicit definition of criteria of equivalence or invariance, with the establishment on a sound theoretical basis of a mapping between domain and range or between subsets of these, and with the development of effective and elegant automatic techniques for obtaining the images of elements of the domain.

Some linguists tend to regard *language** as synonymous with *speech**, a point of view expressed by Bloch and Trager (1942), who define language as "a system of arbitrary vocal symbols by means of which a social group cooperates". From this point of view, writing is only a stepchild, a poor mirror of speech: in the words of Bloch and Trager, it is "either inadequate to the demands of the social organism, or else derive(s) entirely from spoken language and (is) effective only in so far as (it) reflect(s) this". A linguist interested primarily in speech, or working with a language that has no system of writing, will naturally attach prime importance to speech, and will, with good reason, tend to analyze languages and to construct grammars on a purely phonematic basis; but to insist on equating linguistics with the sole study of speech is to be dogmatic and parochial.

The direct study of writing, the direct study of speech, the study of the relation between speech and conventional writing, or the study of the relation between speech and synthetic graphic representations, all have justifiable claims to the status of linguistic studies. Vast amounts of raw material for interlingual translation are in written form; as indicated in Sec. 1.5, the problems of automatic speech recognition are formidable; it is consequently natural that those working on automatic language translation should be interested primarily in written language. For our purposes, therefore, *language** is to be interpreted as *written language**; repetitious usage of the qualifier "written" will be avoided, but only for the sake of simplicity, not as a claim to an exclusive interpretation.

We shall further restrict our attention to those languages, including all the major Western languages, which use linear alphabetic representations. The elements in the domain and in the range of translation are therefore strings of the kind described in Chapter 2. In Chapter 5 and beyond, attention will be focused on translation from Russian to English.

The characterization of the domain and range and the definition of the correspondence between equivalent strings is much more difficult for interlingual translation than for any of the examples given in Sec. 4.2. First, the domain and the range are given, and are not for us to define at will. The relation between signs and the functions they serve is indeed arbitrary, and "foof" and "Grötz" *could* easily play the roles that "dog" and "Hund" play in English and in German. But in fact they do not: the string "dog" is used in English in certain ways and "Hund" in German; "foof" and "Grötz" are neither English nor German; to introduce these new forms into either language would, if desirable, be possible, but it is beyond the normal authority of the translator to do so. Second, equivalence between strings is not freely definable, but must be discovered. *Language, which the individual tends to regard as an artifact, as a collective phenomenon behaves like an element of nature and must be treated as such.*

Ideally, perhaps, equivalent strings should be indiscernible in every respect; that they cannot be so follows from the very fact that they must be representations of different languages. We must settle, somehow, for equivalence in *relevant respects*, and to ask what is a good translation is essentially to ask what is relevant, what must remain invariant under translation.

The answer to this question is difficult to find. The writings of some philosophers succeed in clarifying the question, but offer little practical assistance. Quine's collection *From a Logical Point of View* (1953), especially the lucid and witty essay "Meaning in Linguistics," his more recent essay "Meaning and Translation" in the collection *On Translation* (Brower, 1959), or Goodman's "On Likeness of Meaning" (1952) are among the most helpful. With some exceptions (for instance, Jakobson, 1959; Whatmough, 1956*b*), linguists *qua* linguists seem to regard translation as a purely literary problem; hence little attention has been paid to it from the linguistic point of view, and translators of literary masterpieces are frequently too busy translating to take time out to analyze in a general way what they are doing.

Recently, however, there has been a strong surge of interest in interlingual translation as an area worthy of study. Apart from the literature on automatic translation, four books, in Russian, French, and English, devoted to translation have recently appeared. These are best described in their authors' own words.

Cary (1956) begins *La Traduction dans le Monde Moderne* with the blunt statement: "Il n'existe pas d'ouvrage d'ensemble consacré à la traduction" and continues "le fait peut paraître incroyable. Il surprendra moins si l'on songe que l'époque actuelle est la première à faire systématiquement appel à la traduction sous les formes les plus diverses." He points out, elsewhere in his book, that not even the *Encyclopedia Britannica* had an article on translation. Cary's view is not strictly accurate, as shown by the critical bibliography offered by Morgan in *On Translation*, where works dating from 46 B.C. to 1958 are listed. However, it is clear that a substantial proportion of the works in the bibliography were produced in the last decade, and that, with few exceptions, the major emphasis remains on literary translation. This emphasis remains even in *On Translation*, although philosophy, linguistics, and automatic information processing are represented.

Fedorov (1953), writing in the aftermath of the curious controversy over linguistic gospel which shook Soviet linguistics in 1950, says: "The problems of translation . . . can be fruitfully solved only in the light and on the basis of Stalin's works in linguistics. In the years of the ascendancy of the 'new science of language' the theory of translation was underrated, or rather, under fire. Even the possibility of a scientific linguistic approach to translation was repudiated by the Marrists, despite the self-evident practical importance and scientific interest of this task."

The state of knowledge about translation is best described by Savory (1957), who justifies his effort by calling attention to "first, the fact that translators have freely contradicted one another about every aspect of their art; and secondly, that they have written as if all translation were a conversion of a literary masterpiece in one language into a literary achievement in another."

One outstanding merit of all four books is that they call attention to the extraordinary variety of genres of translation, that is, to the variety of views on what is relevant and deserves to be preserved under translation. Cary distinguishes literary, poetic, technical, commercial, official (military, diplomatic, administrative, judicial) translation, as well as the translation of plays and songs, translation for radio broadcasting and for movies, translation of children's books, and translation at international meetings. Savory makes similar distinctions, and emphasizes, among others, the translation of the classics and of the Bible. The point that different pairs of domain and range require different treatment, to which we shall return later, is also well made.

We have already seen in Chapter 2 how difficult it is to find a single system of notation that will meet several requirements at once. The various notations of Table 5 and the codes of Table 10 illustrate the problem in simple terms. Illustrations of a similar phenomenon in interlingual translation are not difficult to find. Cary points out that, in a description of typing exercises, the French equivalent of the English sentence *the quick brown fox jumps over the lazy dog** should be *Zoé, ma grande fille, veut que je boive ce whisky dont je ne veux pas**. The common denominator of the two sentences is the presence in the representation of each of at least one instance of almost every member of the alphabet of the respective languages. But suppose that this were not the only relevant criterion of correspondence. The French sentence lends itself to breezy comments that would be utterly nonsensical if applied to the inane English. We can imagine the perplexity of a translator faced with an episode of some novel set in a typing school, in which the characters quip about their exercise. The profusion in many grammars of similar carefully selected horrible examples masks the fact that an overwhelming proportion of sentences, especially in scientific and technical writing, present no serious difficulties. Nevertheless, any method of automatic translation that is not mere technical expediency but claims both to make a contribution to linguistics and to follow sound engineering practice must be able to cope with such examples, at the very least in a safe-failing manner.

It is in the translation of poetry that the problem of invariance is most acute. So many things are relevant: the most mechanical, like the number of words, and the most subtle, like an emotion evoked by the juxtaposition of precisely certain words. The notion that translating poetry is impossible doubtless stems from the undeniable difficulty if not impossibility of preserving all relevant aspects at once. The translator of poetry must, for each new poem he faces, decide what it is he wants most to save, and close his eyes to the rest. As a result, poetry lends itself to repeated translation by men who discern untried combinations of relevant features. And the debate on the relative merits of those combinations, as Brower, Cary, and Savory clearly show, can be endless.

In scientific or technical translation, the problem is somewhat less acute, because style is of lesser importance. It seems that an important aspect of meaning can be preserved over quite a range of variation in formal structure. An interesting experiment by van der Pol (1956), in which a passage was translated from English into French by one translator, the French version translated into English by another, this English version into French again and then once more into English, confirms this observation. In van der Pol's words,

The primary conclusion that can be drawn from this test is that the meaning has been retained to a remarkable degree, though by comparison with the original, the style of [the final English] version is entirely corrupted. Thus a person reading the original, and another reading the final text, should be able to agree on the content and the intent of the paper; although they might not be equally assisted in their appreciation of it by the respective styles.

It may be argued, as Whorf (1956) seems to do, that, even if agreement were reached on relevant features to be preserved, what can be said in the domain may be ineffable in the range, and that in such cases translation is truly impossible. This is not the place in which to debate the mystics' claim of ineffability in *any* language (Henle, 1949), but it seems fair to suppose that if something can be said in one language it can be said in any other. As Whatmough (1956a, p. 222) points out, "Those experts who know Hopi or Eskimo or Aranta, and who begin by insisting that this or that feature is

incapable of English expression, always end by explaining it in English."
The equivalent expressions may be formally quite different, for instance, a
word may map into a phrase, or it may be necessary to extend the range by
coining or borrowing new words, but these are practical questions or matters
of style, not questions of possibility. The result may resemble what Fitts
(Brower, 1959, p. 34) expects of a translation of Pushkin by Nabokov:

> He contemplates a line or two of translation accompanied by notes of every
> conceivable kind—exegetical, semantic, aesthetic, metrical, historical, sociological
> —and his printed text would look, I should think, like an eighteenth-century Plato
> on my own shelves: a snippet or two of precious Greek surrounded by a sea of
> commentary. A tireless writer of footnotes, I find this concept endearing; but I am
> not sure that it is anything more. The trouble is that such a translation, though it
> might give the prose "sense" of the original together with an explanation of what-
> ever goes to lift the prose sense above itself and transmute it into a form of art,
> might also provide no evidence beyond the saying so that the art was art in the
> first place.

As an alternative to impossibility, any concept is endearing.

However great the difficulties of translating may be, the difficulties of
evaluating the quality of translations are equally great. To this day, it's every
man for himself. Miller and Beebe-Center (1956) have made a valiant
attempt to develop a reliable, valid, objective, and easily used rating scale,
applicable to translations produced by either men or machines. Their
penetrating analysis of the problem has clarified it considerably, but both
heated arguments and calm investigations can be expected to continue.
Restricting one's attention to scientific and technical writing at least puts one
in the eye of the hurricane.

4.4. The Possibility of Automatic Translation

All the quandaries about the possibility and the quality of translation
discussed in the preceding section are with us, whatever agent we may choose
to do the translating. It then follows that if the possibility of good translation
is granted at all the possibility of good automatic translation can be established
as well, simply by exhibiting an algorithm for translating by means of an
information-processing machine. Such an algorithm is sketched in Fig. 29. The
original strings on which the algorithm is designed to operate consist
of m characters $s_0, s_1, \ldots, s_{m-1}$ followed by a sentinel "$", which is
a member of the domain and range alphabets serving only to mark the ends
of strings. The memory of the machine is assumed to store a table of p
original strings together with their images. The kth original string ($k =
1, 2, \ldots, p$) consists of m_k symbols $s'_{k0}, s'_{k1}, \ldots, s'_{km_k}$ again followed by the
sentinel "$". The kth image string is of the form $s''_{k0} s''_{k1} \cdots s''_{kn_k}$ "$".

The algorithm provides for comparing the input string in turn with each of the original strings stored in memory. If a match is obtained with the kth original, the kth image is printed. The algorithm is fundamentally a pure table look-up. This becomes evident if we note (*a*) that the input string, the

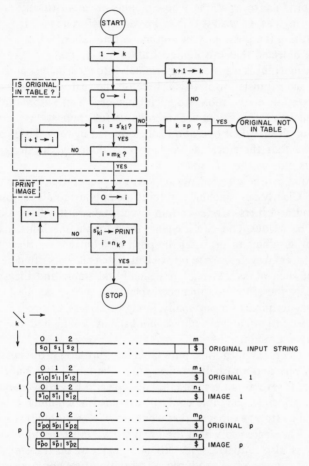

Fig. 29. An algorithm for automatic translation.

kth original stored string, and the kth image could be replaced by the irreducible symbols s, s_k', and s_k'', respectively, and (*b*) that the statements in the boxes enclosed by dashed lines could then be replaced by $s = s_k'$?* and by $s_k'' \rightarrow Print$*. The use of reducible originals and images was postulated for the convenience of a hypothetical user, rather than compelled by the nature of the algorithm.

It follows that a machine programmed to execute this algorithm can translate any text for which an original and image are included in the table stored in memory. Questions regarding the quality of translation must be resolved in the preparation of the images prior to storing them; hence the machine surely can produce translations whose quality, relative to any criteria we choose, is as high as we wish it to be. We can therefore regard the algorithm as demonstrating the possibility of automatic translation.

It may be objected that this demonstration has no real significance, that, although the machine is transforming originals into their images and therefore translating, such translation is trivial because the real work has been done in the preparation of the table stored in memory. Yet, automatic symbol-processing machines have found numerous useful applications, although it can be said of all that their operation is determined by their design and their program, and that therefore the *real* work is done by the designers and programmers, not by the machines.

An objection can be directed not only at the preparation of the table but at its extent. Clearly our machine can translate only originals represented in the table, and no others. Under certain circumstances, this limitation might not in itself be serious. One can imagine, for example, machines installed at international terminals to print, at the push of a button, translations of the stock phrases that travelers use to buy tickets, catch planes, or find hotels. The repertoire of such phrases is relatively small and stable and their use very frequent. It is therefore not inconceivable that such machines might be useful at busy terminals serving many foreign travelers who might not own or might have forgotten their phrase books, but would find the machine available day and night.

For more general applications, pure table look-up soon proves inadequate. If *War and Peace* and an English image are in the table, *War and Peace* can be translated. If several images are tabulated, several English versions can be obtained. In principle, the table could be extended indefinitely, but the cost of the enormous storage capacity that would soon be required sets a practical upper limit. The degree to which the "real work" must be done by human intervention is also embarrassing. No new text can be translated by the machine until someone has produced a translation first and added it to the table. A machine translating whole texts by a pure table look-up algorithm thus appears incapable of matching the performance of a human translator who, when presented with original texts he has never seen before, can translate at least some of them into adequate images.

Simpler instances of this very problem have been analyzed in some detail in Sec. 2.6. It was shown there that replacing pure table look-up of irreducible representations by more sophisticated algorithms defined on reducible representations could yield not only substantial economies in table storage, but also algorithms of greater generality. Example 2-7 demonstrates that the case and number of nouns can be determined without tabulating all the representations of nouns, provided that inflection follows a regular pattern and that this pattern is mirrored in the structure of the representations. The common algorithm for adding decimal numbers enables us to obtain correctly the sum of numbers no one ever before may have added. *To gain similar generality in an algorithm for interlingual translation, the domain and range must be analyzable, not as sets of irreducible texts, but as sets of less numerous, freely combinable elements from which not only extant texts, but texts yet unwritten, may be formed.*

The practical possibility of such analysis is suggested, of course, by the existence of familiar grammars, and by the fact that when we use our own languages and translate from one to another we treat texts as reducible representations. It cannot be hoped, however, that any extant grammar will prove adequate to the needs of automatic translation. Characteristically, grammars intended for use by persons are elliptic in their descriptions of languages; moreover, their outlook tends to be descriptive rather than algorithmic; that is, they describe properties of sentences, but provide no explicit algorithms either for constructing a sentence having prescribed properties (Chomsky, 1957, is a notable exception) or for determining the properties of a given sentence. This is not too serious in most of the usual applications, for the human mind is capable of great feats of understanding, analogical reasoning, and hypothesizing, and can draw on such subjective and ill-defined resources as intuition, "Sprachgefühl", appreciation of shades of meaning, or knowledge of a subject, for *ad hoc* assistance in the resolution of immediate problems. Enough has been said in the first two chapters to indicate why, in marked contrast, the design or program of an automatic translator must provide, in advance and in the most minute, explicit, and systematic detail, for the disposition of every contingency of any consequence which the machine is likely to face in the course of operation. The only resources available to an operating automatic translator are the following objective elements: (*a*) the original texts themselves, regarded strictly as sequences of characters; (*b*) tables provided by the designer or programmer and stored in the internal or auxiliary memory; (*c*) algorithms provided by the design and program for operating on the elements (*a*) and (*b*).

A grammar suitable for automatic translation cannot be based on any other elements. The restriction is not quite as severe as it might seem, because of the provision for tables; it does imply that the thought normally applied by persons to each specific task must be applied once and for all to the compilation of tables and to the specification of rules.

We might add that the machine could have experts at its disposal who, should the machine print "What is the English image of Russian X?" on its output printer, might return the proper image via an input keyboard. It is not uncommon for a mathematical program to be designed to ask in this way for today's date or for the value of some parameter. The procedure must be used with great caution. For one thing, frequent manual intervention materially reduces the operating speed of a machine, robbing it of its most precious asset. More important, it is obvious that the limiting case of this approach is a program which, if the original text is *Doctor Zhivago*, prints "What is the English image of *Doctor Zhivago*?", then idles while the expert supplies the translation to a machine now equivalent to a grossly exaggerated typewriter. Even pure table look-up would be attractive in comparison.

For experimentation and evaluation, a closed loop including both men and machines has marked advantages. An iterative experimental system, in which the results produced by the machine at one stage of experimentation are examined by men to determine desirable modifications in the mode of machine operation at the next stage, is sketched in Sec. 10.5. Manual transcription of texts prior to automatic processing, and some human postediting of machine products will be unavoidable for some time. The ultimate value of automatic machines in translation will depend in part on where the ultimate process lies in the spectrum ranging from the limiting case of the exaggerated typewriter to the production of satisfactorily accurate and readable translations untouched, so to speak, by human hands.

In the next section, we turn our attention to the question of developing grammars capable of assisting in the design of automatic translating machines. The spirit which must guide this development is so close to that of modern structural linguistics as to suggest that the effort has intrinsic linguistic interest as well as practical necessity.

4.5. Grammar and Interlingual Correspondence

A grammar of a language is a model of this language in the sense defined in Sec. 2.5, or, as Chomsky (1956) puts it, it can be viewed as "a theory of the structure of this language". Any model necessarily represents only selected features of the modeled phenomenon, and both the choice of these features and the form of the model depend on the purposes to which the model is to be applied.

Two grammars at least are involved in any interlingual process, one modeling the domain, the other the range. Grammars suitable for translation by means of a pure table look-up algorithm would be of the utmost conceptual simplicity, that is, they would be simple lists of the irreducible elements of the domain and of the range. The domain grammar is identical in form with the range grammar, and the formulation of the translation algorithm is equivalent to the organization of the two individual lists into a table of two columns. More complicated devices are necessary to describe sets of reducible elements.

One possible approach is sketched in Fig. 30 (adapted from Yngve, 1957). The grammar of the domain is supposed to be formulated as a set of rules

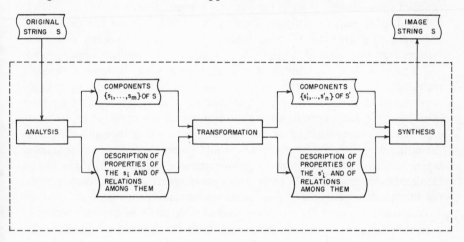

Fig. 30. A translation process.

for analyzing a string S of the domain into a set $\{s_1, \ldots, s_m\}$ of substrings (not necessarily disjoint), and for describing the properties of the elements s_i and significant relations among these elements. Rules are provided for transforming the set $\{s_1, \ldots, s_m\}$ into a set $\{s'_1, \ldots, s'_n\}$ of image elements belonging to the range, and for converting the description of properties and of relations among the original elements into a description of equivalent properties and relations among the elements of the range. The grammar of the range, in turn, is supposed to be formulated as a set of rules for synthesizing an image S' from the set of components and the description of the relations among these.

The nature of the grammar of either domain or range depends to some degree on its purpose. On the one hand, the model may be intended to be explanatory, and therefore required to account for every significant feature of the language it models. On the other hand, for purposes of translation, only those features in which domain and range *differ* need be explicitly provided for in the model. Under these circumstances, the algorithms for analysis, transformation, and synthesis may be viewed as a single grammar, of a kind that Harris (1954) has named *transfer grammar*.

It should be noted that, although algorithms for analysis, transformation, and synthesis, and their associated tables, will initially be developed on the basis of existing grammars, these algorithms and tables may be viewed as grammars in their own right. The study of algorithmic models of language may have a long-range significance for linguistics far transcending the significance of these models as tools for automatic translation.

The development of rigorous, explicit notational systems for describing properties of strings and relations among these suggests to many that such systems may eventually serve the role of synthetic *intermediate languages*, to which properties of many different natural languages may be related. Besides reawaking the old dream of a universal grammar, this idea appears to hold promises of efficient multilingual automatic translation. Translating directly from any of n languages to any other requires of the order of n^2 transfer grammars. For translating first into an intermediate language and then into the ultimate range only $2n$ grammars would be required if the intermediate language were distinct from the n given ones, and only $2(n-1)$ if the intermediate language were one of the n. These promises are superficially attractive, but there is little prospect of their prompt realization.

Traditionally, linguists have distinguished a hierarchy of elements, viewing a given element alternatively as a composite of elements of an alphabet lower in the scale and as an irreducible alphabetic component of elements higher in the scale. One possible hierarchy might comprise distinctive features, phonemes, morphemes, words, phrases, sentences, paragraphs, and so on. Roughly speaking, the study of the word is traditionally called *morphology* when the word is considered to be a reducible string of morphemes, and *syntax* when the word is viewed as a component of higher elements such as phrases or sentences. It is evident that elements at any level, including trivially the lowest and the highest, lend themselves to such alternative consideration, and that the morphology of reducible elements is distinguished from the syntax of their components as analysis is from synthesis.

The choice of a particular level in a hierarchy as the source of alphabetic elements strongly influences the character of the analysis, translation, and synthesis procedures, as we might expect in the light of Chapter 2. If original strings are assumed to be books or journal articles and are themselves taken as alphabetic elements, analysis and synthesis are trivial; the given unit is identical with its one and only component. However, the translation process reduces to the discredited pure table look-up.

At the other extreme we might choose letters as our elementary irreducible components. The set of alphabetic elements is relatively small even if we include in it punctuation marks, brackets, and similar special signs. No serious conceptual difficulties prevent either man or machine from decomposing a given text as a string of elements of this alphabet. However, except for transliteration, there is usually no useful direct correspondence between the alphabets of the domain and of the range. Consequently a tremendous burden is placed on the analysis and synthesis procedures, since from this point of view everything depends on the relations among the now numerous components of the original string. In other words, a dictionary relating Roman letters to Cyrillic letters would contribute little toward translation from Russian into English. Most of the burden would be on "syntactic" analysis and synthesis, that is to say, on the process of describing the relations among the many letters in the string.

Experience suggests that a natural level of analysis would be provided by the selection of *words* as the irreducible components, at least for translation among the major Indo-European languages. Existing word dictionaries attest to the possibility of establishing fruitful correspondences between original and image words, although unfortunately these correspondences are frequently not one-to-one. The number of words is not so large as to lead to immediate rejection of table look-up as a practicable method for obtaining images of words. The fact that we construct sentences as sequences of words and conversely that we interpret them by interpreting their component words and the relations among these suggests that an automatic process along the lines sketched in Fig. 30 may be possible. Interword relations across sentence boundaries are intuitively felt to be very weak. Hence the description of relations among words may possibly be restricted within the relatively narrow confines of sentences.

The printed representation of texts normally lends itself readily to the identification either of texts or, at the other extreme, of characters. Strictly speaking, words are not irreducible relative to the conventional representation. It is possible, nevertheless, to characterize words as irreducible identities even in the ancient texts where spaces and punctuation marks are absent, but where the practiced reader finds no difficulty in isolating individual words. In modern texts, man or machine can consider the representation of a word to be a string of characters between two successive spaces; for many purposes this description is adequate, and justifies viewing words as irreducible elements, although there are some special difficulties in the case of abbreviations, hyphenated words, and the like. For the present, we may assume that words can be identified and treated as irreducible when necessary, and we can examine the consequences of this assumption.

4.6. Word-by-Word Translation

An obvious simplification of the procedure described in Fig. 30 is the elimination of the description of relations among the components of the original string and likewise of the description of relations among their images. This amounts to regarding the translation of a string of words as the string of the translations of these words. Thus, if $S = w_1 w_2 \cdots w_n$ is a string of words, $T(S) = T(w_1 w_2 \cdots w_n) = T(w_1)T(w_2) \cdots T(w_n)$.

A procedure of this sort is commonly called *word-by-word translation* and is known to yield translations which, if not always elegant or even easily readable, may nevertheless be quite intelligible. Word-by-word translation is similar to the direct translation of irreducible texts in that it relies primarily on table look-up. The significant difference, however, is that a set of words is a set of relatively few, freely combinable elements from which not only extant texts but texts yet unwritten may be formed. That is, words provide us with the kind of set that was suggested in Sec. 4.4 as the only reasonable basis for a practicable translation system.

The conceptual simplicity of word-by-word translation suggested that an exploration of the practical problems of achieving it would prove rewarding as a first step toward more sophisticated procedures, a step providing both an indispensable component of any more complex system, and a first approximation to an automatic translator (Oettinger, 1954). Most of the remainder of this book is devoted to a description of theoretical and experimental studies of the lexical problems inherent in translation from Russian to English. The compilation and operation of an automatic dictionary are described in detail, as are the steps taken to provide in the dictionary descriptions of the properties of words not reflected in the structure of their representations, but essential in achieving word-by-word translation and in proceeding beyond it.

4.7. Syntactic and Semantic Problems

Analyzing the relations among the components of any particular text presupposes a prior cataloguing of both the possible significant relations among text components and the properties of representations that reflect these relations. The relations in question are formal ones, as between subject and predicate, or between noun and adjectival modifier. These relations are typically reflected in the order of word representations or in structural similarities of the related representations, for example, in Russian the simultaneous use of plural-indicating desinences in the representations of both a noun and its modifying adjective.

If the same or very similar structural devices are used to represent relations in both domain and range, the transformation of relation representations presents little difficulty. Even if the structural devices are not similar, so long as the underlying relations are similar, there is little need for recourse to nonformal elements, that is, to meaning or reference, at least if fairly good grammars of domain and range are already available.

Because the relation between signs and what they refer to is arbitrary (but fixed by common convention, at least within a single language), and the relation between signs and their representations is equally arbitrary, ultimate recourse to meaning or reference, however distasteful, is unavoidable. However, translating must not be confused with interpreting, and while interpretation may be necessary in developing translation algorithms, or in understanding a translated text, there is every reason to hope that translation algorithms can be based on purely formal elements, and that they can therefore be executed by a machine.

Consider, for example, the string "He got up, dressed, and took a shower". Whatever humor there may be in the situation represented by this string obviously depends on the temporal order of the events described. This temporal order is reflected in the spatial order of the representations of the events described. If we were to seek an equivalent description of the events in another language, our criterion of equivalence would certainly include equivalence of temporal order.

Spatial order of representational elements is not fundamental, as may be illustrated simply by representing the same situation by another English string, "He got up, and took a shower after he dressed". The second string is clearly equivalent to the first in the representation of temporal order, although the significance of spatial order in the two strings is different. The interpretation of spatial order in the second string is affected by lexical properties of the string "after", illustrating how semantic load may be carried by either the configuration or the lexical properties of strings. These lexical properties may be made explicit by marks in a dictionary entry. Once this phenomenon is recognized, there is no fundamental bar to formulating an algorithm for purely formal analysis that will produce different results according to the presence or absence, in a string, of word representations whose dictionary entries are supplied with special marks. The practical difficulties in the large-scale study of this problem are enormous.

The process of analysis, transformation, and synthesis is illustrated in simplified form by the not necessarily typical German-to-English example in Fig. 31. The images s_i' of the elements s_i of the original German string S on line a are shown in line b. Connecting lines denote groups of elements related by their order in the string. The numbered circles denote an equivalent

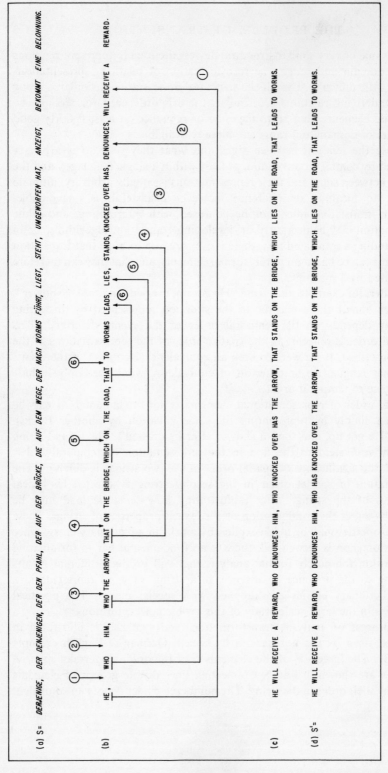

Fig. 31. The English image of a German string.

English order relation. Lines b and c show two intermediate phases of synthesis leading to the image S'. The easiest way for one to follow the process is by beginning with the innermost nested line (6).

Differences in order are only one example of the different devices used in various languages. Relations expressed by prepositions in English may be expressed by desinences in Russian. Reichenbach (1947, pp. 287–298) characterizes the notion of verbal tense in terms of three points in time, the point of speech, the point of the event, and the point of reference. For example, the expression *I had seen John** describes an event that occurred before the point of speech, relative to a point of reference intermediate between the event and speech about it. The expression *I saw John** describes an event whose occurrence in the past coincides with the point of reference. Since it relates formal tenses or circumlocution patterns in natural languages to the possible configurations of the three points, Reichenbach's system facilitates the establishment between pairs of natural languages of correspondences that preserve the configurations of the points. This system may be regarded as a concrete illustration of a partial intermediate language.

The elementary operations in terms of which the complete transformation and synthesis operations of Fig. 30 can be synthesized include insertion, deletion, modification and permutation of strings, and table look-up. Table look-up is unavoidable in obtaining correspondents of words. However, if there is to be hope of effecting transformation and synthesis in any practical and elegant way, the algorithms for these operations cannot depend on viewing individual words or individual contexts as the set of freely combinable elements of which texts are made. The analysis process must yield word *classes* or context *classes*, such that any member of a class is equivalent to any other relative to transformation and synthesis algorithms. Again, the experience of human translators suggests that this is possible. Surely, one and the same set of algorithms should apply to "John came here", "Jim came here", "Roger came here", and so on.

Given any bilingual dictionary and a set of analysis, transformation, and synthesis algorithms, there will remain some classes, some individual strings, that cannot be handled by these algorithms. These are idioms. In Bar Hillel's words (Locke and Booth, 1955, chap. 12), "a given sentence in a language L_1 is *idiomatic with respect to a language L_2, to a given bilingual word dictionary from L_1 to L_2, and to a given list of grammatical rules* if, and only if, none of the sequences of the L_2 correspondents of the sequence of words of the given L_1 sentence is found to be grammatically and semantically a satisfactory translation, after perusal of the applicable grammatical rules". Idioms can therefore be removed by extending the dictionary and the grammatical rules, provided that they either fall into large classes, in other words, are merely heretofore unrecognized patterns, or else, if they are individual cases, are sufficiently few in number to permit their practical treatment by table look-up.

Ultimately, means for selection among multiple correspondents of words, that is, for resolution of problems of homography and polysemanticism, must be incorporated in translation algorithms. In certain instances, for example, prepositions, conjunctions, and other so-called *function* words, formal syntactic properties may provide a clue to proper selection. For the more numerous *meaning* words, the problem is much more difficult.

The foregoing is merely a sketch of some of the syntactic and semantic problems of automatic translation. More elaborate formulations of these problems and suggestions for attacking them, but little by way of satisfactory, tested solutions, may be found in the works mentioned in the bibliographic notes for Chapter 10; a lexical foundation intended to support a practically sound and theoretically interesting syntactic and semantic superstructure is described in the following chapters.

4.8. Bibliographic Notes

General surveys of the problems of translation have been made by Brower (1959), Savory (1957), Cary (1956), and Fedorov (1953). References to publications concerned chiefly with automatic translation are given in Sec. 10.7.

The literature of linguistics is vast and diversified. Carroll (1953) gives "a survey of linguistics and related disciplines in America" which can serve the layman as a very good introduction to many aspects of linguistics. Cherry (1957) deals with diverse aspects of human communication in the light of contributions made by many disciplines. Both of these works have extensive bibliographies. Whatmough (1956a), Bloomfield (1933), and Sapir (1921) are among the basic writers on language who should be read by anyone with a serious interest in linguistics. More specialized works of interest to investigators of automatic translation include those of Hockett (1958), Chomsky (1957), Gleason (1955), Harris (1951), and Bloch and Trager (1942). The journals *Language* and *Word* are among the best sources of current information, and the glossary by Hamp (1957) provides a guide to the terminology of linguistics.

CHAPTER 5

ENTRY KEYS FOR AN AUTOMATIC DICTIONARY

5.1. Inflection, Derivation, Homography, and Polysemanticism

The structure of entries in an automatic dictionary and of their keys is governed both by technological factors of the type discussed in Secs. 1.6 and 1.7 and by the solutions obtained for certain linguistic problems connected with inflection and derivation and with homography and polysemanticism.

Table 15. Paradigm of *студент**

Case	Singular	Plural
Nominative	студент$_{ns}$*	студенты$_{np}$*
Genitive	студента$_{gs}$*	студентов$_{gp}$*
Dative	студенту$_{ds}$*	студентам$_{dp}$*
Accusative	студента$_{as}$*	студентов$_{ap}$*
Instrumental	студентом$_{is}$*	студентами$_{ip}$*
Prepositional	студенте$_{ps}$*	студентах$_{pp}$*

Consider inflection. Everyone knows that English nouns can occur in the singular and in the plural. The noun *dog** has the singular *dog$_s$** and the plural *dogs$_p$**, with the corresponding conventional representations "dog" and "dogs". In a highly inflected language, such as Russian, an even greater multiplicity of variants exist. For example, *студент** (*student**) has the inflected forms given in Table 15. The set of all inflected forms of a word is called the *paradigm* of the word.

The phenomenon of inflection can be approached from two points of view. The first, which may be called the Platonic, synthetic point of view, presupposes the existence of abstract words with well-defined sets of meanings. Inflected forms of these words are generated in accordance with the rules of grammar as required for discourse. Each inflectional variant has the same basic meaning as the abstract word itself, only adapted to a particular function in the sentence. For example, according to Unbegaun (1957, p. 159), the Russian accusative "indicates the object which, in its totality, undergoes the action of the verb", the dative "indicates the remoter object to which the action of the verb applies, though less completely than in the case of the accusative", the main value of the instrumental "is that of instrument, agent, or means", and so forth. This point of view may also be called "classical", being that taught in most schools and that with which, for better or worse, most of us are most familiar; many contemporary structural linguists look askance at it because of its heavy reliance on meaning and on nebulous notions like that of a "remoter object to which the action of the verb applies . . .".

The point of view favored by many structural linguists is nominalistic and analytic. In its extreme form, it presupposes the existence only of large quantities of tokens serving as a *sample* or *corpus* for analysis. These tokens must then be subjected to exhaustive analysis by techniques described by writers like Harris (1951, 1957) and Fries (1952). Ideally, meaning should play no role in these techniques. The end point of the analysis should be *classes* of forms or of words; often these classes turn out to be more or less analogous to paradigms and to parts of speech in the classical sense. In a strict sense, there are no abstract inflected forms, only classes of equivalent tokens; no words, only classes of equivalent classes of equivalent tokens; and no parts of speech, only classes of equivalent classes of equivalent classes of equivalent tokens, with the appropriate definition of equivalence at each level. A program of structural analysis in the strict sense has yet to be carried through to a successful and significant conclusion.

The results, if any, of a nominalistic analysis of language can usually be expressed in Platonic terms. For example, a certain class of tokens can be identified with the abstract inflected form $dogs_p{}^*$, and the class $\{dog_s{}^*, dogs_p{}^*\}$ can be identified with the abstract word dog^* by anyone willing to make the necessary ontological commitments or oblivious of this necessity. Also, the classical view is convenient and the corresponding mode of expression is lucid for problems of synthesis. The structuralists rightly object to attempts to force American Indian or Eskimo languages into a mold that happens to be more or less suitable for Greek or Latin. Their emphasis on searching, without preconceptions, for the best model for each language or language family is eminently sound. However, to insist that the only scientific approach is to start from scratch and let the structure of a language jump out, so to speak, by itself, is to be more royalist than the king. Adopting a working hypothesis, possibly in the form of a structural system expressed in classical terms, and verifying its validity experimentally, is quite different in spirit from insisting that every language must fit the structure of Latin. Indeed, some such hypothesis, as well as covert appeals to meaning, may be detected in many structuralist systems.

In approaching the problems of inflection in relation to automatic translation elements of both points of view can be taken to advantage. Roughly speaking, the nominalistic point of view is most useful as a guide in the linguistic analysis necessary to discover rules for translation, while the Platonic point of view is convenient in describing the application of these rules. The best and most complete existing grammars and dictionaries of Russian are built largely on classical lines. They do not, by any means, provide an adequate basis for an automatic translation system. However, this is not a sufficient reason for rejecting entirely the wealth of experience embodied in them and embarking on a rigorously structural analysis for which, at present, there is little hope of success. In fact, the sources of inspiration matter but little; in any science, the fruits of inspiration must ultimately be judged by combined criteria of practical usefulness and conceptual elegance and simplicity like those by which the theories of the physical sciences are evaluated. As the Russian mathematician Kolmogorov (1956) put it:

It is already obvious that work on automatic translation is profitable . . . for linguists, who are compelled, in the course of this work, to make more precise the linguistic laws they formulate. They acquire here an objective criterion, so important for the development of theory; if some formulation proposed by them, say a set of rules about word order in some language, enables the production of a good, polished translation by machine, there is conclusive evidence that the rules were comprehensively formulated. Here one can't get away with vague phrases masquerading as "laws", as there is a tendency to do in the humanities.

Although inflection has been used as an example in the preceding exposition, the other problems cited in the opening sentence of this chapter, and indeed most problems arising in the study of automatic translation, can be approached in the same spirit. It seems reasonable to start with the assumption that, for Russian and English at least, the classical concepts of inflection and derivation, and of homography and polysemanticism, have some heuristic value and can provide a good starting point for a more rigorous investigation in a nominalistic vein.

Because the idea of inflection is familiar to anyone who knows some Latin, Greek, Russian, or German, or even French or English, little more need be said about it at this point. While inflection is regarded as leaving words invariant—for instance, dog_s* and $dogs_p$* both are forms of the one word dog*—derivation can be said to produce distinct words. Thus, the word-pairs *студент*-*студентка**, *пациент*-*пациентка**, and *демонстрант*-*демонстрантка**, denote, respectively, male and female students, patients, and demonstrators. The relation between inflection and derivation in this case is schematized in Table 16.

The correlation in this example between derivational variants and the gender of nouns or the sex of the referents of these nouns is accidental. Derivation is also used, for example, to turn verbs into nouns. Moreover, in some important cases, like that of Russian adjectives, similar gender modification is subsumed under inflection. For example, the forms

Table 16. Inflection and derivation.

| | Case | Derivation ⟶ | |
		Masculine	Feminine
I		*студент**	*студентка**
n			
f			
l	NS	*студент$_{ns}$**	*студентка$_{ns}$**
e			
c	GS	*студента$_{gs}$**	*студентки$_{gs}$**
t			
i	DS	*студенту$_{ds}$**	*студентке$_{ds}$**
o			
n		·	·
↓		·	·
		·	·

красный$_{mns}$*, *красная$_{fns}$**, *красное$_{nns}$** are commonly treated respectively as masculine, feminine, and neuter nominative singular inflectional variants of one adjective *красный**. One can easily imagine a system where *красный** = of the color red and named by a masculine noun*, *красная** = of the color red and named by a feminine noun*, and *красное** = of the color red and named by a neuter noun* are treated as distinct *derivational* variants. While there may be perfectly sound reasons for preferring one convention to the other (see, for instance, Bloch and Trager, 1942, pp. 54–55), it is essential to note that the question is one of *convention* (modeling), and that a popular convention (grammatical model) may not be consistently formulated or universally accepted.

There is a twilight zone where inflection and derivation seem to raise difficulties for lexicographers. The English gerund ("leading", "hydrogenating") and Russian present active participle ("трудящийся", "читающий") are in this zone. In both languages, these forms are conventionally treated as members of the paradigms of verbs. Certain instances, however, are also treated as independent words. For example, *Webster's New Collegiate*

Dictionary (1953) has an entry for *leading** distinct from that for *lead**. The latter gives "led" and also "leading" as inflected representations of the verb, and we find *leading** described separately as: "*n.* 1. Action of one who leads; guidance. 2. Suggestion; hint.—*adj.* That leads; guiding; direction; foremost." On the other hand, there is no entry *hydrogenating**, but only the entry *hydrogenate**.

In Smirnitskij's *Russko-Anglijskij Slovar'* (1949) we find not one but two entries with key "трудящийся", one describing a participle and adjective translatable as *working**, the other describing a noun declined like an adjective and translatable as *worker**. There is, of course, an independent entry for the verb *трудиться** = *to work**, *to toil**, *to labor**. On the other hand, there is but one entry for the verb *читать**, *to read**. There is no distinct entry *читающий**.

The difficulty in distinguishing between inflection and derivation for lexicographic purposes is intimately related to the general problem of *homography*. Homographs are loosely defined as a set of distinct words written alike. In the terms of this definition, a set of distinct words $w_1{}^*$, $w_2{}^*$, $w_3{}^*$ form a homographic set if they all happen to have the representation "w". Such a point of view is not readily reconciled with an extreme structuralist approach. It requires more than collecting a set of tokens for "w" to arrive at the conclusion that these are not tokens for a single word w^*, but that some really stand for $w_1{}^*$, others for $w_2{}^*$, and so on. The arguments presented in favor of distinguishing several homographs tend to rely on phonetic distinctions, on historical data, on functional criteria, or on some intuitive measure of "distance between meanings". A single meaning, or a tight cluster of meanings, is attributed to a single word; meanings that are far apart are attributed to two or more words. Intermediate cases are handled by distinguishing several clusters, but attributing them all to one word.

For example, *Webster's* uses "lead" in two major headings. Of these, the first subsumes three main parts, one describing a noun, the second an adjective, the third a transitive verb. The noun part is itself subdivided into five parts, two of which are still further subdivided, and the verb part also has five subdivisions. The second major heading subsumes four parts, describing a transitive verb, an intransitive verb, a noun, and an adjective, respectively. The noun part is subdivided into ten parts. The meanings under the first major heading all have some relation to metallic lead, while those under the second derive from the concepts of guiding, preceding, directing, and so forth. In this case, it happens that the distinction between the two main headings coincides also with a phonematic distinction—the words are homographic but not homophonic—although this criterion is by no means consistently used. The boundary between *polysemanticism*, attribution of several meanings to one word, and *homography*, the distribution of several meanings among several words of identical spelling, is very hazy indeed. The problem of defining this boundary belongs to lexicography in general (Abaev, 1957), but the definition of a good solution depends to some extent on the type of dictionary under consideration. A further analysis of this problem, as it affects automatic translation, is given in Chapter 9.

No attempt will be made here to treat the problems of inflection, derivation, homography, and polysemanticism in full generality. They will be considered in the context of a study of automatic translation from printed Russian into English. The solutions adopted or contemplated in the course of experimental work performed at the Harvard Computation Laboratory will be emphasized, although some attention is given to alternative approaches that must be evaluated when the design of machines especially adapted to automatic translation is under consideration.

5.2. Paradigm and Reduced Paradigm

The fact that Russian is a highly inflected language is an important factor in determining the organization of an automatic dictionary. The best treatment of the inflectional system of a given language depends also on whether this language is the domain or the range of translation. When a language is considered as the domain of translation, the problem of analysis of textual material dominates. When a language is considered as the range, the emphasis is on synthesis. It is natural therefore that the point of view adopted here should emphasize the analysis of Russian. The problem of synthesizing Russian is receiving careful consideration on the part of Russian scholars, and a number of American investigators have already given considerable thought to the problem of English synthesis (see Sec. 10.7).

Our investigation is based on a tentative acceptance of the Platonic description of the Russian inflectional system, or at least of its broad outlines. To the extent that orthography, rather than phonematics, is primordial in the context of automatic translation, many of the defects ascribed by structuralists to the classical grammars either are less serious than in other contexts, or else must be accepted, since they cannot be remedied by appeal to phonematic criteria. To the extent that a machine is limited to operations on the tokens for a given text, the classical framework must be altered to eliminate appeals to intuition, or, more specifically, appeals to anything that cannot be reduced either to well-defined practical operations on textual tokens or to look-up in finite tables. This elimination and a replacement by more suitable elements are most readily performed in the nominalistic spirit.

As a first step in this direction the concept of the paradigm must be examined somewhat more closely. From the point of view of school grammars, each Russian noun has twelve variant forms, six in the singular and six in the plural, as illustrated in Table 15. If, however, one considers the representations that actually occur in Russian texts, frequently only fewer than twelve can be distinguished (Table 17).

From a purely structuralist point of view, therefore, the classical twelve inflectional variants cannot be taken for granted. By the kind of exhaustive

Table 17. Representations of *студент**
(reduced paradigm).

(1) "студент"	(6)	"студенты"
(2) "студента"	(7)	"студентов"
(3) "студенту"	(8)	"студентам"
(4) "студентом"	(9)	"студентами"
(5) "студенте"	(10)	"студентах"

contextual analysis that structuralists propose but have not yet been able to carry out in full for any major language, one might be able to define an invariable correspondence between some representations, for instance, "студент" and "студентах", and something equivalent to the nominative singular and prepositional plural, respectively. The status of the representations "студента" and "студентов" is not so clear. On the sole evidence of isolated nouns like *студент**, it would be difficult to justify postulating a distinction between genitive and accusative. In English, for example, one can argue for the existence of cases, but the morphology of modern English strings reflects so little of any case distinctions that a grammar built on an elaborate case structure would be a rather poor model of the language. Because one could arrive at a useful distinction between Russian genitive and accusative in other ways, for example, by examining the behavior of a multitude of nouns other than those of the *студент** type, this model is accepted even for *студент**.

The intermediate point of view adopted here as a working hypothesis is that there exist distinct inflectional variants $студента_{gs}$* and $студента_{as}$*, and $студентов_{gp}$* and $студентов_{ap}$* (Table 15), represented homographically by the strings "студента" and "студентов", respectively. This condition, which will be called *internal homography**, reflects the fact that one cannot, by inspection of the string "студента" alone, determine whether it stands for $студента_{gs}$* or $студента_{as}$*. The possibility remains open, however, that inspection of neighboring strings can lead to a specific choice. It is of course this possibility that leads to postulating two distinct inflectional variants, an act that otherwise would be wanton distinction of indiscernibles.

As a consequence, it becomes useful to distinguish between what we have called the paradigm of a word, as illustrated in Table 15, and what we shall call the *reduced paradigm* of the word, as illustrated in Table 17. Note that the paradigm is a set of (abstract) forms, while the reduced paradigm is a set of strings (representations). If these two sets were in one-to-one correspondence, and if we were not concerned with coding problems, this distinction would be hairsplitting. In a context where inflection is discussed informally and no one cares whether words are represented in the Cyrillic alphabet, Morse code, or some binary code, what is meant by /the inflected forms of *студент**/ is probably crystal clear. Because of internal homography, and because we are interested, on the one hand, in properties such as case and number that are invariant under code transformation, and, on the other hand, in properties of representations, such as word length and reducibility, the distinction will be valuable in many contexts.

5.3. Canonical Strings; Direct and Inverse Inflection

In ordinary dictionaries, the paradigm of a word is conventionally represented by a standard or *canonical* form which is a homomorphic image (Sec. 2.2) of the paradigm; the nominative singular is a common canonical form for nouns, the masculine nominative singular for adjectives, and the infinitive for verbs. This lexicographic device presupposes on the part of a person using the dictionary the ability to transform an arbitrary inflected form of a word into the canonical form used as the key to the dictionary entry for the word. As conventionally practiced, selecting and applying the proper transformation requires a fairly thorough acquaintance with the inflection system of the language in question, as well as a certain amount of imagination. Under unusual circumstances more than one entry may be given for a single word: *child** is represented by both $child_s$* and $children_p$* in *Webster's*.

The usefulness of canonical entries in English and Russian dictionaries is based on the fact that, for a large proportion of words, all members of a reduced paradigm may be generated from the string representing the canonical form by a relatively simple algorithm, and conversely. Given the canonical string "dog", a string for the plural of *dog** is made simply by adding "s" to the right-hand end of the canonical string. Conversely, given the string "dogs", the canonical string is obtained simply by deleting "s". In practice, the matter is slightly more complicated, for we must learn that "loss" is a canonical string, but not "los", and that the plural-forming desinence of "losses" is "es", not "s". A transformation generating one or more members of a reduced paradigm, given the canonical string, will be called *direct inflection** or just *inflection** as it usually is. The "inverse" transformation producing the canonical string from a member of the reduced paradigm will be called *inverse inflection**. It will become apparent subsequently that inverse inflection is not the inverse of direct inflection in the strict sense of Sec. 2.3, but the terminology is sufficiently precise for our purposes.

It is important to note that inflection, direct or inverse, is not necessarily isomorphic with the addition or deletion of desinences. Many inflected languages use other representational techniques, of which traces may be found even in Russian and in English. The nominative singular of *человек** is "человек", but the nominative plural is "люди"; and English has the pair "man"-"men". That most Russian representations are reducible over the ordinary alphabet into initial substrings identifying words and terminal substrings identifying case and number is in a sense accidental. Other modes of representation are possible and are used. However, granted the occurrence of the accident in a particular language, one can make the most of it in the formulation of algorithms for inflection.

An automatic dictionary may provide for the look-up of strings occurring in texts by either: (*a*) providing a distinct complete entry for each distinct string (*paradigm dictionary*); or (*b*) providing a single canonical entry, to which the distinct strings may be referred (*canonical dictionary*).

The first of these alternatives entails using a pure table look-up algorithm, with the usual advantage of great conceptual simplicity, and the usual disadvantage due to limitations on the size of tables that can be used practically. The second entails using an algorithm capable of relating an arbitrary string to the proper canonical string used as the key to the corresponding entry. A variety of compromises between these two extremes can be made (Secs. 9.6, 10.3).

The experimental work at the Harvard Computation Laboratory has been based chiefly on the use of canonical entries. The algorithms for direct and inverse inflection developed for this purpose are described in detail in the following sections of this chapter. The system is capable of being shifted toward greater use of direct table look-up, to take advantage of the expected development of cheap, high-capacity storage devices and to supplement other algorithms in situations where only table look-up can help (Secs. 9.6, 10.3).

A glance at the reduced paradigm of *студент**, given in Table 17, shows that each member may be factored into two substrings, an initial string "студент" and a terminal string selected from among the null string " # " and the strings "а", "у", "ом", "е", "ы", "ов", "ам", "ами", and "ах". The fact that "студент" is present as an initial substring in every string of the reduced paradigm and that the same set of terminal substrings occurs in a multitude of other paradigms when these are similarly factored suggests that "студент" be used as a canonical string for *студент**. The identity of the initial string "студент" with the string for the nominative

Table 18. Reduced paradigm of *дама**.

(1)	"дама"	(6)	"дам"
(2)	"дамы"	(7)	"дамам"
(3)	"даме"	(8)	"дамами"
(4)	"даму"	(9)	"дамах"
(5)	"дамой"		

singular minus its terminal null string " # " is one reason why the nominative singular is conveniently taken as a canonical form. In this case, direct inflection can be effected by concatenating appropriate terminal strings with the canonical string, and inverse inflection can be effected by factoring strings into two substrings, of which one is the canonical string, the other a terminal string.

Other words, for example, *дама**, have reduced paradigms with somewhat different properties. It is evident from Table 18 that although the substring "дам" plays the same role as "студент", it is used in conjunction with a different set of terminal strings, namely, "а", "ы", "е", "у", "ой", " # ", "ам", "ами", "ах". Moreover, the string "дам" stands for the genitive plural of *дама**, not for the nominative singular. The reasoning adopted for *студент** would in this case suggest the use of the genitive plural representation as a canonical form. In conventional dictionaries it is nevertheless the string "дама" that is used as the key to the entry for *дама**. Direct inflection in this case must be effected by first deleting the terminal substring from the conventional canonical string, and then joining the remaining initial string with one of the set of permissible terminal strings. For inverse inflection, the procedure is reversed. This procedure appears somewhat inefficient because of the necessity of both deleting and adding a terminal string. Strictly speaking, this is true also in the case of *студент**, where the null string must be either added or deleted. However, in that case we can have the best of both worlds: on the one hand, postulating a null string simplifies the conceptual framework of inflection; on the other hand, the addition or deletion of a null string is a null operation which costs nothing to realize since in practice nothing need be done.

In the reduced paradigm of *знание** (Table 19), there is no string such as "студент" or "дам" which both stands alone (with a null terminal string) and serves as a substring of the remaining strings of the reduced paradigm.

Table 19. Reduced paradigm of *знание**.

(1)	"знание"	(6)	"знаний"
(2)	"знания"	(7)	"знаниям"
(3)	"знанию"	(8)	"знаниями"
(4)	"знанием"	(9)	"знаниях"
(5)	"знании"		

There is, however, a string "знани" which is a substring common to all members of the reduced paradigm, and which therefore appears suitable as a canonical string.

It becomes clear from an examination of a large number of reduced paradigms that the presence of a substring common to all members of the paradigm is a very frequent although by no means universal phenomenon. Moreover, the distinct terminal substrings joined with the common substrings are not unique with the paradigm of each distinct word. Indeed, the set of all words may be partitioned into large subsets, each word in a subset having a paradigm formed with the same characteristic set of terminal substrings. The total number of distinct terminal substrings is well under 100.

Loosely speaking, the initial substrings and terminal substrings may profitably be identified with classical *stems* and *desinences*, respectively. It will be shown, in the next section and in Chapters 9 and 10, that the rigorous definition for Russian of initial strings and terminal strings that can easily be identified by a machine does not coincide exactly with the classical idea of stems and desinences. Distinct terms, such as *generating stem**, *canonical stem**, *first-order affix**, will therefore be used to avoid confusion with classical stems and desinences whenever the need for distinction arises.

The fact that a relatively small number of terminal strings enters into the formation of a very large number of paradigms in turn influences the choice of the canonical string for certain paradigms, such as that of *окно** given in Table 20. Although "ок" is the common initial substring, it is not the commonly accepted stem, nor do the terminal strings "но", "на", and so on, recur in a manner that would justify their inclusion among the desinences. In this case, it turns out to be simpler to adopt both "окн" and "окон" as canonical stems. This question is considered further in Sec. 9.3.

Something close to but not identical with the classical stem is chosen as a canonical string because it can usually be obtained from arbitrary members of a reduced paradigm by algorithms simpler than those that would be

Table 20. Reduced paradigm of *окно**.

(1)	"окно"	(6)	"окон"
(2)	"окна"	(7)	"окнам"
(3)	"окну"	(8)	"окнами"
(4)	"окном"	(9)	"окнах"
(5)	"окне"		

necessary if the classical canonical string were retained. This should not be surprising. The classical canonical strings are used in conventional dictionaries because they suit the capabilities, or at least the habits, of persons. There is no reason to expect that what is good for man is good for a machine, or vice versa.

To the extent that terminal strings correspond to desinences, data regarding the syntactic role of a word in a sentence can readily be made available as a by-product of dictionary look-up (Sec. 10.1). In the following section, an algorithm defining canonical stems will be described.

5.4. A Definition of Russian Affixes and Canonical Stems

In considering translation from Russian into English, both inverse inflection of Russian to gain access to a dictionary and direct inflection of English to produce grammatically correct English strings are of great importance. In the initial stages of research on Russian-English automatic translation, the inverse inflection of Russian took priority, since little else of a concrete nature could be accomplished without an operating automatic dictionary.

Because inverse inflection precedes dictionary look-up, the inverse inflection algorithm must operate on strings as they occur in texts, initially without benefit of information that can be made available within the dictionary for use in later stages of the translation process. In theory, inverse inflection of a given string can be made a function not only of this string, but also of neighboring strings, that is to say, a function of the context of the string. Certain ambiguous situations could be resolved in this way. The problem is enormously simplified, however, if inverse inflection is considered simply as a function of the given string alone. As it happens, the results achieved in this way are quite satisfactory both practically and aesthetically, and adjustments based on context can easily be made at a later stage of the translation process.

A starting point for the definition of Russian affixes developed in this section is given by the list of substrings shown in Table 21. Such a list could have been obtained by a structural analysis of some large sample of words, alphabetized on the letters from right to left, and lined up on the right in the manner illustrated in Table 22. It proved more expedient to include in the list of Table 21 the desinences described in conventional grammars, allowing for modifications of the list in the light of later experience with large text samples.

The list of Table 21 is divided into two parts, to reflect an important characteristic of Russian verbal strings. Many Russian verbs occur in pairs, of which one member is an ordinary verb, the other a verb usually having the same or related meanings but with a reflexive, reciprocal, or passive connotation; in some instances, the second verb has altogether different meanings. Strings for the second type of verb are distinguished from those for the first kind by the presence of a terminal string "ся" or "сь" in place of the null string. These terminal strings actually play a derivational rather than an inflectional role. They must be taken into account in considering inflection because when they are present inflection is represented in a medial rather than in a terminal position in the string. Because of their special role these terminal strings are distinguished and called *reflexive substrings**. They are listed in Table 21*a*. The initial string preceding a reflexive substring is called a *stem of order zero**. For many words, the stem of order zero is distinguished from an ordinary word string only by the absence of the phantom null reflexive substring.

The inflectional substrings listed in Table 21*b* include the strings commonly recognized as desinences. The initial string preceding an inflectional substring is a *stem of order one.*

Should we want to use stems of order one as canonical stems it would be necessary to specify how they might be obtained from strings occurring in texts and then matched against dictionary keys to obtain access to the proper entries. One method of obtaining such canonical stems would be to factor each text string into its component reflexive substring, inflectional substring, and stem of order one. The stem of order one might then be matched against

Table 21. The sets of substrings.

a. Reflexive

1	#
2	сь
3	ся

b. Inflectional

1	#	30	о
		31	его
2	а	32	ого
3	в	33	ат
4	ев	34	ет
5	ов	35	ит
		36	ут
6	е	37	ют
7	ее	38	ят
8	ие		
9	ое	39	у
10	ете	40	ему
11	ите	41	ому
12	ые		
		42	ах
13	и	43	их
14	ами	44	ых
15	ими	45	ях
16	ыми		
17	ями	46	ы
18	вши		
		47	ь
19	й	48	ть
20	ей	49	ешь
21	ий	50	ишь
22	ой		
23	ый	51	ю
		52	ую
24	ам	53	ью
25	ем	54	юю
26	им		
27	ом	55	я
28	ым	56	ая
29	ям	57	яя

dictionary keys in the manner of Exercise 1-3. The problem of identifying the terminal characters of a string with substrings of Table 21 must now be examined.

Let α_i be the ith substring of Table 21a or 21b. The problem is how to instruct a machine to make a proper factoring of text strings. The following set of instructions defines one possible way:

(1) Examine the terminal substrings of the given text string. Factor the

longest terminal substring identical with some α_j of Table 21a. The factored terminal string is a reflexive substring; the remaining initial string is a stem of order zero.

(2) Examine the terminal substrings of the stem of order zero. Factor the longest terminal substring identical with some α_j of Table 21b. The factored terminal substring is an inflectional substring; the remaining initial string is a stem of order one.

Table 22. Section of an end-alphabetized word list.

гармоникам	заключаем
графикам	произведем
точкам	приведем
узлам	будем
правилам	имеем
формулам	покажем
проблемам	скажем
системам	можем
схемам	условием
нормам	следствием
нам	действием
величинам	присутствием
ионам	запаздыванием
законам	указанием
электронам	желанием
изотопам	испытанием
приборам	подтверждением
кулонметрам	изображением
вопросам	движением
массам	снижением
ординатам	предложением
результатам	изложением
расчетам	положением
частотам	напряжением
лицам	сопротивлением
называем	проявлением
предполагаем	выделением
желаем	газовыделением
принимаем	уравнением
получаем	изменением
случаем	применением

These instructions are not in the form of a program for a machine, but the reader who has done Exercise 2-3 will surely see how they could be programmed.

Consider the result of applying these instructions to the string "знаний". According to the first instruction, the reflexive substring would be found to be "#", and the stem of order zero would be "знаний". The second instruction provides for the separation of the inflectional substring "ий", leaving the first order stem "знан". The same instructions applied to all the members of the reduced paradigm of *знание** (Table 19) would give the results shown in Table 23. The net effect would be to produce, not the single canonical stem "знани" which we were led to expect from an examination of Table 19, but also a string "знан".

If we keep the instructions as they are, we must either divide the reduced paradigm of *знание** into two sets, one characterized by the canonical

Table 23. Tentative inverse inflection of *знание**.

(1)	знан-ие	(6)	знан-ий
(2)	знани-я	(7)	знани-ям
(3)	знани-ю	(8)	знани-ями
(4)	знани-ем	(9)	знани-ях
(5)	знани-и		

string "знан", the other by the canonical string "знани", or else abandon strict structuralism and admit that certain forms of one and the same paradigm may belong to one canonical string, and others to another, even though the basis for grouping all these forms into a single paradigm is now somewhat shaky.

The example of *окно** (Table 20) suggests that admitting the possibility of more than one canonical string being associated with a reduced paradigm is the most practical way out of certain inevitable difficulties. From this point of view, it would be consistent to admit the same possibility for *знание**. Nevertheless, the fact that words with reduced paradigms like that of *знание** are highly abundant in technical Russian makes having two entries for each of these words seem not only uneconomical but also conceptually inelegant.

Classical grammars, and extensive lists like that of Table 22 (Bielfeldt, 1958; Vasmer, Greve, and Kroesche, 1957), indicate that the substrings "ие" and "ий", which are responsible for the factoring of the initial substring "знан", are considered to be adjectival desinences, and indeed yield reasonable stems when separated from the adjectival strings where they occur. The substrings in Table 21*b* intended for use with "знание" and "знаний" are "е" and "й". The fact that each of the latter substrings is embedded also in longer substrings (7–12, and 20–23, respectively, in Table 21*b*) explains why instructions 1 and 2 do not yield a satisfactory definition of canonical stems.

A definition of canonical stems closer to that of classical stems must determine whether a given substring or a second one of which the first is a terminal substring is to be factored. For the substring "й", conditions for factoring may be stated as follows:

> If a word ends in "й", then "й" is factorable provided that the penultimate character is neither "е" nor "и" nor "о" nor "ы"; if the penultimate character is "и", "й" is factorable only if either the third character from the end is "в", "т", or "ц" or the third character from the end is "н" and the fourth character from the end is "а", "е", or "я".

Complex statements such as this are more readily expressible in the notations of the propositional calculus or of switching algebra (Quine, 1950; Caldwell, 1958). Let x_j be the proposition *the character x is in the jth position from the right-hand end of a given string*. Thus, the proposition "й" *is the last character in a string* may be written simply as $й_1$. In this notation, the condition for factoring "й", $F(й)$, may be expressed as follows:

$$F(й) = й_1\{e_2'и_2'o_2'ы_2' + и_2[в_3 + н_3(а_4 + e_4 + я_4) + т_3 + ц_3]\}.$$

This expression is to be interpreted as asserting that "й" is a factor of a string whenever the proposition on the right-hand side is true, but not otherwise; the corresponding expressions for most substrings are far less complicated. The complete set of expressions defining conditions for factoring substrings (Oettinger, 1954) is given in Table 24.

Substrings into which word strings are factored according to the set of defining expressions $F(\alpha_j)$ given in Table 24 will be called *canonical stems**, *first-order affixes**, and *affixes of order zero**. Canonical stems correspond to classical stems, and first-order affixes to desinences, far more closely than stems of order one correspond to classical stems or inflectional substrings to desinences. The reader may verify this in at least one paradigm, by evaluating the relevant expressions of Table 24 for the members of the reduced paradigm of *знание** (Table 19). A list of strings factored according to the definitions of Table 24 is shown in Table 25. The factoring of most of the strings in Table 25 corresponds to the factoring that persons with a good knowledge of Russian would give, although some exceptions are evident. An analysis of the significance of these exceptions and of the effectiveness of the definitions of Table 24 is given in Chapter 9.

5.5. An Inverse Inflection Algorithm for Russian

The definitions of affixes given in Table 24 are in a form that lends itself readily to the formulation of elegant and efficient algorithms for the automatic factoring of strings. One such algorithm is sketched in flow-chart form in Fig. 32.

Table 24. Definition of affixes.

j	α_j	$F(\alpha_j)$
		a. Zero order
1	#	$я'_1ь'_1 + с'_2$
2	сь	$ь_1с_2$
3	ся	$я_1с_2$
		b. First order
1	#	$\displaystyle\prod_{\alpha_j^1 \neq \#} [F(\alpha_j^1)]'$
2	а	$а_1$
3	в	$в_1е'_2о'_2(т'_2 + с'_3)$
4	ев	$в_1е_2$
5	ов	$в_1о_2$
6	е	$е_1\{е'_2и'_2о'_2ы'_2(т'_2 + е'_3и'_3) + и_2[в_3 + н_3(а_4 + е_4 + я_4) + т_3]\}$
7	ее	$е_1е_2$
8	ие	$е_1и_2в'_3(н'_2 + а'_4е'_4я'_4)т'_3$
9	ое	$е_1о_2$
10	ете	$е_1т_2е_3$
11	ите	$е_1т_2и_3$
12	ые	$е_1ы_2$
13	и	$и_1(м'_2 + а'_3и'_3ы'_3я'_3)(ш'_2 + в'_3)$
14	ами	$и_1м_2а_3$
15	ими	$и_1м_2и_3$
16	ыми	$и_1м_2ы_3$
17	ями	$и_1м_2я_3$
18	вши	$и_1ш_2в_3$
19	й	$й_1\{е'_2и'_2о'_2ы'_2 + и_2[в_3 + н_3(а_4 + е_4 + я_4) + т_3 + ц_3]\}$
20	ей	$й_1е_2$

Table 24.   (*Continued*)

j	α_j	$F(\alpha_j)$
21	ий	$й_1и_2в_3'(н_3' + a_4'e_4'я_4')т_3'ц_3'$
22	ой	$й_1о_2$
23	ый	$й_1ы_2$
24	ам	$м_1a_2$
25	ем	$м_1е_2$
26	им	$м_1и_2$
27	ом	$м_1о_2$
28	ым	$м_1ы_2$
29	ям	$м_1я_2$
30	о	$о_1(г_2' + о_3'е_3')$
31	его	$о_1г_2е_3$
32	ого	$о_1г_2о_3$
33	ат	$т_1a_2$
34	ет	$т_1е_2$
35	ит	$т_1и_2$
36	ут	$т_1у_2$
37	ют	$т_1ю_2$
38	ят	$т_1я_2$
39	у	$у_1(м_2' + е_3'о_3')$
40	ему	$у_1м_2е_3$
41	ому	$у_1м_2о_3$
42	ах	$х_1a_2$
43	их	$х_1и_2$
44	ых	$х_1ы_2$
45	ях	$х_1я_2$

Table 24. (*Continued*)

j	α_j	$F(\alpha_j)$
46	ы	$ы_1$
47	ь	$ь_1[c_2't_2'(ш_2' + e_3'и_3') + t_2c_3o_4]$
48	ть	$ь_1т_2(c_3' + o_4')$
49	ешь	$ь_1ш_2e_3$
50	ишь	$ь_1ш_2и_3$
51	ю	$ю_1y_2'ь_2'ю_2'$
52	ую	$ю_1y_2$
53	ью	$ю_1ь_2$
54	юю	$ю_1ю_2$
55	я	$я_1a_2'c_2'я_2'$
56	ая	$я_1a_2$
57	яя	$я_1я_2$

Table 25. A sample of Russian affixes and canonical stems
as defined by Table 24.

являющ-ее	нов-ый
главн-ым	мог-ут
источник-ом	устройств-о
исключа-ет-ся	приспособлени-е
исключ-ая	больш-ие
благодар-я	расстояни-й
красн-ая	измерени-и
метод-е	пр-и
леж-ат	действующ-ей
аппар-ат	длительност-ью
аппарат-а	сущност-ь

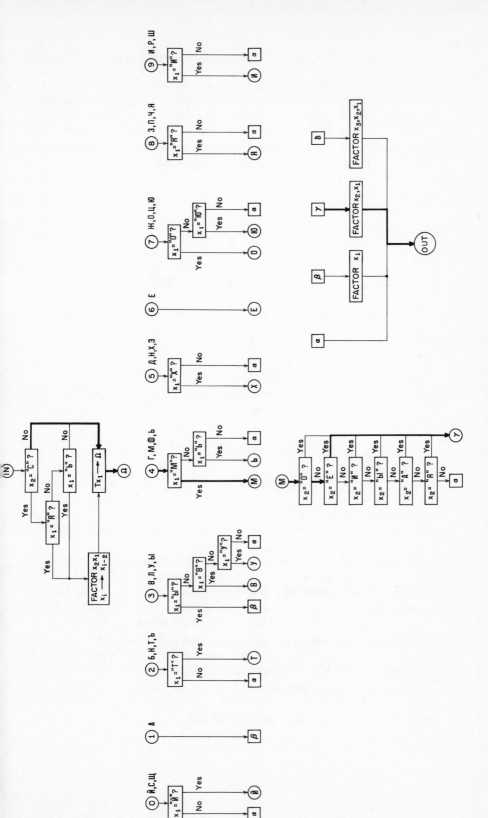

Fig. 32. An inverse-inflection algorithm for Russian.

The first portion of the algorithm factors affixes of order zero, if any. This portion, like the remainder of the algorithm, is designed by interpreting the expressions of Table 24 as defining conditions for selecting alternative *branches* of a *tree*. The origin of the botanical terminology is evident from an examination of Fig. 33, which represents the topological structure of the flow chart of Fig. 32; *root** might have been more apt in this case than

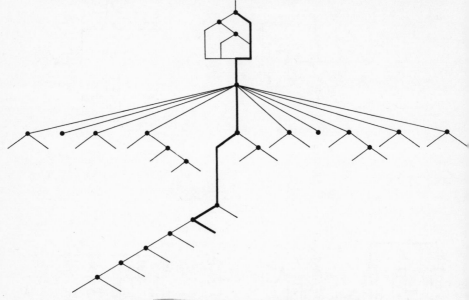

Fig. 33. Tree structure of the inverse-inflection algorithm.

branch*, but the custom has been set. Treelike structures of this kind are widely used as models, not only of programs, but also of switching networks in telephony and in railroading, for obvious reasons.

All branches of the zeroth-order tree lead to a node of ten branches, each of which corresponds to a set of possible terminal characters of stems of order zero. The sentence $Tx_1 \to \Omega$ defines the selection of one of these ten branches by providing for the setting of the variable connector Ω. The transformation T is the "multiplication by 1" operation described in Sec. 2.7. A program corresponding to $Tx_1 \to \Omega$ is given in Example 2-9.

It is important to notice that, while the program corresponding to the entire structure of Fig. 33 may consist of many instructions, only a small subset of these is actually executed in any particular instance. The path followed when the terminal string of a word representation is "ем" is illustrated by the heavy lines on Figs. 32 and 33.

With reference to Fig. 32, we note that when a representation ends in "ем", $x_2 \neq$ "с", hence there is no affix of order zero. Table 11 shows that T"м" $= 4$, hence Ω is set equal to 4. The tree for distinguishing "м" from "ь" and from "г" and "ф" corresponds to the secondary selection program to which the U instruction in register *104* (Example 2-9) should transfer control. Once it has been established that $x_1 =$ "м", entries 24 through 29 of Table 24*b* indicate that further selection depends only on x_2. One possible mode of selection on x_2 is shown in Fig. 32.

> *Exercise 5-1.* What simplifications, if any, could you make in the tree for selection on x_2, given $x_1 =$ "м", under the following circumstances: (*a*) the values of x_2 that are significant in selection all occur on distinct lines of Table 11; (*b*) the representations of Cyrillic characters are reducible, and a specific component of the representation has one value if the character is a vowel, another if it is a consonant. How valuable would such a reducible structure be in selections following values of x_1 other than "м"?

Separate output branches are not necessary for every affix in Table 24, since it is sufficient to distinguish whether the affix of order one consists of 0, 1, 2, or 3 letters. Accordingly, once an affix has been identified, control is transferred to one of the connectors α, β, γ, or δ. Figure 33s hows only certain paths in their entirety, among them those which lead to α or β following selection on x_1 alone. and the paths for $x_1 =$ "м". The majority of the paths not shown are similar in structure to that for $x_1 =$ "м", but those for $x_1 =$ "е", "и", "й", or "ь" are somewhat more complex.

> *Exercise 5-2.* Given that $x_1 =$ "й", draw a flow chart for the subsequent selections. Use the data of Table 26 as a guide in designing the tree.

Once the algorithm described in this section has been applied to a Russian word string, the first-order stem can be used as a key for dictionary look-up. The detailed look-up procedure depends on the kind of information storage medium used to store the dictionary. Whatever medium is used, a word occurring in a text is presumed to be found in the dictionary *only if its stem matches a dictionary stem identically*, and only one attempt at matching need be made for each stem. The affixes of orders zero and one are saved, to be interpreted after look-up with the aid of grammatical data found in the dictionary (Chapter 10).

Alternative definitions of desinence-like terminal substrings and of corresponding alternative factoring and search algorithms have been given in the literature. At least one system (Ramo-Wooldridge, 1958, pp. 27–35) is based in part on the idea, considered in Sec. 5.4, of factoring the longest terminal substrings occurring in word representations and also present in a table. This system has some properties, good and bad, akin to those of the system adopted and described in this section, but seems far more cumbersome and prone to error. A major difference between the system that was finally adopted and the alternatives examined in Sec. 5.4 lies in the degree to which affixes, as factored by the systems, correspond to "classical" desinences, and in the related question of stem definition, as illustrated in Table 23.

Table 26. First-order affixes ranked by frequencies
(based on a 6000-word sample of texts).

	$-\alpha$	$f(\%)$	Σf		$-\alpha$	$f(\%)$	Σf
1	#	12.25	12.25	29	ым	0.89	91.99
2	а	9.49	21.74	30	ит	0.89	92.88
3	е	7.17	28.91	31	ами	0.82	93.70
4	и	6.74	35.65	32	ю	0.74	94.44
5	я	5.84	41.49	33	ыми	0.69	95.13
6	о	5.40	46.89	34	ее	0.61	95.74
7	ы	4.93	51.82	35	ие	0.61	96.35
8	ой	4.49	56.31	36	ах	0.56	96.91
9	ет	4.00	60.31	37	ий	0.50	97.41
10	ого	3.05	63.36	38	ому	0.47	97.88
11	ть	2.44	65.80	39	ью	0.45	98.33
12	ом	2.38	68.18	40	в	0.26	98.59
13	ых	2.29	70.47	41	ях	0.22	98.81
14	ов	2.05	72.52	42	ями	0.20	99.01
15	ем	2.03	74.55	43	ам	0.19	99.20
16	у	1.82	76.37	44	ят	0.17	99.37
17	их	1.71	78.08	45	ими	0.15	99.52
18	ая	1.43	79.51	46	ут	0.13	99.65
19	ей	1.34	80.85	47	ему	0.09	99.74
20	ь	1.34	82.19	48	вши	0.06	99.80
21	ые	1.30	83.49	49	ям	0.06	99.86
22	ый	1.19	84.68	50	ев	0.04	99.90
23	им	1.19	85.87	51	ат	0.04	99.94
24	ют	1.17	87.04	52	юю	0.04	99.98
25	й	1.08	88.12	53	яя	0.02	100.00
26	ое	1.06	89.18	54	ете	0.00	—
27	его	0.94	90.17	55	ите	0.00	—
28	ую	0.93	91.10	56	ешь	0.00	—
				57	ишь	0.00	—

Other proposed systems depend, as far as one can make out from their descriptions, on programming a machine "to select from the dictionary the longest entry which is a part of the input word" (Booth, Brandwood, and Cleave, 1958, p. 67), thereby abandoning the idea of entering the dictionary by identity match only. At least one variant of this approach is based on ordering a dictionary, not in the straight alphabetic order of stems, but so that "a short stem is listed . . . after the longer stems in which it is contained" (Kulagina and Mel'chuk, 1956, p. 112). Many of these systems also appear to require uneconomical and inelegant repeated attempts at matching each stem.

The factoring rules defined in Table 24 may also be expressed by an algorithm that can be executed directly by the input mechanism of the translating system. The form of such an algorithm has been considered by Cohn (1958).

5.6. Direct Inflection

Direct inflection is of importance in two phases of automatic translation. One kind of direct-inflection algorithm may serve in the synthesis of sentences belonging to the *range* of translation to generate properly inflected strings from the canonical strings given in the dictionary. For example, if English is the range, and a Russian noun occurs in the plural, its English correspondent must be formed accordingly. Thus, if *dog** is represented in a dictionary by the canonical string "dog", and if the original string is in the plural, means must be provided to supply "dogs" rather than "dog" as the image string. The direct-inflection algorithm must, in this case, be capable either of *forming* "dogs" from "dog" if only the latter form is explicitly given in the dictionary, or else of *selecting* the proper one of "dog" and "dogs", should it prove more advantageous to store both forms explicitly in a table. A direct-inflection algorithm for this purpose may therefore be either a string-generating algorithm, capable of producing a specific single string on demand, or else a table look-up algorithm capable of finding the proper single string in a table containing all members of a reduced paradigm. In the latter case the string-generating algorithm may be used to generate the table during dictionary compilation.

A second type of direct-inflection algorithm is useful in the compilation of a dictionary, to aid in producing the *domain* strings to be used as keys in dictionary look-up. As pointed out in Sec. 5.3, a dictionary may be a paradigm dictionary, where each word is represented by a *set* of entries, one for each member of its reduced paradigm, or else it may be a canonical dictionary, where each word is represented by a canonical entry to which any member of its reduced paradigm will be referred.

Consider first the paradigm dictionary. How are the keys for the entries to be obtained? One unattractive possibility is to process an amount of text sufficiently large to guarantee that at least one instance of every member of the reduced paradigm of every word of interest will be found. Another possibility is to list only the classical canonical string for every word of interest, and to use the list as a point of departure for generating the other members of the reduced paradigms of these words.

Generating paradigms would be staggering as a purely manual process, if a dictionary of many thousand words is contemplated (University of

Washington, 1958). The standard Russian noun paradigm has 12 members, the adjective paradigm has 24 or 28, and verbs have well over 100, if participles are counted. The membership of the reduced paradigms is somewhat less, but still enormous. An automatic method of direct inflection in the form of a string generator is therefore essential.

With a canonical dictionary, a method of direct inflection hardly seems necessary, especially if the classical canonical strings themselves were to be used as keys, or if the inflectional system of the domain language were completely free of irregularities. Unfortunately, the classical canonical strings are not well suited for use in an automatic dictionary, for the reasons given in Sec. 5.3, and the inflectional system of Russian is far from being wholly regular.

Unless it can be guaranteed that the canonical strings used as keys are precisely those that will result from the application of the inverse-inflection algorithm to strings occurring in texts, there is a danger of either looking up the wrong entry or missing an entry which is in the dictionary. *Complete consistency in the operation of an automatic dictionary can be guaranteed, whatever algorithm may be used for inverse inflection, if only this inverse-inflection algorithm lends itself to being used also to generate the canonical strings.* If the complete reduced paradigm of every word to be represented in the dictionary is made available by direct inflection, the inverse-inflection algorithm can be applied to each member during the process of compilation, thus generating canonical strings consistent with those that will be produced by the application of the same algorithm to strings occurring in texts.

In the case of a word such as *окно** the procedure described in the preceding paragraph guarantees that the word will be represented by the two stems "окн" and "окон", and not by "окн" alone. Likewise, if the longest-terminal-string-identical-with-an-entry-in-the-affix-table method of inverse inflection were used, the procedure would guarantee that *знание** (Table 23) would be represented by both "знан" and "знани". If this procedure were to be carried out manually, there might be no need for explicitly generating every member of every reduced paradigm. A person who knows Russian fairly well will be aware of the existence of the string "окон" and can act accordingly, without seeing all other strings of Table 20 written out. However, if only because of the great tedium of handling thousands of words in this way, the probability of mistakes is extremely high, and catching mistakes is extremely difficult. To be aware of all subtle variations on this basic theme, the person would have to know Russian more than fairly well, and consequently be disinclined to subject himself to a tedious, repetitious task that might well be performed automatically by a machine.

For these reasons an algorithm for automatic direct inflection was felt to be necessary and such an algorithm was developed. This algorithm is intended primarily for dictionary compilation, not for synthesis, although the classification system on which it is based could be used to develop a synthesis algorithm useful for translating into Russian. Synthesis algorithms for English, useful for translating into English, are also being developed.

In formulating a direct-inflection algorithm for compilation purposes, every effort was made to reduce the need for manual intervention to a minimum. Where intervention was felt to be unavoidable, the procedures developed were such as to maximize speed while minimizing tedium, the probability of mistakes, and the need for a profound knowledge of Russian. Auxiliary procedures for checking manual work also had to be developed. The direct-inflection algorithm is based on a morphological classification system partitioning the set of all Russian words into classes each of which contains only words with like inflection patterns; this system is described in Chapter 6.

CHAPTER 6

MORPHOLOGICAL AND FUNCTIONAL
CLASSIFICATIONS OF RUSSIAN WORDS

6.1. A Morphological Classification of Russian Words

A new system of classification of Russian words is described in this chapter. This system was developed to enable the automatic direct inflection of any word after it had been identified as belonging to a specific class whose members are inflected according to a well-defined pattern (Magassy, 1956, 1957; Matejka, 1957; Landau, 1955). Classification systems given in existing grammars were found to be incomplete and often incompatible with our requirements. Our system of classification was based on the assumption that the identification of the inflectional pattern of a word should remain, at least initially, a manual function while the actual generation of distinct inflected forms could be an almost completely automatic process; ease and accuracy of identification could therefore be promoted by using any readily obtainable data meaningful to a person, while generation of inflected forms had to be based strictly on explicit *orthographic* data recognizable by a machine.

Some of the criteria used in defining classes are illustrated in Fig. 34. The first column of Fig. 34 describes the paradigms of the Russian words *дама**, *комната**, and *граница**. The corresponding *generated paradigms* are in the second column. It will be noted that *дама** and *комната** have been included in the same class, even though they differ in the accusative plural, because their reduced paradigms can be generated by identical procedures.

The different correspondences between the generated strings "дамы", "комнаты", and members of the paradigms of *дама** and *комната**, respectively, are due to the fact that one of these strings represents an animate noun, the other an inanimate one. Nothing in the morphology of either word reflects this functional difference; a distinction can be made only *after* dictionary look-up provided that animate and inanimate nouns are identified by appropriate marks in their dictionary entries. It should be noted that while the animate/inanimate distinction will enable a machine to avoid relating "дамы" spuriously with a hypothetical $дамы_{ap}*$, it does not remove internal homography (Sec. 5.2), for example, that associating "дамы" with both $дамы_{gs}*$ and $дамы_{np}*$. The problem of functional analysis is considered further in Sec. 6.2.

154

The justification for including *граница** in the same class as *дама**
and *комната** is twofold: first, differentiating *граница** from *комната**
is inefficient at this stage and is accomplished more readily at the later stage
described in Sec. 6.2. Second, the only price to be paid for simplifying initial
classification in this way is the creation of generated paradigms each of which
may include not only the reduced paradigm of a member of a class, but also
one or more *vacuous strings* that may not belong to the reduced paradigm

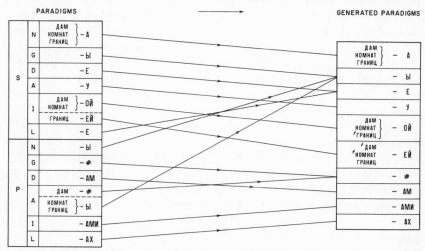

*STRINGS BELONGING TO A GENERATED PARADIGM, BUT NOT TO A REDUCED PARADIGM (VACUOUS STRINGS)

Fig. 34. Noun paradigms (examples drawn from class *N4*).

of this word (for instance, "границой", "комнатей", "дамей"). This
addition is harmless. If a paradigm dictionary is being compiled, the vacuous
strings will never be consulted in the absence of error, and therefore can be
eliminated once it is noted that their frequency of use is zero. In a canonical-
stem dictionary a vacuous string often leads to a stem identical to those
obtained from the other strings and therefore leaves no traces. Should a
vacuous string lead to a distinct stem of its own, this stem can be removed
in the manner described in Sec. 7.8. Moreover, should a string like "дамей"
occur in a text owing to a typographical mistake, the mistake will usually be
detected by the methods outlined in Sec. 10.1.

The generation of vacuous strings is very much like a procedure used
quite frequently in approximating mathematical functions over a given range;
any convenient function may be used that suitably approximates the desired
function in the specified range; its behavior outside of this range is of no
consequence. Other expedients of this kind are also proving their worth

in terms of greater systematization than would be possible without them. As those experienced in automatic data processing well know, the handling of a single exceptional case often proves more costly and more time-consuming than the routine handling of a few extra elements.

The definitions of inflectional classes and the rules for forming inflected strings for words belonging to these classes are illustrated in Table 27. The characteristics identifying a class are given under *class identification* and the process of inflection is described under *generation rules*. It will be noted that all structural operations to be performed in the process of automatic inflection, or the phenomena bearing upon them, are described strictly in terms of orthography. For example, in the class *N4*, the generated paradigm is described as consisting of the classical canonical string, or *word* for short, plus several other strings generated by joining specified strings to a generating stem (GS). The rule of formation for the generating stem itself is given in such terms as *word minus last letter**, with *minus** represented by the usual "—".

The *generating* stems defined under *generation rules* are not necessarily identical with the *canonical* stems that will be used as keys for entries in the dictionary. The latter are obtained by applying the inverse-inflection algorithm to the strings in the generated paradigm, and the resulting canonical stems are not always identical with the generating stems. For example, the past passive participial strings of some verbs are constructed most readily by adding "енный" to a generating stem. The canonical stem of the corresponding paradigm includes the string "енн".

Because assigning words to classes is intended, for the present, to be a manual task, any classification criterion that is easily recognized by persons can be used under *class identification*. For example, stress distinctions, which cannot be used in defining generation rules, do serve as a means of class identification. In addition, significant examples, lists of exceptions, and so forth, have been given wherever possible.

This system for generating inflected strings also accounts for a large number of words whose formation is "irregular". For example, the class *N4.31* comprises words in which the character "o" is introduced in one string. The generation rules for this class (Table 27) therefore specify that an "o" must be inserted between the last and penultimate characters of the generating stem before the form *g* is constructed. The class *N4.31* is further distinguished from the class *N4* by the use of "и" instead of "ы" in the inflected string *b*. Whenever the generation rules for one class deviate only slightly from those for another class, they have been stated as exceptions to the rules for this other class. Considerable economies in programming inflections are achieved as a result. The complete classification system (Magassy, 1957) is given in the appendix to this chapter.

Table 27. Illustration of class identification and generation rules.

Class	Examples	Class identification	Generation rules	
			Generating stem	Inflected strings
N4	дама лампа игла служба	A class ending in a preceded by any consonant *except*: 1. г, ж, к, х, ч, ш, щ; 2. ц, whenever preceded by another consonant; 3. the majority of cases when the consonant is л, н, preceded by another consonant; 4. some cases when the consonant is м, р, preceded by another consonant; 5. a few cases when the consonant is б, preceded by another consonant	word – last letter	*a.* word *f.* GS + ей *b.* GS + ы *g.* + # *c.* + e *h.* + ам *d.* + y *i.* + ами *e.* + ой *j.* + ах
N4.31	выплавка кишка	A "reappearing о" class embracing the nouns ending in a preceded by: 1. any consonant (*except* й, ж, ш, ч, ц), not followed by ь, +к; 2. ж, ш, ч, +к, whenever the stress falls upon the ultima of the word	word – last letter, but an o must be inserted between the penult and the ultima of the generating stem in *g*	As in *N4*, except и for ы in *b*

One feature of the generation rules for verbal paradigms deserves special mention. A given verb may have up to four participial forms associated with it, namely, present and past active participles, and present and past passive participles. As indicated in Sec. 5.1, these forms belong to a zone where inflection and derivation are difficult to distinguish. Conventionally, participles are regarded as belonging to the paradigms of verbs, and this convention was followed in making the tables in the appendix.

It is not sufficient, however, to generate a single string for each participle, since participles themselves are inflected like adjectives. The complete generation rules for participles therefore call for two major steps. In the first step, the classical canonical form of each participle is generated as one of the members of the reduced paradigm of the verb. At the same time, the appropriate adjectival class mark is generated automatically. This step may be viewed as the automatic *derivation* of an adjectival form. In the second step, the generated paradigm of each participle is created according to the rules for the class of adjectives to which it has been assigned by the first step.

For example, the conventional participial canonical forms "делаюший", "делавший", "делаемый", and "деланный", are associated with *делать**, a verb of class *VI*. These forms are automatically assigned to the classes *A4*, *A4*, *A3*, and *A1*, respectively.

The system of morphological classification comprises 8 classes of adjectives, 38 classes of nouns, and 46 classes of verbs. Invariable words are assigned to a special class, labeled *I*. The system of classification is sufficiently comprehensive to include all but a few unproductive classes with highly atypical paradigms, listed in the appendix to this chapter; it is completed by the definition of classes, labeled *N99.99*, *A99.99*, and *V99.99*, to which all words not falling into any of the other classes are assigned. These words must be inflected by hand. Of a total of 7600 words classified at one stage of our work, all but 24 were assigned to genuine classes. The distribution of these words among the major groups is indicated in Table 28.

Class definitions given as illustrated in Table 27 are not in a form that readily lends itself to rapid classification. To make classifying as easy as possible, the rules were expressed in a synoptic tabular form (Matejka, 1957) illustrated in Table 29 by the classification table for adjectives. The vertical line in Table 29 divides desinences on the right from significant terminal stem characters on the left. The special characters "C", "V", and "C$_ь$C" are interpreted as follows: "C" denotes any consonant not specified earlier in the group between horizontal lines, and "V" likewise for vowels. The combination "C$_ь$C" signifies any consonant, followed by the soft sign or not, and not specified earlier within the group. For example, an adjective ending in "ий" preceded by "к" may be assigned to any of the classes *A6*, *A7*, or *A8* depending on what precedes "к". If the character preceding "к" is one of the three indicated next to *A7*, the adjective is assigned to this class; if the character preceding "к" represents any consonant not in the list belonging to *A7*, the adjective is assigned to the class *A6*; finally, if the character preceding "к" represents any vowel, the adjective is assigned to the class *A8*.

Vowel changes are marked by the character ">". On the left side of the character is the vowel before the change, and on the right side the vowel

Table 28. Sample distribution of words among major groups.

Group	Number	Percent
adjectives	2477	32.59
nouns	3972	52.26
invariables	74	0.97
unclassified nouns and adjectives (*N99.99, A99.99*)	13	0.17
verbs	1053	13.86
unclassified verbs (*V99.99*)	11	0.14
	7600	99.99

after the change. In the third group in Table 29, for example, adjectives ending in "ый" preceded by "л" or "р" are normally assigned to the class *A3*, unless "е" is inserted as the second character from the end in the masculine predicative form of the adjective.

The complete set of synoptic tables is given in the appendix to this chapter. By means of these tables, assigning to classes becomes a routine task which can be done by a person with relatively little knowledge of Russian, although occasionally dictionaries must be consulted in the process. Experience has shown that the amount of dictionary consultation at this stage is negligible. With this one exception, all inevitable dictionary consultation is concentrated at a single stage of the compilation process, namely, when English correspondents and grammatical codes are adjoined to the canonical stems (Sec. 7.8). The distribution of an initial group of 7600 words among the several classes is shown in Table 30. The grouping of invariables includes a variety of words such as indeclinable nouns, prepositions, conjunctions, interjections, particles, and certain adverbs and comparative adjective forms. For obvious reasons, invariables need not be considered in greater detail in

Table 29. Synoptic classification table for adjectives.

ADJECTIVES

1

$$\begin{bmatrix} ш & ий \\ ж & ий \\ ч & ий \end{bmatrix} \quad A4$$

$$н \mid ий \qquad A5$$

$$\begin{bmatrix} г & ий \\ х & ий \end{bmatrix} \quad A8$$

$$C \rightarrow \begin{bmatrix} жк & ий \\ йк & ий \\ ьк & ий \end{bmatrix} \quad A7$$

$$C \longrightarrow к \mid ий \qquad A6$$

$$V \longrightarrow к \mid ий \qquad A8$$

2

$$\begin{bmatrix} ж & ой \\ к & ой \\ ш & ой \\ г & ой \\ х & ой \end{bmatrix} \quad A8$$

$$C_ьC \longrightarrow н \mid ой \qquad A2$$

$$V \longrightarrow н \mid ой \qquad A3$$

$$C \longrightarrow \quad ой \qquad A3$$

3

$$нн \mid ый \qquad A1$$

$$C_ьC \longrightarrow н \mid ый \qquad A2$$

$$V \longrightarrow н \mid ый \qquad A3$$

$$\begin{bmatrix} л & ый \\ р & ый \end{bmatrix} \quad A3, \text{ but } A2 \text{ if masculine predicate: } \# > e_2$$

$$C \longrightarrow \quad ый \qquad A3$$

Table 30. Distribution of words among classes.

Class	Number	Percent	Class	Number	Percent
A1	262	3.45	V1	431	5.67
A2	1425	18.75	V2	24	0.32
A3	450	5.92	V2.01	4	0.05
A4	24	0.32	V3	232	3.05
A5	25	0.33	V4	144	1.90
A6	243	3.20	V4.01	42	0.55
A7	31	0.41	V4.02	8	0.11
A8	17	0.22	V4.1	29	0.38
			V4.11	1	0.01
N1	1332	17.52	V4.2	30	0.40
N1.1	223	2.93	V4.21	6	0.03
N1.2	14	0.18	V5	14	0.18
N1.3	68	0.90	V5.1	1	0.01
N1.4	4	0.05	V5.2	3	0.04
N2	41	0.54	V5.3	1	0.01
N2.1	1	0.01	V5.4	4	0.05
N3	259	3.41	V5.41	1	0.01
N3.05	10	0.13	V6	11	0.15
N3.1	29	0.38	V6.1	3	0.04
N4	260	3.42	V6.2	13	0.17
N4.05	6	0.08	V7	0	0
N4.06	0	0	V8	6	0.08
N4.1	63	0.83	V8.1	2	0.03
N4.3	105	1.38	V8.2	11	0.15
N4.31	280	3.69	V9	2	0.03
N5	4	0.05	V9.1	2	0.03
N5.05	4	0.05	V10	0	0
N5.1	2	0.03	V10.01	1	0.01
N5.15	1	0.01	V10.1	1	0.01
N5.2	1	0.01	V10.2	1	0.01
N5.3	1	0.01	V10.3	1	0.01
N6	267	3.51	V10.4	0	0
N6.1	1	0.01	V11	2	0.03
N7	254	3.34	V11.1	0	0
N8	49	0.65	V12	7	0.09
N8.1	1	0.01	V13	4	0.05
N8.15	7	0.09	V14	6	0.08
N8.3	4	0.05	V15	0	0
N8.4	2	0.03	V15.1	0	0
N9	2	0.03	V15.2	0	0
N9.2	0	0	V16	2	0.03
N10	666	8.76	V17	0	0
N11	6	0.08	V18	2	0.03
N11.1	1	0.01	V19	1	0.01
N11.2	0	0	V20	0	0
N12	1	0.01	V21	0	0
N13	3	0.04	Unclassified	11	0.15
Invariables (I)	74	0.97			
Unclassified	13	0.17	Total	7600	99.99

Table 31. Correlation of conventional and morphological classes.

Conventional classes	Morphological classification
Nouns:	
1. declinable	nouns
2. indeclinable	invariables
3. substantivized adjectives	adjectives
Adjectives:	
1. positive degree	
a. declinable	adjectives
b. indeclinable	invariables
2. comparative degree flectional	
a. regular—indeclinable	invariables
b. irregular—indeclinable	invariables
c. irregular—declinable	adjectives
3. comparative degree analytical	
a. first term	invariables
b. second term	adjectives
4. superlative degree flectional	adjectives
5. superlative degree analytical	
a. inflectional terms	adjectives
b. noninflectional terms	invariables
Verbs:	
1. full system (synthetic)	verbs
2. separate participles	adjectives
3. analytical future	verbs
4. conditional, subjunctive	
a. inflectional terms	verbs
b. noninflectional terms	invariables
Pronouns	pronouns
Numerals:	
1. cardinal	nouns
2. ordinal	adjectives
Adverbs	invariables
Prepositions	invariables
Conjunctions	invariables
Interjections	invariables
Particles	invariables
Impersonal predicatives	invariables
Abbreviations	invariables

the present context. A table correlating conventional groupings with the classification system described in this section is given in Table 31.

6.2. A Functional Classification of Russian Words

The classification system described in Sec. 6.1 was designed to lead toward an algorithm for automatic direct inflection, and therefore could be based adequately on purely morphological criteria. It was acknowledged in Sec. 6.1 that different functions, of "дамы" and "комнаты", for example, could be distinguished by machine (Sec. 10.1) only if distinct marks were provided in the dictionary entries for the corresponding words. The distinction between the reduced paradigms of *дама** and *комната** on the one hand, and of *граница** on the other, must likewise be provided for.

By sufficient extension of the classification system of Sec. 6.1, all distinctions of this kind could have been taken into account at a single stage of classification (Sec. 10.3). It was, however, felt desirable to segregate routine clerical operations applicable to large classes of words, readily identified by the structure of their canonical strings, from the detailed consideration of each word as an individual. Such individual treatment is made inevitable by the need to assign English correspondents to each Russian word; it seemed therefore that other nonroutine adjustments to individual words could conveniently be made at the same time. Experience now indicates that it will be possible in the future and more efficient to do all manual clerical operations together at a stage corresponding to the present classification stage.

The morphological classification system was refined by a system of functional classification (Matejka, 1958). Significant grammatical properties of words, reflected in the written representation of these words either not at all or else not explicitly, are specified by including appropriate notations in the dictionary entries for these words at the time their English correspondents are assigned. The animate/inanimate distinction for nouns, the perfective/ imperfective distinction for verbs, and the use of adjectival forms as nouns (for example, *портной**) are among the features accounted for in this way.

6.3. The Classification of Nominal Forms

Table 32 defines a subclassification system for nominal forms, which both accounts for the animate/inanimate distinction and resolves morphological classes of the *комната**/*граница** type into disjoint subclasses. Applications of this system are described in Sec. 10.1. An *A* is placed in a suitable place in the dictionary entry if the noun is animate, an *I* if it is inanimate. One of *1*, *2*, *3*, or *4* is joined to *A* or *I* according to the significance of certain critical desinences. Thus, in the class *N4*, animate nouns for which "ой" denotes the instrumental singular (*Is* in Table 32) are marked by *A2*, and inanimate nouns with the same interpretation of "ой" are marked by *I2*. If "ой" does not naturally occur in the reduced paradigm of a noun in this class, as indicated by ϕ in Table 32, the second code character is *1*. In this system, *учительница**, *граница**, *дама**, and *комната**, all of which belong to the class *N4*, are assigned to the subclasses *A1*, *I1*, *A2*, and *I2*, respectively.

Words belonging to the class *N8.3* are subclassified according to the functions of the desinences "и" and "#" (null desinence). For example, *яблоко**, where "и" is employed in both the nominative and the accusative plural, and "#" is employed in the genitive plural, would be marked by *I3*. Twenty-five of the major classes, including the extensively used *N4.3*, *N4.31*, *N7*, and *N10*, are sufficiently homogeneous to require no more than the simple animate/inanimate subclassification, as indicated by the *x* marks in Table 32. In fact, in the case of *N10*, even this distinction is, strictly speaking, superfluous. However, routine work, both manual and automatic, tends to proceed more smoothly in the absence of exceptions. It therefore simplifies matters to use *I1* even though it is redundant in this case.

One machine word (Sec. 1.3) of 12 characters is provided in the dictionary entry for a Russian word (Sec. 7.8) to represent the grammatical characteristics of the Russian word. For reasons given in Sec. 6.5, the machine word is called the *organized word**. The interpretation attached to characters in each of the 12 positions is defined in Table 33 in a tabular form designed for quick reference. Examples of usage are also given in Table 33. A more detailed explanation of the significance of the characters that may occur in each position follows.

First Position: *N* denotes words functioning as nouns, including those originally classified as invariable, for instance *реле** or *какаду** (Example 14).

Second Position: *D* denotes declinable nouns, *U* indeclinable ones. The character *K* is provided for the sake of completeness, to cover the exotic situations where a noun may be declinable in some contexts, indeclinable in others. Some proper names provide an example: Вижу Отто Гуревича/ Елисавету Гуревич.

Third and Fourth Positions: These positions are normally reserved for subclassification marks determined in accordance with Table 32. *K* in position 3 denotes nouns which may be animate in some contexts but inanimate in others such as *гений** (ценить гений художника/ценить гения) or *бактерия** (Example 13). *0* is in the fourth position for indeclinable nouns. *Z* in the fourth position marks exceptions for which no provision has been made in Table 32.

Table 32. Subclassification of nominal forms.

Class	Critical desinence	Subclassification marks[1]							
		I1	A1	I2	A2	I3	A3	I4	A4
N1	ы	NpAp	Np	φ	φ				
N1.1	и	NpAp	Np	φ	φ				
N1.2	ы	NpAp	Np	φ	φ				
N1.3	None	x	x						
N1.4	None	x	x						
N2	и	NpAp	Np	PsNpAp	PsNp	φ	φ		
N2.1	None	x	x						
N3	и	NpAp	Np	φ	φ				
N3.05	None	x	x						
N3.1	{ и	NpAp	Np	NpAp	Np	φ	φ		
	ем	φ	φ	Is	Is	Is	Is		
N4	ой	φ	φ	Is	Is				
N4.05	None	x	x						
N4.06	None	x	x						
N4.1	ей	φ	φ	IsGp	IsGpAp	Gp	GpAp	Is	Is
N4.3	None	x	x						
N4.31	None	x	x						
N5	ей	IsGp	IsGpAp	Is	Is				
N5.05	None	x	x						
N5.1	ь	φ	φ	Gp	GpAp				
N5.15	None	x	x						
N5.2	None	x	x						
N5.3	None	x	x						
N6	ьми	φ	φ	Ip	Ip				
N6.1	None	x	x						
N7	None	x	x						
N8	None	x	x						
N8.1	None	x	x						
N8.15	None	x	x						
N8.3	{ и	NpAp	Np	φ	φ	NpAp	Np		
	#	φ	φ	φ	φ	Gp	GpAp		
N8.4	None	x	x						
N9	ев	Gp	GpAp	φ	φ				
N9.2	None	x	x						
N10	None	x	x						
N11	None	x	x						
N11.1	None	x	x						
N11.2	None	x	x						
N12	None	x	x						
N13	None	x	x						

[1] Use *S* in position 7 of the organized word to signal any special usage not covered by the table.

Fifth Position: *M*, *F*, and *N* denote masculine, feminine, and neuter gender, respectively. Some rare cases are covered by *C* and *K*. For example _сирота_* (_orphan_*) would be marked by *C* to denote the possibility of different agreement (наш слепой сирота/наша слепая сирота). The

Table 33. Coded grammatical properties of nouns and examples of usage.

Character 1:	NOUN = *N*
Character 2:	DECLINABLE = *D*; INDECLINABLE = *U*; OSCILLATION = *K*
Character 3:	ANIMATE = *A*; INANIMATE = *I*; OSCILLATION = *K*
Character 4:	SUBCLASSES = 1, 2, 3, 4; INDECLINABLE = 0; NO PROVISION = *Z*
Character 5:	GENDER: Masculine = *M*; Feminine = *F*; Neuter = *N*; Common = *C*; Oscillation = *K*
Character 6:	NUMBER: Singular, Plural = 0; Singular Only = 1; Plural Only = 2; Change of meaning in plural = 3
Character 7:	SPECIAL INFLECTED STRINGS: If there are = *S*; If not = 0
Character 8:	SPECIAL CANONICAL FORMS: Canonical form created for dictionary purposes = *X*; Deficient canonical form = *L*; None = 0
Characters 9–12:	NOT USED = △

Example	Class	Noun	Position 1	2	3	4	5	6	7	8	9	10	11	12
1	*N1*	взор	*N*	*D*	*I*	*1*	*M*	*0*	*0*	*0*	△	△	△	△
2	*N1*	адрес	*N*	*D*	*I*	*2*	*M*	*3*	*S*	*0*	△	△	△	△
3	*N1*	бензин	*N*	*D*	*I*	*1*	*M*	*1*	*S*	*0*	△	△	△	△
4	*N1*	лес	*N*	*D*	*I*	*2*	*M*	*3*	*S*	*0*	△	△	△	△
5	*N4*	леса	*N*	*D*	*I*	*2*	*F*	*0*	*0*	*0*	△	△	△	△
6	*N4*	ножница	*N*	*D*	*I*	*2*	*F*	*2*	*0*	*X*	△	△	△	△
7	*N1.1*	человек	*N*	*D*	*A*	*1*	*M*	*1*	*S*	*0*	△	△	△	△
8	*N6*	людь	*N*	*D*	*A*	*2*	*C*	*2*	*0*	*X*	△	△	△	△
9	*N1*	зуб	*N*	*D*	*I*	*1*	*M*	*0*	*0*	*0*	△	△	△	△
10	*N11.1*	зубье	*N*	*D*	*I*	*1*	*M*	*2*	*S*	*X*	△	△	△	△
11	*N1*	месяц	*N*	*D*	*I*	*1*	*M*	*0*	*S*	*L*	△	△	△	△
12	*N9*	месяце	*N*	*D*	*I*	*1*	*M*	*0*	*S*	*X*	△	△	△	△
13	*N7*	бактерия	*N*	*D*	*K*	*1*	*F*	*0*	*0*	*0*	△	△	△	△
14	*I*	какаду	*N*	*U*	*A*	*0*	*M*	*0*	*0*	*0*	△	△	△	△
15	*N1*	солдат	*N*	*D*	*A*	*1*	*M*	*0*	*S*	*0*	△	△	△	△

word _боа_* might be treated as two entries, $боа_1$* ↔ _boa_* coded *M* and $боа_2$* ↔ _scarf_* coded *N*, or as a single entry coded *K*, depending on one's views regarding homography versus polysemanticism.

Sixth Position: *0* denotes nouns used in the singular or plural with the same meaning. *1* denotes nouns used in the singular only (Examples 3, 7), *2* those used only in the plural (Examples 6, 8, 10), and *3* those which undergo a change of meaning in the plural (Examples 2, 4).

Seventh Position: S denotes the existence of special forms, such as alternative genitive or prepositional singular representations in "y", or genitive plurals in "#" in class *N1*. Otherwise *0* is used.

Eighth Position: X is used with an entry for which a synthetic canonical

Table 34. An artifice for producing generated paradigms of irregular words.

англичанини	*N1*	*A1*	+ англичана	*N4*	*A2*
теленок	*N1.3*	*A1*	+ телато	*N8*	*A1*
брат	*N1*	*A1*	+ братье	*N11.2*	*A1*
друг	*N1*	*A1*	+ друзье	*N11.2*	*A1*
сын	*N1*	*A1*	+ сыновье	*N11*	*A1*
крыло	*N8*	*I1*	+ крылье	*N11.2*	*I1*
колено	*N8*	*I1*	+ колень	*N6*	*I1*
небо	*N8*	*I1*	+ небесо	*N8*	*I1*
церков	*N6*	*I1*	+ церквь	*N6*	*I1*
мать	*N6*	*A1*	+ матерь	*N6*	*A1*
дочь	*N6*	*A2*	+ дочерь	*N6*	*A2*
господин	*N1*	*A1*	+ господо	*N8*	*A1*
судно	*N8*	*I1*	+ судо	*N8.3*	*I1*
сосед	*N1*	*A1*	+ соседь	*N3*	*A1*
путь	*N6*	*I1*	+ путь	*N3*	*I1*
месяц	*N1*	*I1*	+ месяце	*N9*	*I1*
лист	*N1*	*I1*	+ листье	*N11.2*	*I1*
око	*N8*	*I1*	+ очь	*N6.1*	*I1*
etc.					

representation has been created to enable the use of the regular direct-inflection algorithm (Examples 6, 8, 10, 12). For some kinds of irregularities, the reduced paradigm of a word may be contained in the set of inflected strings produced by the direct-inflection algorithm acting on two canonical strings, one of which alone would not produce the complete reduced paradigm, the other of which may not exist at all. The deficient word is marked by *L*, the other by *X* (Examples 11 and 12). How this artifice may be applied to a number of irregular patterns is illustrated in Table 34. *0* is normally used in this position.

Positions 9–12: Not in use at the time of writing.

Just as, in Sec. 6.1, the system of morphological classification was completed by the definition of the classes *N99.99*, *A99.99*, and *V99.99* so, in the functional system, similar escape provisions are necessary. The quality of the system depends, of course, on where the overwhelming majority of words fall. Positions 6 through 8 provide the necessary escape facilities. They are supplemented by allowing the use of comments in the body of the dictionary entries. The presence of characters other than *0* is usually due to deep-seated problems of morphology or semantics; the use of special characters in positions 6 through 8 does not pretend to solve these problems fully, but at

AVT	OBLOKIROVK	NO4.31		GUBK	NO4.31
AVTO	REGULIROVK	NO4.31	D	EBLOKIROVK	NO4.31
AV	TOUSTANOVK	NO4.31		DOBAVK	NO4.31
A	RANZHIROVK	NO4.31		DOVODK	NO4.31
	ARK	NO4.31		DOGADK	NO4.31
	ARMIROVK	NO4.31		DODELK	NO4.31
B	ALANSIROVK	NO4.31		DoZIROVK	NO4.31
	BANK	NO4.31		DOSK	NO4.31
	BIRK	NO4.31		DYRK	NO4.31
	BLOKIROVK	NO4.31		ZAVARK	NO4.31
BO	MBARDIROVK	NO4.31		ZAVODK	NO4.31
	BOROZDK	NO4.31		ZAVJAZK	NO4.31
	BRAKOVK	NO4.31		ZAGOTOVK	NO4.31
	BROVK	NO4.31		ZAGRUZK	NO4.31
	BUDK	NO4.31		ZADAVK	NO4.31
	BULAVK	NO4.31		ZADELK	NO4.31
	BUSINK	NO4.31		ZAKOROTK	NO4.31
VAL•TSOVK		NO4.31		ZAKRASK	NO4.31
	VETRJANK	NO4.31		ZALIVK	NO4.31
	VILK	NO4.31		ZAMAZK	NO4.31
	VOLYNK	NO4.31		ZAMETK	NO4.31
	VPISK	NO4.31		ZAPISK	NO4.31
	VPRAVK	NO4.31		ZAPLAVK	NO4.31
	VREMJANK	NO4.31		ZAPRAVK	NO4.31
	VSTAVK	NO4.31		ZARUBK	NO4.31
	VTULK	NO4.31		ZARJADK	NO4.31
	VYBIVK	NO4.31		ZASVETK	NO4.31
	VYBORK	NO4.31		ZASYPK	NO4.31
	VYVARK	NO4.31		ZATSEPK	NO4.31
	VYVERK	NO4.31		ZACHISTK	NO4.31
	VYVESK	NO4.31		ZASHCHELK	NO4.31
	VYDUMK	NO4.31	ZVU	KOPRISTAVK	NO4.31
	VYEMK	NO4.31		IGOLK	NO4.31
	VYKLADK	NO4.31		IZOLIROVK	NO4.31
	VYNOSK	NO4.31	INS	TRUMENTOVK	NO4.31
	VYPISK	NO4.31		ISKORK	NO4.31
	VYPLAVK	NO4.31		NAGRUZK	NO4.31
	VJAZK	NO4.31		OSHIBK	NO4.31
	GOLOVK	NO4.31		RAZVERTK	NO4.31
	GORELK	NO4.31		RAZRABOTK	NO4.31
	GRADUIROVK	NO4.31		REGULIROVK	NO4.31
GR	AMPLASTINK	NO4.31		SETK	NO4.31
GR	AMPRISTAVK	NO4.31		TRUBK	NO4.31
	GREBENK	NO4.31		USTANOVK	NO4.31

Fig. 35. Some canonical stems of the class *N4.31*.

least calls them to the attention of man or machine. This apparatus is used only in rare instances, for each of which individual attention is essential anyway.

For certain words, as already mentioned in the description of Table 32, the morphological and functional classification is redundant in whole or in part. It might be suggested that a word such as *здание** is explicitly and fully classified by the last three characters of its canonical stem. A glance at Fig. 35 strongly suggests the possibility of factoring canonical stems into stems and affixes of higher order which could assist, at least in some cases, in explicitly supplying the necessary classification, both morphological and functional.

This suggestion has much to recommend it, but its acceptance at the early stages of investigation would have entailed grave dangers, owing to the practical impossibility of accurately accounting for exceptions to patterns. Having a large dictionary and a large corpus of texts available on magnetic tape (or some equivalent medium) will enable the easy discovery of patterns through the automatic production of lists like that of which Fig. 35 is a fragment. Exceptions are also readily checked on such lists. These possibilities are a result of the research described in this book, and have only recently become available for the exploitation they deserve.

6.4. The Classification of Adjectival Forms

Specifying grammatical properties is somewhat simpler for adjectival forms than it is for either nominal or verbal forms. One machine word is again provided to represent grammatical characteristics, and a detailed explanation of the significance of the characters that may occur in each position follows.

First Position: *A* is used for the majority of words of adjectival form, namely, those that also function as adjectives. For words, such as *портной**, which function exclusively as nouns, *N* is used. Some words, of which *постоянный** is an example, function as adjectives in some contexts and as nouns in others, and are marked by *K*.

Second Position: This position is significant only for forms marked by *N* or *K*, and is used according to the conventions described for nominal forms.

Third Position: *0* is used unless *N* or *K* is used in the first position and the conventions for nominal forms are followed.

Fourth Position: Adjectives represented by both long (attributive) and short (predicative) forms are marked by *0*. Adjectives with long forms only are marked by *1*, and those with short forms only by *2*.

Fifth and Sixth Positions: As for the third position.

Seventh through Twelfth Positions: Not in use at the time of writing.

Adjectival forms arising from verbs, that is to say, capable of functioning as participles, are identified by additional marks. These are described in Sec. 6.5 together with the marks for verbal forms.

6.5. The Classification of Verbal Forms

The grammatical properties of verbs, like those of adjectives and nouns, are explicitly expressed by a string of code characters. It is difficult to ensure the systematic and consistent encoding of the complex properties of verbs, especially since there is much room for disagreement. Embarking on a detailed study of Russian verbs prior to dictionary compilation did not seem attractive, since the careful examination of verbs in context would have been very difficult. Using initially an existing and convenient reference work and introducing corrections and refinements once the dictionary is in operation and texts are available in a form highly suitable for contextual studies (Sec. 8.5) seemed the better choice. Accordingly, the initial grammatical description of the verbs included in the automatic dictionary file is based, wherever possible, on that given in Daum and Schenk's *Die Russischen Verben* (1954; denoted subsequently by the abbreviated form "D.S."); information in this work is presented in a tabular form highly suitable for coding purposes.

As is done for nouns and adjectives, escape facilities are provided in the form of room for comments, available in the body of dictionary entries.

In what follows, therefore, notations such as "несов. in D.S." refer to marks in Daum and Schenk used to determine the choice of this or that code character. The interpretation of characters in each of the 12 positions of the grammatical code word is given in tabular form in Table 35. A more detailed explanation follows; usage is illustrated in Table 36.

First Position: V denotes words functioning as verbs.

Second Position: N and S denote imperfective and perfective aspect respectively. Some verbs, such as *задвигать**, where questions of homography versus polysemanticism arise, as well as some others, if treated as polysemantic words require the notation K denoting a dependence of aspect on the sense intended in any given context. For certain verbs M, denoting iterative action (imperfective), and U, marking momentary action (perfective), are used instead of N and S.

Third Position: R marks verbs with infinitive representation ending in "ся" or "сь", 0 marks all others.

Fourth Position: Normal verbs used in all persons are marked by 0. If the third person only is used, the mark 3 applies, and if the second and third persons are used, 2 applies. Impersonal usage is marked by U, and K (with

Table 35. Coded grammatical properties of verbs.

Character 1:	verb	$= V$
Character 2:	(a) imperfective aspect	$= N$ (несов. in D.S.)
	(b) perfective aspect	$= S$ (сов. in D.S.)
	(c) momentary action (perf.)	$= U$ (однокр. in D.S.)
	(d) iterative action (imperf.)	$= M$ (многокр. in D.S.)
	(e) oscillating aspect	$= K$ (несов. и сов. in D.S.)
Character 3:	(a) verbs ending in "ся/сь"	$= R$
	(b) otherwise	$= 0$
Character 4:	(a) 3d person only	$= 3$ ("1. u. 2. Pers. ungebr." in D.S.)
	(b) 2d and 3d person only	$= 2$ ("1. Pers. ungebr." in D.S.)
	(c) impersonal use	$= U$ ("unper." in D.S.)
	(d) oscillating usage	$= K$ (two entries marked by subscripts in D.S.)
	(e) otherwise	$= 0$
Character 5:	first government	$=$ letter ⎫ see Table 37
Character 6:	first government	$=$ number ⎭
Character 7:	second government	$=$ letter ⎫ see Table 37
Character 8:	second government	$=$ number ⎭
Character 9:	(a) two variants of any government	$= K^1$ (usually marked by semicolon or "u." in D.S.)
	(b) more than two variants of any government	$= A^1$
	(c) otherwise	$= 0$
Character 10:	(a) idiomatic use only	$= T$ (the idiom only in D.S.)
	(b) idiomatic use also	$= Y$ (the idiom quoted besides other usages in D.S.)
	(c) otherwise	$= 0$
Characters 11 and 12:	not used	$= \triangle$

[1] If Character 9 is K or A, then the other variants of government must be expressed in the semiorganized words (Table 38).

appropriate comment) covers the situation where two or more possibilities seem to apply simultaneously.

Russian verbs enter in a *governing* relation with other words or phrases. The *first government*, usually of a "direct object", is frequently expressed by the accusative case (for instance, "я вижу книгу"); the *second government*, usually of an "indirect object", is frequently expressed by the dative case, as by "нам" in "ваш отец показывал нам сад", where "сад", the object of the first government, stands for an accusative form. Unfortunately, the classical notion of a direct or indirect object is not invariant under a transformation from Russian to English or vice versa or, from another point of view, the dative case, for example, does not invariably denote an indirect object. Moreover, certain verbs require the use of specific prepositions in addition to some definite case. To provide a starting point for further investigations of Russian verbs at a later date, as well as useful if not always

Table 36. Examples of usage.

Example	Organized word 1 2 3 4 5 6 7 8 9 10 11 12	First semiorganized word 1 2 3 4 5 6 7 8 9 10 11 12	Second semiorganized word 1 2 3 4 5 6 7 8 9 10 11 12
1. советаться	V N R O J 1 0 0 0 0 △ △	△ △ △ △ △ △ △ △ △ △ △ △	N 3 N 4
consult, talk over			
2. разломать	V S O O P 7 0 0 0 0 △ △	△ △ △ △ △ △ △ △ △ △ △ △	M 1 N 1 N 3
break, destroy			
3. разломаться	V S R 3 0 0 0 0 0 0 △ △	△ △ △ △ △ △ △ △ △ △ △ △	M 1 N 1 N 3 N 4
break, be broken			
4. зевнуть	V U 0 0 0 0 0 0 0 0 △ △	△ △ △ △ △ △ △ △ △ △ △ △	M 1 N 1 N 3
yawn, gape			
5. знавать	V M 0 0 P 3 0 0 K 0 △ △	△ △ △ △ △ △ △ △ △ △ F 2	
know, inform			

rigorously systematic or even correct information about Russian verbs, the following conventions were adopted: governing relations are referred to simply by "first government" and "second government", where "first" and "second" refer to the order given by Daum and Schenk (D.S.). For example, "применять что к кому–чему" is said to have first government "что" and second government "к кому–чему". The first government is marked by characters in positions 5 and 6, and the second government is marked by characters in positions 7 and 8.

   Fifth and Sixth Positions: First government *without* preposition is indicated in fifth and sixth position by characters chosen as indicated in Table 37*a*. First government *with* preposition is marked in the same positions by characters chosen as specified in Table 37*b*.

   Seventh and Eighth Positions: The *second* government is characterized in these positions by characters chosen as specified in Table 37.

   Ninth Position: Alternative first or second governments may exist. This is often associated with a problem of homography versus polysemanticism, or simply with optional usage. The presence of such an alternative is marked by *K*, its absence by *0*. When more than two variants are possible, *A* replaces *K*.

   Tenth Position: Verbs entering in idiomatic constructions are marked by *T* if they occur only in such constructions, and by *Y* if normal usages exist. Normally, *0* is in this position.

   Positions 11 and 12: Not in use at the time of writing.

   The machine word whose structure has been described in the preceding paragraphs is called the *organized word** because each character position has a definite role, and the significance of a character usually depends on its position. Additional machine words, where characters may occupy any position without change in significance, are provided for verbs. These machine words are called *semiorganized**. The functions of the semiorganized words are specified in Table 38.

   Whenever *K* or *A* occurs in the ninth position of the organized word, indicating the presence of variant governments, the nature of these variants is marked in the first semiorganized word, as specified in Table 38*a*. For purposes of long-term research on the structure of Russian, information regarding the pattern of association of participles and gerunds with verbs is of some value. This information is represented by marks in the second semiorganized word, chosen as specified in Table 38*b*. Since most verbs have most of the potential participles and gerunds, it is economical to give a positive indication of the absence of some forms, rather than of the presence of the others. Examples of the use of the semiorganized words are given in Table 36.

   A third semiorganized word is used with adjectival forms serving as participles. The notations *Q0*, *Q1*, *Q2*, or *Q3*, when present in this word,

Table 37. Marking systems for governing properties of verbs.[1]

a. Government without preposition
(in D.S.)

P1	кого–чего
P2	кому–чему
P3	кого–что
P4	кем–чем
P5	чего
P6	чему
P7	что
P8	чем
P9	"mit Inf."
00	no information

b. Government with preposition

	(in D.S.)		(in D.S.)
C1	без кого–чего	F6	перед кем–чем
C2	без чего	F7	перед чем
C3	в кого–что	F8	по кому–чему
C4	во что	F9	по чему
C5	в ком–чем	G1	по кого–что
C6	в чем	G2	по что
C7	для кого–чего	G3	по ком–чем
C8	для чего	G4	по чем
C9	до кого–чего	G5	под кого–что
D1	до чего	G6	под что
D2	за кого–что	G7	под кем–чем
D3	за что	G8	под чем
D4	за кем–чем	G9	при ком–чем
D5	за чем	H1	при чем
D6	из кого–чего	H2	про кого–что
D7	из чего	H3	про что
D8	к кому–чему	H4	ради кого–чего
D9	к чему	H5	ради чего
E1	между кого–чего	H6	с кого–чего
E2	между чего	H7	с чего
E3	между кем–чем	H8	с кого–что
E4	между чем	H9	с что
E5	на кого–что	J1	с кем–чем
E6	на что	J2	с чем
E7	на ком–чем	J3	сквозь кого–что
E8	над чем	J4	сквозь что
E9	над кем–чем	J5	у кого–чего
F1	над чем	J6	у чего
F2	о ком–чем	J7	через кого–что
F3	о чем	J8	через что
F4	от кого–чего	00	no information
F5	от чего		

[1] The dash indicates that the government admits both animate and inanimate forms. Lack of the dash means that the government generally admits only inanimate forms.

Table 38. Notation for verbs in the semiorganized words.

a. Variant governments

If K or A is used in position 9 of the organized word, then the other variants of government must be expressed in the first semiorganized word according to the marking system given in Table 37.[1]

b. Absence of participles and gerunds (second semiorganized word)

		(in D.S.)	Example
Present gerund	M1	g. pr. a.	делая
Past gerund	M2	g. pt. a.	делав
Present participle active	N1	p. pr. a.	делающий
Past participle active	N2	p. pt. a.	делавший
Present participle passive	N3	p. pr. p.	делаемый
Past participle passive	N4	p. pt. p.	деланный

If D.S. refers to the set of paradigms by:

1	1a	1b	1c
2	2a	2b	2c
3	3a	3b	3c
4	4a	4b	4c
5	5a	5b	5c
		6b	6c
		7b	7c

then:

g. pr. a.		M1	M1
g. pt. a.			
p. pr. a.		N1	N1
p. pt. a.			
p. pr. p.	N3	N3	N3
p. pt. p.	N4		N4

Generally,[2] if not specified otherwise in D.S.:

(1)	imperfective	M2
(2)	ся/сь (imperfective):	M2N3N4
(3)	perfective	M1N1N3
(4)	ся/сь (perfective):	M1N1N3N4

[1] Do not use 00 in semiorganized words.

[2] The irregular occurrence of participles and gerunds can be noted in the space provided in each dictionary entry for comments.

indicate that the adjectival form may serve as a present active participle, present passive participle, past active participle, or past passive participle, respectively. The third semiorganized word is also used for verbal forms in a manner described in Sec. 7.8 (Table 40).

6.6. Other Forms

While nouns, verbs, and adjectives account for the vast majority of the words in any Russian or English dictionary, the invariable forms account for much of the bulk of any text (Fig. 42) and play syntactic roles of considerable importance. Other forms, such as pronouns and abbreviations, also must be taken into account.

The coding system in the organized word now provides for the identification of numerals, pronouns, abbreviations, adverbs, prepositions, conjunctions, and particles by an appropriate character in the first position. A complete coding system is still under development at the time of writing.

The further treatment of adverbs is not expected to raise any serious difficulties other than certain problems of homography (Sec. 9.7). The syntactic role of abbreviations, like that of mathematical formulas, still requires extensive special study. Russian cardinal numerals have certain peculiar syntactic properties which are already fairly well understood. The remaining word classes—pronouns, prepositions, and so forth—are each characterized by a small, highly individualistic membership, requiring careful attention in the course of syntactic studies.

APPENDIX TO CHAPTER 6

Class Identification and Generation Rules

Words with Atypical Paradigms

Synoptic Classification Table for Nouns and Adjectives

Synoptic Classification Table for Verbs

APPENDIX

Class Identification and Generation Rules

Class	Examples	Class identification	Generation rules — Generating stem	Generation rules — Inflected strings

N1	студент
	кузнец
	
	(See also *N1.2, N1.4*)
		A class of nouns ending in any consonant not followed by ь, *except:*
		
		1. г, ж, й, к, х, ч, ш, щ;
		
		2. ц, whenever any of the following conditions is present:
		  a. ц is preceded by one vowel, or a consonant + е;
		  b. ц is preceded by a pair of consonants + е, and the word carries a stem stress;
		  c. in some cases, if ц is preceded by a pair of consonants + е, and the word carries an end stress
			word itself	*a.* word
				b. GS + a
				c. + y
				d. + ом
				e. + e
				f. GS + ы
				g. + ов
				h. + ам
				i. + ами
				j. + ах

N1.1	градусник (See also *N1.3*)	A class of nouns ending in к, г, х	As in *N1*, except и for ы in *f*	As in *N1*
N1.2	лев лоб вымысел посол заем хребет боец столбец (See also *N1*)	A "fugitive е, о" class, embracing: 1. two-syllable, or longer, nouns ending in a stressed ец preceded by one consonant or vowel; 2. some of the two-syllable, or longer, nouns ending in a stressed ец preceded by a pair of consonants;	1. whenever the penult е is preceded by л, replace the е by a ь; 2. whenever the penult е is preceded by a vowel, replace the е by an й;	As in *N1*

3. in all other cases.

3. some of the one-syllable nouns containing e, o and ending as in N1, Class I.D. 1

4. some of the two-syllable and three-syllable nouns ending in:
   a. ел, ол;
   b. ер, ор;
   c. ем;
   d. ет, especially whenever the paradigm reveals a consistent ultima stress

word — penultimate letter

As in N1.2, except и for ы in f

N1.3

мох
беспорядок
кулечек
барашек

(See also N1.1)

A "fugitive o, e" class, embracing:

1. some of the one-syllable nouns containing o, e and ending as in N1.1;

2. some of the two-syllable or longer nouns ending in ек, ок;

3. all nouns ending in:
   a. ечек;
   b. шек, шок

N1.4

палец
стахановец
шлиссельбуржец

(See also N1)

A "fugitive e" class ending in an unstressed ец preceded by:

1. one consonant or vowel;

2. a pair of consonants

As in N1.2

N2

строй
лишай
гений
музей

(See also N2.1)

A class embracing:

1. the nouns ending in о, а, и, у } + й

2. some nouns ending in е

word — last letter

As in N1, except ем for ом in d and ев for ов in g

a. word	f. GS + и
b. GS + я	g. + ев
c. + ю	h. + ям
d. + ем	i. + ями
e. + е	j. + ях

Class Identification and Generation Rules (*continued*)

Class	Examples	Class identification	Generating stem	Inflected strings
			Generation rules	
N2.1	ручей (See also N2)	A class constituted by a majority of nouns ending in й preceded by e	1. remove the last two letters from the end of the word; 2. add ь to the remainder	As in N2
N3	(masculine) словарь (See also N3.05)	A class ending in a consonant + ь		As in N2, except ей for ев in g

N3.05	(masculine)
	день
	стебель
	камень
	деготь
	
	(See also N3)
		A "fugitive e, o" class embracing:
		1. some of the one-syllable nouns containing e, o and ending as in N3;
		2. some of the two-syllable and three-syllable nouns ending in:
		  a. ель, оль;
		  b. ень;
		3. two-syllable and three-syllable nouns ending in оть
			word — the ultimate and the antepenultimate letters	As in N2, except ей for ев in g

N3.1	нож товарищ	A class ending in ж, ш, ч, щ	As in N1, except и for ы in f, ей for ов in g, and added: k. GS + ем.	
N4	дама лампа игла служба (See also N4.05)	A class ending in a preceded by any consonant *except*: 1. г, ж, к, х, ч, ш, щ; 2. ц, whenever preceded by another consonant;	word — last letter	a. word b. GS + ы c. + e d. + y e. + ой f. GS + ей g. + # h. + ам i. + ами j. + ах

3. the majority of cases when the consonant is л, н, preceded by another consonant;

4. some cases when the consonant is м, р, preceded by another consonant;

5. a few cases when the consonant is б, preceded by another consonant

N4.05 ветла
гривна
дверца
овца
тюрьма
судьба

(See also N4)

A "reappearing e" class embracing: word — last letter, but: As in N4

1. most nouns ending in a preceded by a consonant (*except* ж, ч, ш, if the word has an ultima stress) + л, н;

1. if the penult of the generating stem is either й or ь, replace it by e in g;

2. the nouns ending in a preceded by a consonant + ц;

2. in all other instances, insert e between the penult and the ultima of the generating stem in g

3. some nouns ending in a preceded by a consonant + р, м;

4. a few nouns ending in a preceded by a consonant + б

N4.06 княжна́

A "reappearing o" class embracing: As in N4.31   As in N4

1. the nouns ending in a preceded by ж, ш, ч + л, н, *except* whenever the stress does not fall upon the last syllable of the word;

2. some nouns ending in a preceded by ж, ш, ч + р, м

Class Identification and Generation Rules (*continued*)

Class	Examples	Class identification	Generation rules	
			Generating stem	Inflected strings
N4.1	бумага свеча	A class ending in a preceded by:	As in *N4*, except и for ы in *b*	
		1. a vowel + г, к, х;		
		2. ж, ш, ч, щ		
N4.3	копейка люлька выдержка (See also *N4.31*)	A "reappearing e" class embracing:	As in *N4.05*, except и for ы in *b*	
		1. the nouns ending in a preceded by: *a.* й *b.* a consonant + ь } + к;		
		2. the nouns ending in a preceded by ж, ш, ч + к, with ultima unstressed;		
		3. some nouns ending in a preceded by a consonant + г		
N4.31	выплавка кишка́ (See also *N4.3*)	A "reappearing o" class embracing the nouns ending in a preceded by:	word — last letter, but an o must be inserted between the penult and the ultima of the generating stem in *g*	As in *N4*, except и for ы in *b*
		1. any consonant (*except* й, ж, ш, ч, ц), not followed by ь, + к;		
		2. ж, ш, ч + к, whenever the stress falls upon the ultima of the word		

			a. word
			b. GS + и
			c. + e
			d. + ю
			e. + ей
			f. GS + ь
			g. + ям
			h. + ями
			i. + ях

N5 — няня, доля, клешня (See also *N5.1*)

A class embracing:
1. the nouns ending in я preceded by a vowel + a consonant;
2. some nouns ending in я preceded by a consonant + л, н, р

word — last letter

N5.05 — струя

A class ending in any vowel (*except и*) + я

As in *N5*, except й for ь in *f*

N5.1 — земля, готовальня, двойня (See also *N5*)

A "reappearing e" class ending in я preceded by any consonant (*except* г, к, х) + л, н

word — last letter, but:
1. if the penult of the generating stem is either ь or й, replace it by e in *f* and *j*;
2. in all other instances, insert e between the penult and the ultima of the generating stem in *f* and *j*

As in *N5*, but add: *j*. GS + #

N5.15 — кухня

A "reappearing o" class ending in я preceded by г, к, х + л, н

word — last letter, but an o must be inserted between the penult and the ultima of the generating stem in *f* and *j*

As in *N5*, but add: *j*. GS + #

N5.2 — статья

A class ending in я preceded by a consonant + ь, the stress falling upon the last syllable of the word

As in *N5*, but in *f* replace the last letter of the generating stem (= ь) by e

As in *N5*, except й for ь in *f*

Class Identification and Generation Rules (continued)

Class	Examples	Class identification	Generation rules	
			Generating stem	Inflected strings
N5.3	гостья	A class ending in я preceded by a consonant + ь, the last syllable of the word unstressed	As in N5, but in f replace the last letter of the generating stem (= ь) by и	As in N5, except й for ь in f
N6	(feminine) лошадь	A class constituted by nouns ending in any consonant (except ж, ш, ч) + ь, with the exclusion of мать, любовь, церковь	word – last letter	a. word b. GS + и c. + ью d. + ей e. GS + ям f. + ями g. + ьми h. + ях
N6.1	(feminine) ночь	A class constituted by nouns ending in ж, ш, ч + ь, with the exclusion of ложь, рожь, вошь, дочь	As in N6, except ам for ям in e, ами for ями in f, and ах for ях in h	
N7	профессия	A class ending in ия	As in N5, except й for ь in f, and omission of c	
N8	вещество войско русло село (See also N8.1, N8.15, N8.3, N8.4)	A class embracing the majority of nouns ending in о preceded by: 1. a pair or cluster of consonants, with any penultimate consonant, and any ultimate consonant except: a. ц, ч; b. к (however, a few nouns, like войско, forming the nominative plural case with an a	word – last letter	a. word b. GS + a c. + у d. + ом e. + e f. GS + # g. + ам h. + ами i. + ах

instead of an и desinence, and the genitive plural case with a # desinence instead of employing an ов desinence or inserting a е between the penultimate and ultimate letters of the stem, belong to class *N8*; the shift of the stress from the stems in the singular to the desinences in the plural numbers may be a possible differentia of this type of noun);

c. л, н, м, р (however, it is becoming proper usage to form the genitive plural cases (= generating stems) of a few nouns, like русло, ремесло, трпло, etc., without any vowel insertion between the penultimate and ultimate consonants of the generating stem; such nouns belong to class *N8*);

2. a vowel + any consonant, *except* к, ч

A "reappearing o" class embracing:

1. most nouns ending in о preceded by г, к + л, н, м, р;

2. a few nouns ending in о preceded by з + л

word — last letter, but an o must be inserted between the penultimate and ultimate letters of the generating stem in *f*

As in *N8*

N8.1   стекло
волокно
ело

(See also *N8*)

Class Identification and Generation Rules (continued)

Class	Examples	Class identification	Generation rules	
			Generating stem	Inflected strings
N8.15	число бревно письмо ядро кольцо (See also N.8)	A "reappearing e" class constituted by the majority of nouns ending in o preceded by any consonant (except г, к, з) + л, н, м, р, ц	word — last letter, but: 1. if the penult of the generating stem is ь, replace it by e in f; 2. if the penult of the generating stem is й, replace it by и in f; 3. in all other instances, insert e between the penult and the ultima of the generating stem in f	As in N8
N8.3	очко́ плечико яблоко (See also N8)	A class embracing the nouns ending in o preceded by: 1. a consonant + к, whenever the word carries an end stress; 2. a vowel + к, with the exclusion of the noun око		As in N8, except for addition of: j. GS + и; k. GS + ов

N8.4	колечко (See also N8)	A "reappearing e" class ending in о preceded by a consonant + к (a steady nondesinential stress offers a hint that a noun ending thus should be assigned to this class, rather than to N8.3)	word — last letter, but a e must be inserted between the penult and the ultima of the generating stem in f	As in N8, but add: j. GS + и
N9	коленце вместилище (See also N9.2)	A class constituted by: 1. most nouns ending in e preceded by т, д, н, ф, в + ц; 2. nouns ending in e preceded by a vowel + ж, ц, ч, щ	As in N8, except ем for ом in d, and ев for e in e	
N9.2	зеркальце полотенце (See also N9)	A "reappearing e" class embracing: 1. some nouns ending in e preceded by т, д, н, ф, в + ц; 2. most nouns ending in e preceded by any other consonant + ц	word — last letter, but: 1. if the penult of the generating stem is either ь or й, replace it by e in f; 2. in all other instances, insert e between the penult and the ultima of the generating stem in f	As in N8, except ем for ом in d and elimination of e
N10	напряжение	A class ending in ие	word — last letter	a. word b. GS + я c. + ю d. + ем e. + и f. GS + й g. + ям h. + ями i. + ях

Class Identification and Generation Rules (*continued*)

Class	Examples	Class identification	Generation rules	
			Generating stem	Inflected strings
N11	питье (See also *N11.1, N11.2*)	A class of nouns: 1. ending in е; 2. forming the genitive (accusative) plural case with ей	word — last letter, but the last letter of the generating stem (= ь) must be replaced by е in *f*	As in *N10*, except for elimination of *e*
N11.1	копье (See also *N11*)	A class of nouns: 1. ending in ье; 2. forming the genitive (accusative) plural case with ий	word — last letter, but the last letter of the generating stem (= ь) must be replaced by и in *f*	As in *N10*, except for elimination of *e*
N11.2	подмастерье (See also *N11*)	A class of nouns: 1. ending in ье; 2. forming the genitive (accusative) plural case with ьев	As in *N10*, except ев for й in *f*, and elimination of *e*	

N12 (neuter)
бремя
время
вымя
знамя
имя
пламя
племя
семя
стремя
темя

A class ending in мя

word – last letter

a. word
b. GS + ени
c. + енем
d. + ена

e. GS + ен
f. + енам
g. + енами
h. + енах

N13 горе (the plural number is not used)
поле
море

A class ending in e preceded by л, р

As in *N10*, except ей for й in *f*, and elimination of *e*

Class Identification and Generation Rules (*continued*)

Class	Examples	Class identification	Generation rules	
			Generating stem	Inflected strings
A1	бедственный безвременный деланный	A class ending in ый preceded by нн	word — last three letters	a. word b. GS + ного k. GS + ных c. + ному l. + ными d. + ным m. + # e. + ном n. + ен f. + ная o. + на g. + ной p. + но h. + ную q. + ны i. + ное r. + а j. + ные s. + о t. + ы
A2	больной буйный видный острый	A "reappearing e" class embracing: 1. the adjectives ending in ой preceded by any consonant + н; 2. the adjectives ending in ый preceded by any consonant (except н) + н; 3. a few adjectives ending in ый preceded by a consonant + р, л (this group consists of острый, хитрый, светлый, кислый, теплый)	word — last two letters, but the following modifications must be introduced in *m*: a. if the penult of the generating stem is either й¹ or ь, replace it by е; b. in all other cases,² insert e between the penult and the ultima of the generating stem	a. word b. GS + ого i. GS + ое c. + ому j. + ые d. + ым k. + ых e. + ом l. + ыми f. + ая m. + # g. + ой n. + а h. + ую o. + о p. + ы

¹ Except in the adjective достойный, whose masculine predicative form, that is, *m*, requires an й > и modification.

² Except in the adjectives полный and смешной whose masculine predicative forms require an insertion of o (= полон, смешон).

A3

новый
молодой
делаемый
столкнутый

A class of adjectives ending in: word — last two letters   As in A2

1. ый preceded by:
    a. a vowel + any consonant *except* ц;
    b. a consonant + any consonant *except* н, and in some cases (cf. *A2*), р, л;

2. ой preceded by any consonant except г, ж, к, х, ш, щ, and (if preceded by another consonant) н

A4

хороший
делающий
делавший

A class ending in ий preceded by ж, ч, ш, щ   word — last two letters

a. word
b. GS + его
c.   + ему
d.   + им
e.   + ем
f.   + ая
g.   + ей
h.   + ую
i.   + ее

j. GS + ие
k.   + их
l.   + ими
m.   #
n.   + а
o.   + о
p.   + е
q.   + и

A5

синий

A class ending in ий preceded by н   word — last two letters

1. *a–e, g, i–l* as in *A4*;

2. *f.* GS + яя
    h.   + юю
    m.   + ь
    n.   + я
    o.   + е
    p.   + и

Class Identification and Generation Rules (*continued*)

Class	Examples	Class identification	Generation rules — Generating stem	Generation rules — Inflected strings
A6	броский	A "reappearing o" class ending in ий preceded by any consonant (*except* ж, й), not followed by ь, + к	word – last two letters, but an o must be inserted between the penult and the ultima of the generating stem in m	1. a–c, e–i, m–o, as in *A2*; 2. d, j–l as in *A4*; 3. to be added: p. GS + и
A7	тяжкий бойкий горький	A "reappearing e" class ending in ий preceded by: 1. a consonant followed by ь + к; 2. й + к; 3. ж + к	word – last two letters, but the following modifications must be introduced in m: *a.* if the penult of the generating stem is either ь or й, replace it by e; *b.* in all other cases, insert e between the penult and the ultima of the generating stem	1. a–c, e–i, m–o, as in *A2*; 2. d, j–l, as in *A4*; 3. to be added: p. GS + и
A8	дикий ветхий чужой	A class ending in: 1. ий preceded by: *a.* a vowel + к; *b.* г,[1] х; 2. ой preceded by г, ж, к, х, ш	word – last two letters	1. a–c, e–i, m–o, as in *A2*; 2. d, j–l, as in *A4*; 3. to be added: p. GS + и

[1] The adjective долгий is an exception since its masculine predicative form requires a vowel insertion (= долог).

Verbs

V1

делать
гулять
белеть

A class embracing the verbs ending in ать forming the ять } third person еть } plural with { ают нют еют

word — last two letters

a. word	i. GS + ла
b. GS + ю	j. + ло
c. + ешь	k. + ли
d. + ет	l. + й
e. + ем	m. + йте
f. + ете	n. + я
g. + ют	o. + в
h. + л	p. + вши

q. GS + ющий } → A4
r. + вший }
s. + емый → A3
t. + нный → A1

V2

толкнуть
аукнуть
тонуть

A class

1. ending in:
   a. нуть preceded by one or more consonants;
   b. a stressed нуть preceded by a vowel;
2. preserving ну in the masculine past tense forms;
3. forming the imperative forms with и, ите

word — last three letters

a. word	i. GS + ула
b. GS + у	j. + уло
c. + ешь	k. + ули
d. + ет	l. + и
e. + ем	m. + ите
f. + ете	n. + ув
g. + ут	o. + увши
h. + ул	

p. GS + увший → A4
q. + утый → A2

V2.01

сунуть
двйнуть

A class

1. ending in an unstressed нуть preceded by a vowel;
2. preserving ну in the masculine past tense forms;
3. forming the imperative forms with ь, ьте

As in V2, except ь for и and ьте for ите in *l* and *m*, respectively

Class Identification and Generation Rules (*continued*)

Class	Examples	Class identification	Generation rules	
			Generating stem	Inflected strings
V3	формулировать, межевать	A class embracing the verbs ending in овать) and forming the third person евать) plural with уют	word — last five letters	a. word
				b. GS + ую
				c. + уешь
				d. + ует
				e. + уем
				f. + уете
				g. + уют
				h. + вал
				i. + e, o + вала
				j. вало
				k. вали
				l. + уй
				m. + уйте
				n. + уя
				o. + вав
				p. + e, o + вавши
				q. + ующий
				r. + e, o + вавший → A4
				s. + уемый → A3
				t. + e, o + ванный → A1

Desinences containing ов, ев: the choice of a vowel (either e or o) to precede the в, or the second в from the end, of a desinence must be made in accordance with the fifth letter from the end of the infinitive

V4

говори́ть
по́мнить
пропусти́ть
буди́ть

A class

1. ending in
   a. a stressed ить preceded by any vowel or consonant *except* т, д, whenever these alternate with щ, жд, and ж, ш, ч;
   b. ить preceded by a pair or cluster of consonants, with any end consonant *except* one of those listed above;

2. forming the imperative forms with и, ите;

3. forming the third person plural with ят

word — last three letters, but whenever the last letter of the generating stem is one of the labials or dentals (or an ultima in a cluster of dentals) listed below, the following modifications must be introduced in *b* and *s*:

1. the labials

$$\left. \begin{array}{l} \text{б} \\ \text{п} \\ \text{м} \\ \text{в} \\ \text{ф} \end{array} \right\rangle \begin{array}{l} \text{бл} \\ \text{пл} \\ \text{мл} \\ \text{вл} \\ \text{фл;} \end{array}$$

2. the dentals

$$\left. \begin{array}{l} \text{с} \\ \text{з} \\ \text{т} \\ \text{д} \end{array} \right\rangle \begin{array}{l} \text{ш} \\ \text{ж} \\ \text{ч} \\ \text{ж;} \end{array}$$

and the clusters of dentals

$$\left. \begin{array}{l} \text{ст} \\ \text{зд} \end{array} \right\rangle \left\{ \begin{array}{l} \text{щ} \\ \text{жд}^1 \end{array} \right.$$

and the ю desinence in *b* must be replaced by у

a. word
b. GS + ю, у
c.   + ишь
d.   + ит
e.   + им
f.   + ите
g.   + ят
h.   + ил
i.   + ила
j.   + ило
k.   + или
l.   + и
m.   + я
n.   + ив
o.   + ивши
p.   + ящий ⎫ → *A4*
q.   + ивший ⎬
r.   + имый → *A3*
s.   + енный → *A1*

V4.01   ста́вить

A class

1. ending in an unstressed ить preceded by any single consonant *except* т, д, whenever these alternate with щ, жд, and ж, ш, ч;

As in *V4*, except ь for и in *l* and the addition of:

t. GS + ьте

¹ The жд alternant appears only in *s*, that is, in the past passive participles.

Class Identification and Generation Rules (*continued*)

Class	Examples	Class identification	Generation rules — Generating stem	Generation rules — Inflected strings
		2. forming the imperative forms with ь, ьте; 3. forming the third person plural with ят		
V4.02	успокоить	A class 1. ending in an unstressed ить preceded by any vowel; 2. forming the imperative forms with й, йте; 3. forming the third person plural with ят	word — last three letters	As in *V4*, except: 1. й for и in *l*; 2. elimination of the y variant in *b*; 3. the addition of: *t.* GS + йте

V4.1	возвратить
	утвердить
		A class
		
		1. ending in
		  a. a stressed ить preceded by т, д alternating with щ, ж/жд, respectively;
		  b. ить preceded by a pair or cluster of consonants, with т, д (alternating as above) as the end consonant;
		
		2. forming the imperative forms with и, ите;
		
		3. forming the third person plural with ят
			word — last three letters, but the following modifications of the last letter of the generating stem must be introduced:
			
			1. т > щ in *b*, *t*;
			
			2. д > ж in *b*;
			
			3. д > жд in *s*
				As in *V4*, except for the dropping of the ю variant in *b*

V4.11	насы́тить прину́дить	**A class** 1. ending in an unstressed ить preceded by a single т, д alternating with щ, ж/жд; 2. forming the imperative forms with ь, ьте; 3. forming the third person plural with ят	As in *V4.1*, except ь for и in *l* and the addition of: *t.* GS + ьте
V4.2	учи́ть верши́ть	**A class** 1. ending in 　*a.* a stressed ить preceded by ж, ш, ч; 　*b.* ить preceded by a pair or cluster of consonants, with ж, ш, ч as the end consonant; 2. forming the imperative forms with и, ите; 3. forming the third person plural with ат	word — last three letters 1. *a, c–f, h–l, n–o, q–s* as in *V4*: 2. *b.* GS + y 　*g.*　+ ат 　*m.*　+ а 　*p.*　+ ащий → *A4*
V4.21	раскула́чить	**A class** 1. ending in an unstressed ить preceded by a single ж, ч, ш; 2. forming the imperative forms with ь, ьте; 3. forming the third person plural with ат	As in *V4.2*, except ь for и in *l* and the addition of: *t.* GS + ьте

Class Identification and Generation Rules (*continued*)

Class	Examples	Class identification	Generating stem	Inflected strings

V5	писать
	выхлопотать
	плакать
	колебать
		A class
		1. ending in ать preceded by
		    a. с, з, к, г, х, б, м, п;
		    b. т, д, *except* whenever these consonants alternate with щ, жд;
		2. forming the third person plural with ут
			word — last three letters, but the following modification must be introduced in *b–g*, *l–p*,[1] *s*, and, in the case of 1 below, *u*:
			
			1. the labials
			  б } { бл
			  п } > { пл
			  м } { мл;
			
			2. the dentals
			  с } { ш
			  з } > { ж
			  т } { ч
			  д } { ж;
			
			3. the gutturals
			  к } { ч
			  г } > { ж
			  х } { ш;
			
			4. the dental-gutteral cluster ск } > щ
			5. the dental cluster ст }
				a. word
				b. GS + у
				c. + ешь
				d. + ет
				e. + ем
				f. + ете
				g. + ут
				h. + ал
				i. + ала
				j. + ало
				k. + али
				l. + и
				m. + ите
				n. + ь
				o. + ьте
				p. + я, а[1]
				q. + ав
				r. + авши
				s. + ущий } → *A4*
				t. + авший → *A3*
				u. + емый → *A1*
				v. + анный → *A1*

[1] Whenever the generating stem ends in ж, ш, щ, ч, the proper desinence for *p* is а.

V5.1	клеветать	A class 1. ending in атъ preceded by т, д alternating with щ, жд; 2. forming the third person plural with ут	word – last three letters, but a modification $\left.\begin{array}{l}т \\ д\end{array}\right\} > \left\{\begin{array}{l}щ \\ жд\end{array}\right.$ must be introduced in b–g, l–p, and s	As in *V5*
V5.2	наврать ждать ткать	A class 1. ending in атъ preceded by any nonalternating consonant, *except* ж, ш, ч, щ; 2. forming the third person plural with ут	word – last three letters	As in *V5*, except for the elimination of n–p and u
V5.3	веять	A class 1. ending in ятъ preceded by any vowel; 2. forming the third person plural with ют	word – last three letters	*a.* word *b.* GS + ю *c.* + ешь *d.* + ет *e.* + ем *f.* + ете *g.* + ют *h.* + ял *i.* GS + яла *j.* + яло *k.* + яли *l.* + й *m.* + йте *n.* + я *o.* + яв *p.* + явши *q.* GS + ющий $\Big\}$ → *A4* *r.* + явший *s.* + янный → *A1*
V5.4	брать, драть (and derivatives)	A class ending in атъ and distinguished by a #–e alternation	word – last three letters, but a e must be inserted between the penult (= б, п) and the ultima (= р) of the generating stem in b–g, l–m, p, and s	As in *V5*, except for the elimination of n, o, and u, and the dropping of the variant in p

Class Identification and Generation Rules (*continued*)

Class	Examples	Class identification	Generation rules	
			Generating stem	Inflected strings
V5.41	звать (and derivatives)	A class ending in ать and distinguished by a ≠–о alternation	word — last three letters, but an о must be inserted between the penult (= з) and the ultima (= в) of the generating stem in b–g, l–m, p, s, and u	As in V5, except омый for емый in u, the dropping of the a variant in p, and the elimination of n, o
V6	бренчать брозжать бурчать верещать визжать ворчать держать дребезжать дрожать дышать жужжать журчать звучать кричать лежать молчать мчать мычать пищать рычать слышать стучать торчать трещать урчать	A class 1. ending in ать preceded by ж, ш, ч, щ; 2. forming the third person plural with ат	word — last three letters	a. word b. GS + у c. + ишь d. + ит e. + им f. + ите g. + ат h. + ал i. + ала j. + ало k. + али l. + ь m. + ьте n. + и o. + а p. + ав q. + авши r. + ащий } → A4 s. + авший → A3 t. + имый → A3 u. + анный → A1

V6.1   бояться
       стоять
       (and derivatives)

A class

1. ending in ять preceded by о;

2. forming the third person plural with ят

word — last three letters

a. word		i. GS + яла	
b. GS + ю		j. + яло	
c. + ишь		k. + яли	
d. + ит		l. + й	
e. + им		m. + йте	
f. + ят		n. + я	
g. + ят		o. + явши	
h. + яли			

p. GS + ящий }
q. + явший } → A4

V6.2   предвидеть
       терпеть
       смотреть

A class

1. ending in еть preceded by any consonant;

2. forming the third person plural with ят

word — last three letters, but whenever the last letter of the generating stem is one of the labials or dentals listed below, the following modifications must be introduced in b and t:

a. word		i. GS + ела	
b. GS + ю, y		j. + ело	
c. + ишь		k. + ели	
d. + ит		l. + и	
e. + им		m. + ь	
f. + ят		n. + ьте	
g. + ят		o. + я	
h. + ел		p. + ев	

q. GS + ящий }
r. + евший } → A4
s. + имый → A3
t. + енный → A1

1. the labials
   б }   бл
   м } > мл
   п }   пл;

2. the dentals
   с }   ш
   т } > ч   } and the
   д }   ж;   } ю desinence in b must be replaced by y

3. the cluster of dentals
   ст > щ

Class Identification and Generation Rules (continued)

Class	Examples	Class identification	Generation rules — Generating stem	Inflected strings
V7	мерзнуть возникнуть гаснуть	A class 1. ending in нуть; 2. losing ну in the masculine past tense forms	word — last four letters	a. word b. GS + ну j. GS + ло c. + нешь k. + ли d. + нет l. + ни e. + нем m. + ните f. + нете n. + нь g. + нут o. + ньте h. + # p. + нув i. + ла q. + нувши r. + ши s. GS + нущий ⎫ t. + нувший ⎬ → A4 u. + ший ⎭
V8	нести пасти трясти везти ползти грызть лезть (and derivatives) and the verb спасти	A class ending in сти, зти, and зть	word — last two letters	a. word b. GS + у j. GS + ло c. + ешь k. + ли d. + ет l. + и e. + ем m. + ите f. + ете n. + ь g. + ут o. + ьте h. + # p. + я i. + ла q. + ши r. GS + ущий ⎫ → A4 s. + ший ⎭ t. + омый → A3 u. + еный → A1

V8.1

гнести
мести
цвести
обрести (обрести) (no past tense forms), (and derivatives) and the verb рассвести (only third person singular future and neuter past tense forms)

A class

1. ending in сти, сть;

2. forming the first person singular with ту

word — last three letters

a. word
b. GS + ту   i. GS + ла
c. + тешь   j. + ло
d. + тет   k. + ли
e. + тем   l. + ти
f. + тете   m. + тите
g. + тут   n. + тя
h. +л   o. + тши

p. GS + тущий ⎱ → A4
q. + тший ⎰
r. + томый → A3
s. + тенный → A1

V8.11

вычесть
зачесть
учесть
перечесть
протесть
derived from the seldom used parent verb честь

A class

1. ending in сть;

2. forming the first person singular with ту;

3. losing the last letter of the generating stem in some inflected forms

word — last three letters, but the last letter of the generating stem (= е) must be dropped in b–g, i–n, and s

As in V8.1, except for the elimination of p and r

V8.2

блюсти
брести
вести
грясти (no past tense forms),
класть,
красть,
пасть,
прясть
(and derivatives)

A class

1. ending in сти, сть;

2. forming the first person singular with ду

word — last three letters

a. word
b. GS + ду   i. GS + ла
c. + дешь   j. + ло
d. + дет   k. + ли
e. + дем   l. + ди
f. + дете   m. + дите
g. + дут   n. + дя
h. +л   o. + дши

p. GS + дущий ⎱ → A4
q. + дший ⎰
r. + домый → A3
s. + денный → A1

Class Identification and Generation Rules (*continued*)

Class	Examples	Class identification	Generation rules	
			Generating stem	Inflected strings
V9	беречь запрячь мочь пренебречь стеречь стричь (and derivatives)	A class 1. ending in чь; 2. forming the first person singular with гу	word — last two letters	*a.* word *i.* GS + гла *b.* GS + гу *j.* + гло *c.* + жешь *k.* + гли *d.* + жет *l.* + ги *e.* + жем *m.* + гите *f.* + жете *n.* + жа *g.* + гут *o.* + гши *h.* + г *p.* GS + гущий ⎫ → *A4* *q.* + гший ⎬ *r.* + женный → *A1*
V9.1	влечь волочь облечь обречь печь сечь течь (and derivatives)	A class 1. ending in чь; 2. forming the first person singular with ку	word — last two letters	*a.* word *i.* GS + кла *b.* GS + ку *j.* + кло *c.* + чешь *k.* + кли *d.* + чет *l.* + ки *e.* + чем *m.* + ките *f.* + чете *n.* + ча *g.* + кут *o.* + кши *h.* + к *p.* GS + кущий ⎫ → *A4* *q.* + кший ⎬ *r.* + ченный → *A1*

V10	жать₁ мять (and derivatives) and the verbs начать распять	A class 1. ending in ать, ять; 2. distinguished by an а, я–н alternation	word – last two letters, but the last letter of the generating stem (= а, я) must be replaced by н in *b–g*, *l–m*, and *p*

a. word
b. GS + у
c. + ешь
d. + ет
e. + ем
f. + ете
g. + ут
h. + л
i. GS + ла
j. + ло
k. + ли
l. + и
m. + ите
n. + в
o. + вши
p. GS + ущий ⎱ → A4
q. + вший ⎰
r. + тый → A3

V10.01	жать₂ (and derivatives)	A class ending in ать and distinguished by an а–м alternation	As in *V10*, but the alternant н must be replaced by м
V10.1	обнять отнять снять	A class ending in нять preceded by a consonant, the ня alternating with ним	As in *V10*, but: 1. the alternant н must be replaced by им; 2. *p* must be eliminated
V10.2	донять занять перенять понять пронять унять	A class ending in нять preceded by a vowel, the ня alternating with йм	word – last two letters, but the last two letters of the generating stem (= ня) must be replaced by йм in *b–g* and *l–m*
V10.3	взять(ся)	As in *V10*, except for the elimination of *o–p*	

a. word
b. возьму
c. возьмешь
d. возьмет
e. возьмем
f. возьмете
g. возьмут
h. взял
i. взяла
j. взяло
k. взяли
l. возьми
m. возьмите
n. взяв
o. взявши
p. взявший → A4
q. взятый → A3

Class Identification and Generation Rules (*continued*)

Class	Examples	Class identification	Generation rules Generating stem	Inflected strings
V10.4	принять(ся)		word – last two letters, but the last two letters of the generating stem (= ня) must be replaced by м in *b–g* and *l–m*	As in *V10*, except for the elimination of *p*
V11	терёть простерёть мерёть перёть	A class 1. ending in еть preceded by ер; 2. distinguished by an ер–р alternation	word – last three letters, but the penult of the generating stem (= e) must be eliminated in *b–g*, *l–m*, and *o*	*a.* word *b.* GS + y *c.* + ешь *d.* + ет *e.* + ем *f.* + ете *g.* + ут *h.* GS + # *i.* + ла *j.* + ло *k.* + ли *l.* + и *m.* + ите *n.* + ши *o.* GS + ущий ⎫ → *A4* *p.* + ший ⎭ *q.* + тый → *A3*
V11.1	бороть(ся) колоть полоть пороть (and derivatives)	A class ending in оть preceded by ор, ол	word – last three letters	*a.* word *b.* GS + ю *c.* + ешь *d.* + ет *e.* + ем *f.* + ете *g.* + ют *h.* + ол *i.* GS + ола *j.* + оло *k.* + оли *l.* + и *m.* + ите *n.* + я *o.* + ов *p.* + овши *q.* GS + ющий ⎫ → *A4* *r.* + овший ⎭ *s.* + отый → *A3*

V12

1. давать } and other verbs derived from these by means of prefixation
2. the bases -знавать -ставать

A class ending in авать and distinguished by a ва–"j" alternation

word — last four letters

a. word
b. GS + ю
c. + ешь
d. + ет
e. + ем
f. + ете
g. + ют
h. + вал

i. GS + вала
j. + вало
k. + вали
l. + вай
m. + вайте
n. + ван
o. + вав

p. GS + ющий ⎫ → A4
q. + вавший ⎬
r. + ваемый → A3

V13

бить
вить
лить
пить
шить
(and derivatives)

A class ending in ить and distinguished by an и–"j" alternation

word — last two letters, but the following modifications of the last letter of the generating stem must be introduced:

1. и > ь in b–g and q;
2. и > е in l–m

a. word
b. GS + ю
c. + ешь
d. + ет
e. + ем
f. + ете
g. + ют
h. + л

i. GS + ла
j. + ло
k. + ли
l. + й
m. + йте
n. + я
o. + в
p. + вши

q. GS + ющий ⎫ → A4
r. + вший ⎬
s. + тый → A3

V14

быть
крыть
мыть
ныть
рыть
петь
(and derivatives)

A class

1. ending in ыть, еть;
2. in which the letter preceding ть (= ы, е) alternates with о

word — last two letters, but the last letter of the generating stem (= ы, е) must be replaced by о in b–g, l–n, and q

As in V13

V15

обуть
разуть
дуть
(and derivatives)

A class ending in уть

word — last two letters

As in V13

Class Identification and Generation Rules (*continued*)

Class	Examples	Class identification	Generation rules	
			Generating stem	Inflected strings
V15.1	жить плыть слыть (and derivatives)	A class 1. ending in ить, ыть; 2. distinguished by и–ив, ы–ыв alternations	word – last two letters, but а в must be added, in b–g, l–m, p, and s, to the last letter (= и, ы) of the generating stem	As in *V10*, except for the addition of: *s.* GS + я
V15.2	застрять деть стать (and derivatives)	A class 1. ending in еть, ать, and ять; 2. distinguished by е–ен, а–ан, and я–ян alternations	word – last two letters, but an н must be added, in b–g and l–m, to the last letter (= е, а, я) of the generating stem	As in *V10*, except ь for и in l, ьте for ите in m, and the elimination of p
V16	идти (and derivatives, e.g., дти, йти, перейти)	A class of the verbs ending in дти, йти	1. word – last two letters, but а д must be affixed to it in the case of the verbs ending in йти; 2. ше but: *a.* the last letter (= e) must be dropped in i–k of all verbs; *b.* in the case of the verbs ending in йти, the portion of the word preceding йти must be prefixed	*a.* word *b.* GS₁ + у *c.* + ешь *d.* + ет *e.* + ем *f.* + ете *g.* + ут *h.* GS₂ + л *i.* GS₂ + ла *j.* + ло *k.* + ли *l.* GS₁ + и *m.* + ите *n.* + учи *o.* + я *p.* GS₂ + дши *q.* GS₁ + ущий → A4 *r.* GS₂ + дший → A4 *s.* GS₁ + енный → A1

V17 ехать (and derivatives)

A class ending in ать and distinguished by an еха–ед alternation

word – last three letters, but the ultima of the generating stem (= х) must be replaced by а д in b–g, l, and o

a. word		h. GS + ал	
b. GS + у		i. + ала	
c. + ешь		j. + ало	
d. + ет		k. + али	
e. + ем		l. + учи	
f. + ете		m. + ав	
g. + ут		n. + авши	
o. GS + ущий			
p. + авший	} → A4		

V18 дать (and derivatives)

A class ending in ать

word – last two letters

a. word	i. GS + ла	
b. GS + м	j. + ло	
c. + шь	k. + ли	
d. + ст	l. + й	
e. + дим	m. + йте	
f. + дите	n. + в	
g. + дут	o. + вши	
h. + л		
p. GS + вший → A4		
q. + нный → A1		

V19 бежать (and derivatives)

A class ending in ать, characterized by a г–ж alternation as well as a conjugational miscegenation

word – last three letters, but the ultima of the generating stem (= ж) must be replaced by а г in b, g, l–m, and p

a. word		i. GS + ала	
b. GS + у		j. + ало	
c. + ишь		k. + али	
d. + ит		l. + и	
e. + им		m. + ите	
f. + ите		n. + а	
g. + ут		o. + ав	
h. + ал			
p. GS + ущий			
q. + авший	} → A4		

Class Identification and Generation Rules (*continued*)

Class	Examples	Class identification	Generation rules	
			Generating stem	Inflected strings
V20	хотеть (and derivatives)	A class ending in еть, characterized by a т–ч alternation as well as a conjugational miscegenation	word – last three letters, but the ultima of the generating stem (= т) must be replaced by а ч in *b–d*	*a.* word *h.* GS + ел *b.* GS + y *i.* + ела *c.* + ешь *j.* + ело *d.* + ет *k.* + ели *e.* + им *l.* + и *f.* + ите *m.* + ев *g.* + ят *n.* GS + ящий } → *A4* *o.* + евший }
V21	быть			*a.* быть *j.* был *b.* есть *k.* была *c.* суть *l.* было *d.* буду *m.* были *e.* будешь *n.* будь *f.* будет *o.* будьте *g.* будем *p.* будучи *h.* будете *q.* бывши *i.* будут *r.* бывший → *A4*

Words with Atypical Paradigms (Hand-inflected)

a. Nouns (*N99.99*)
1. крестьянин, англичанин, гражданин, etc.;
2. теленок, галчонок, опенок, etc.;
3. брат, друг, сын, etc.;
4. звено, дерево, дно, etc.;
5. колено (joint), колено (link), око, ухо, etc.;
6. небо, чудо;
7. ложь, рожь, любовь, церковь;
8. мать, дочь;
9. (family names) Некрасов, Пушкин, Карамзин, etc.;
10. господин, русин, грузин, etc.;
11. судно;
12. сосед, черт;
13. путь;
14. дитя;
15. плечо.

b. Adjectives (*A99.99*)
1. possessive, ending in
    (i) "ов" ("ев");
    (ii) "ин";
    (iii) "ий";
2. qualitative, ending in "ц" + "ый";

c. Verbs (*V99.99*)
1. стлать, слать, лгать;
2. сесть, грести, скрести, класть, расти;
3. жечь, лечь, толочь;
4. изъять, объять, подъять;
5. молоть;
6. гнить, почить;
7. брить, ошибиться, реветь;
8. all verbs with varyingly expandable prefixes (such as "с" → "со", "в" → "во", etc.).

Synoptic Classification Table for Nouns and Adjectives

1. A vertical line divides the desinence from the terminal portion of the stem.
2. Horizontal lines divide each column into groups of related classes.
3. V stands for any vowel not specified earlier within the group of related classes.
4. C stands for any consonant not specified earlier within the group of related classes.
5. C$_ь$C stands for any consonant either followed or not followed by "ь" and not specified earlier within the group of related classes.
6. The accent marks indicate stress, wherever distinctive.
7. Subscripts, when used, indicate the position within the stem (1 = last character, 2 = penultimate).
8. Vowel changes within a stem are marked by ">" and are located by the subscripts. The vowel before the change is given on the left side of ">", the vowel after the change on the right side.
9. Unless specified otherwise, the vowel changes affect all cases except nominative singular (*Ns*) and accusative singular (*As*) (inanimate).
10. Gp means that the vowel change occurs in genitive plural only.
11. Np: a means that the nominative plural has the desinence "a".
12. # indicates that a vowel has disappeared or will reappear.
13. Only the combination of consonants, vowels, or both, relevant for the classification of the whole word is indicated.

Nouns

Masculine

$\begin{bmatrix} \text{ечек} \\ \text{шек} \\ \text{шок} \end{bmatrix}$ *N1.3*

[éц]] *N1.2*
[ец]] *N1.4* $\Big\}$ but *N1* if not $e_2 > \#$

$\begin{bmatrix} \text{ж} \\ \text{ч} \\ \text{ш} \\ \text{щ} \end{bmatrix}$ *N3.1*

$\begin{bmatrix} \text{г} \\ \text{к} \\ \text{х} \end{bmatrix}$ *N1.1* but *N1.3* if $e_2/o_2 > \#$

C → *N1* but *N1.2* if $e_2 > \text{ь}_2$, $e_2 > \text{й}_2$, $e_2/o_2 > \#$

$\begin{bmatrix} \text{а} & \text{й} \\ \text{и} & \text{й} \\ \text{о} & \text{й} \\ \text{у} & \text{й} \end{bmatrix}$ *N2*

$\begin{bmatrix} \text{е} & \text{й} \end{bmatrix}$ *N2* but *N2.1* if $e_1 > \text{ь}_1$

$\begin{bmatrix} \text{ел} & \text{ь} \\ \text{ол} & \text{ь} \\ \text{ен} & \text{ь} \\ \text{он} & \text{ь} \\ \text{от} & \text{ь} \end{bmatrix}$ *N3* but *N3.05* if $e_2/o_2 > \#$

C → ь *N3* but *N3.05* if $e_2/o_2 > \#$

Feminine

$\begin{bmatrix} \text{ж} & \text{ь} \\ \text{ч} & \text{ь} \\ \text{ш} & \text{ь} \end{bmatrix}$ *N6.1*

C → ь *N6*

$\begin{bmatrix} \text{жн} & \text{á} \\ \text{шн} & \text{á} \end{bmatrix}$ *N4.06*

$\text{C}_\text{ь}\text{C} \begin{bmatrix} \text{л} & \text{а} \\ \text{н} & \text{а} \\ \text{ц} & \text{а} \end{bmatrix}$ *N4.05* but *N4* if not Gp: $\text{ь}_2 > e_2$ nor $\text{й}_2 > e_2$ nor $\# > e_2$

$\text{C}_\text{ь}\text{C} \begin{bmatrix} \text{м} & \text{а} \\ \text{р} & \text{а} \\ \text{б} & \text{а} \end{bmatrix}$ *N4.05* but *N4* if not Gp: $\text{ь}_2 > e_2$ nor $\text{й}_2 > e_2$ nor $\# > e_2$

$\begin{bmatrix} \text{жк} & \text{а} \\ \text{чк} & \text{а} \\ \text{шк} & \text{а} \end{bmatrix}$ *N4.3* but *N4.31* if final stress

$\text{C} \rightarrow \text{ьк} \begin{bmatrix} \text{йк} & \text{а} \\ \text{ьк} & \text{а} \end{bmatrix}$ *N4.3*

C → к а *N4.31*
V → к а *N4.1*
$\text{C}_\text{ь}\text{C}$ г а *N4.3*
V → г а *N4.1*

$\begin{bmatrix} \text{х} & \text{а} \\ \text{ж} & \text{а} \\ \text{ч} & \text{а} \\ \text{ш} & \text{а} \\ \text{щ} & \text{а} \end{bmatrix}$ *N4.1*

C → а *N4* but *N4.05* if Gp: $\text{ь}_2 > e_2$ or $\text{й}_2 > e_2$ or $\# > e_2$

$\begin{bmatrix} \text{л} & \text{е} \\ \text{р} & \text{е} \end{bmatrix}$ *N13*

$\begin{bmatrix} \text{ь} & \text{е} \end{bmatrix}$ *N11* but *N11.1* if Gp: $\text{ь}_1 > \text{и}_1$ or *N11.2* if Gp: ев

C → е *N9* but *N9.2* if Gp: $\text{ь}_2 > e_2$ or $\text{й}_2 > e_2$ or $\# > e_2$

и е *N10*

	Adjectives
⎡ гл | о ⎤ ⎢ кл | о ⎥ ⎣ зл | о ⎦	⎡ ж | ий ⎤ ⎢ ч | ий ⎥ A4 ⎢ ш | ий ⎥ ⎣ щ | ий ⎦
гм | о км | о *N8.1* but *N8* if not Gp: $\# > о_2$	н | ий *A5*
гн | о кн | о	⎡ г | ий ⎤ *A8* ⎣ х | ий ⎦
гр | о кр | о	жк | ий ⎤ йк | ий ⎥ *A7* С → ьк | ий ⎦
$C_ьC$ ⎡ л | о ⎤ *N8.15* but *N8* if ⎢ м | о ⎥ Gp: $ь_2 > е_2$ nor ⎢ н | о ⎥ $й_2 > и_2$ nor ⎣ ц | о ⎦ $\# > е_2$	С → к | ий *A6* V → к | ий *A8*
$C_ьC$ [к | о] *N8.4* ⎤ ⎬ but *N8* if $C_ьC$ [к | ó] *N8.3* ⎦ Np: а	⎡ ж | ой ⎤ ⎢ к | ой ⎥ ⎢ ш | ой ⎥ *A8* ⎢ г | ой ⎥ ⎣ х | ой ⎦
У → к | о *N8.3* С → | о *N8*	$C_ьC$ н | ой *A2* V → н | ой *A3* С → | ой *A3*
⎡ гл | я ⎤ ⎢ кл | я ⎥ ⎣ хл | я ⎦	
гн | я кн | я *N5.15* хн | я	нн | ый *A1* $C_ьC$ н | ый *A2* V → н | ый *A3*
гр | я кр | я хр | я	⎡ л | ый ⎤ *A3* but *A2* if ⎣ р | ый ⎦ masc. pred.: $\# > е_2$
$C_ьC$ ⎡ л | я ⎤ *N5.1* but *N5* if ⎢ н | я ⎥ Gp: $ь_2 > е_2$ nor ⎣ р | я ⎦ $й_2 > е_2$ nor $\# > е_2$	С → | ый *A3*
м | я *N12* С → ь | й *N5.2* С → ь | я *N5.3* С → | я *N5* и | я *N7* V → | я *N5.05*	

Synoptic Classification Table for Verbs

The right-hand side of the double column gives the infinitive; the left-hand side gives the third person plural and, if helpful, the first person singular, the imperative singular, the past passive participle, and the past active participle. Light vertical lines divide desinences from infinitive stems.

формулируют	формулир	овать	*V3*		колют	кол	оть	*V11.1*
межуют	меж	евать	*V3*		белеют	бел	еть	*V1*
					поют	п	еть	*V14*
делают	дел	ать	*V1*		трут	тер	еть	*V11*
слышат	слыш	ать	*V6*					
звучат	звуч	ать	*V6*		денут	д	еть	*V15.2*
держат	держ	ать	*V6*					
верещат	верещ	ать	*V6*		видят	вид	еть	*V6.2*
пишут	пис	ать	*V5*		хотят	хот	еть	*V20*
хлопочут	хлопот	ать	*V5*					
мажут	маз	ать	*V5*		будут	б	ыть	*V21*
гложут	глод	ать	*V5*					
плачут	плак	ать	*V5*		плывут	пл	ыть	*V15.1*
движут	двиг	ать	*V5*					
машут	мах	ать	*V5*		воют	в	ыть	*V14*
дремлют	дрем	ать	*V5*					
треплют	треп	ать	*V5*		несут	не	сти	*V8*
колеблют	колеб	ать	*V5*		лезут	ле	зть	*V8*
клевешут	клевет	ать	*V5.1*		гнетут	гне	сти	*V8.1*
ждут	жд	ать	*V5.2*					
					вычтут	выче	сть	*V8.11*
жнут	ж	ать	*V10*					
					ведут	ве	сти	*V8.2*
жмут	ж	ать	*V10.01*		прядут	пря	сть	*V8.2*
бегут	беж	ать	*V19*					
					идут	и	дти	*V16*
дерут	др	ать	*V5.4*		перейдут	пере	йти	*V16*
зовут	зв	ать	*V5.41*					
					берегут	бере	чь	*V9*
дают	дав	ать	*V12*		могут	мо	чь	*V9*
					запрягут	запря	чь	*V9*
дадут	д	ать	*V18*		стригут	стри	чь	*V9*
станут	ст	ать	*V15.2*		влекут	вле	чь	*V9.1*
едут	ех	ать	*V17*		волокут	воло	чь	*V9.1*

гуляют	гул	ять	*V1*
стоят	сто	ять	*V6.1*
веют	ве	ять	*V5.3*
мнут	м	ять	*V10*
обнимут	обн	ять	*V10.1*
доймут	дон	ять	*V10.2*
возьмут	вз	ять	*V10.3*
примут	прин	ять	*V10.4*
застрянут	застр	ять	*V15.2*

толкнут толкни толкнул	толкн	уть	*V2*
сунут сунь сунул	сун	уть	*V2.01*
гаснут гас	гасн	уть	*V7*

голосят голошу голоси	голос	ить	*V4*
возят вожу вози	воз	ить	*V4*
платят плачу плати	плат	ить	*V4*
плодят пложу плоди пложен	плод	ить	*V4*
любят люблю люби	люб	ить	*V4*
простят прошу прости	прост	ить	*V4*
ставят ставлю ставь	став	ить	*V4.01*
успокоят успокой	успоко	ить	*V4.02*
возвратят возврашу возврати	возврат	ить	*V4.1*
предупредят предупреди предупрежден	предупред	ить	*V4.1*
насытят насышу насыть	насыт	ить	*V4.11*
учат учи	уч	ить	*V4.2*
раскулачат раскулачь	раскулач	ить	*V4.21*
вьют	в	ить	*V13*
живут	ж	ить	*V15.1*

CHAPTER 7

DICTIONARY COMPILATION

7.1. The Compilation Process

Lexicography is an ancient art, demanding of its practitioners either superhuman patience and longevity, or else a willingness to accept the words of predecessors without firsthand reference to actual usage. The problem of compiling an automatic dictionary is significant far beyond the immediate problems of automatic translation. The automatic compilation of dictionary files and the use of these files by automatic translating machines promise revolutionary advances in lexicography in terms of increased timeliness and accuracy of dictionaries. If provision is made for some kind of automatic or semiautomatic feedback from the texts processed, an operating automatic dictionary can be continuously improved and augmented so that it will accurately reflect current syntactic and semantic usage.

The compilation of a dictionary still presents vast information-handling problems, so that the use from the very beginning of automatic machines and associated techniques is of great importance in the effective solution of these problems. Research workers with both adequate qualifications in linguistics and experience in the design and operation of automatic information-processing machines are rather scarce. Careful planning is therefore essential in order to enable the performance of large-scale routine tasks by a team of clerical and technical personnel assisted by automatic machines. In the initial stages, old-fashioned reliance on existing dictionaries is still a definite help.

Some features of the compilation process described in this chapter are not intrinsic, but were determined in part by the accidental order in which ideas arose and in part by specific characteristics of the Univac system. An excessively detailed description of the process would therefore have only ephemeral value. Some detail, pedestrian though it may seem, is essential nevertheless. Many who have little acquaintance with the problems of automatic information processing tend either to view automatic machines as miraculous devices capable of performing untold wonders or else to shun these machines as products of the devil; the close examination of a concrete application should help toward a more realistic appraisal. Many techniques used in dictionary compilation have a broader range of applications in

216

structural and mathematical linguistic studies, and therefore deserve careful
description. Moreover, there is little in the literature by way of description
of techniques of large-scale information processing. With the exception of
one or two recent books cited in Sec. 1.8, what is available consists mainly
of manufacturers' reports. These are understandably sanguine about the

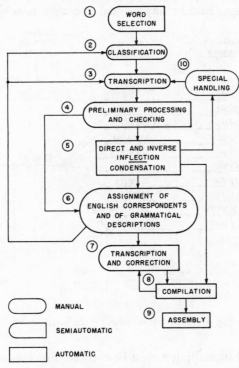

Fig. 36. The compilation process.

capabilities of machines, and tend to gloss over many serious difficulties.
Our description will therefore aim at sufficient generality to assist those
working with machines other than the Univac or those planning the design
of new ones, and at sufficient detail to maintain contact with reality whenever
similar detailed problems can be expected to arise with any machine. Certain
significant questions, not considered in this chapter in order to avoid repeated
digressions, are analyzed in Chapter 9.

Figure 36 shows the broad outlines of the compilation process (Giuliano,
1959a, 1957a,b). Some steps of this process are illustrated with the word
заем* in Fig. 37. The first step is the selection of the Russian words to be

included in the initial dictionary or added to an existing one. Eventually, this selection can be made a by-product of the operation of an automatic translator, by providing for the listing of words not found in the dictionary (Fig. 73). Second, the selected Russian words must be classified according to the system described in Sec. 6.1, to enable their automatic inflection at a later stage. The next step is transcription of the words and their class marks onto magnetic tape. Because any manual operation is subject to mistakes,

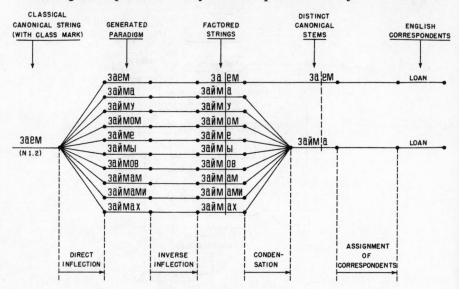

Fig. 37. Compilation steps for the word *заем**.

classification and transcription must be checked. For reasons given in Secs. 7.6 and 7.8, action on mistakes detected at step 4 is postponed to step 6. In the fifth step, the generated paradigm of each word is produced automatically; each member of the generated paradigm is factored by the machine as stem + affix; the resulting set of stems is condensed to a set including only distinct stems. The fifth step is fully automatic.

The sixth step comprises the assigning of appropriate English correspondents and grammatical marks to each stem. Material found to be in error at this step or at step 4 is deleted, and returned to step 2 or step 3 for another try. The bulk of the material is transcribed, checked for mistakes, and corrected where necessary. The new group of dictionary entries is assembled with other groups or with an existing dictionary in step 9. Some of the entries in such a group are shown in Fig. 38.

Fig. 38. Sample dictionary entries.

Word						
NABEGAL	V01.00	019-0224P07 DASHED AGAIN ST	1 COVERED	2 FELL UPON %	AOA1A2	P7 B3
	ENTRY 011097 VKOOE5OOKO					
NABEGAN	A01.00	019Q0224P007 DASHED AGAIN ST	1 COVERED	2 FALLEN UPON%	AOA1A2	P7 Q3
	ENTRY 011098 ADOOOO					
NABEGANI	N10.00F600	019-0202P09 TAILING %	A1			QO
	ENTRY 011099 ND1NOOO					
NABEGANN	A01.00FT10	019Q0224P010 DASHED AGAIN ST	1 COVERED	2 FALLEN UPON%	AOA1A2	P7 Q3
	ENTRY 011100 ADOOOO					
NABEGAJUSHCH	A04.00F910	019A0224P010 DASHING AGAI NST	1 COVERING	2 FALLING UPON	AOA1A2	QO
	ENTRY 011101 ADO100					
NABIVK	N04.31F100	019-0203P007 PACKING	1 FILLER	2 PRINTING %	AOA1	
	ENTRY 011102 ND1FOOO					
NABIVOK	N04,31	019-0203P007 PACKING	1 FILLER	2 PRINTING %	AOA1	
	ENTRY 011103 ND1FOOO					
NABIR	V01.00 1YO	019-0226P007 DIALING	1 SETTING UP	2 COLLECTING %	AOA1A2	P7 B5
	ENTRY 011104 VNOOP1OOKO					
NABIRA	A03.00 600	019C0226P008 DIALED	1 SET UP	2 COLLECTED %	AOA1A2	Q1
	ENTRY 011105 ADOOOO					
NABIRA	V01.00FKUO	019-0226P008 TO DIAL	1 TO SET UP	2 TO COLLECT %	AOA1A2	P7 B0K0B1B4B6
	ENTRY 011106 VNOOP1OOKO					
NABIRAVSH	A04.00F910	019B0226P010 DIALING	1 SETTING UP	2 COLLECTING %	AOA1A2	P7 Q2
	ENTRY 011107 ADO100					
NABIRAEM	A03.00FT10	019C0226P010 DIALED	1 SET UP	2 COLLECTED %	AOA1A2	Q1
	ENTRY 011108 ADOOOO					
NABIRAJT	V01.00 600	019-0226P009 DIAL	1 SET UP	2 COLLECT %	AOA1A2	P7 B4
	ENTRY 011109 VNOOP1OOKO					
NABIRAL	V01.00	019-0226P007 DIALED	1 SET UP	2 COLLECTED %	AOA1A2	P7 B3
	ENTRY 011110 VNOOP1OOKO					

Fig. 38. Sample dictionary entries.

7.2. The Structure of Items

Among the practical problems that arise in the planning of compilation procedures, the selection of a good format for the representations of Russian words and other information elements is of paramount importance. The representational unit associated with a Russian word is called an *item*, in keeping with terminology widely used in information processing. An item, in general, is a token irreducible relative to some conventional set of filing operations (Sec. 1.7). For example, cards in a library card catalog and folders in a correspondence file are commonly treated as items. Because our dictionary is to be stored on magnetic tapes, the following is concerned chiefly with magnetic-tape files.

The structure of an item must meet several requirements. By definition, it should be irreducible relative to a set of filing operations. The description of Univac characteristics in Secs. 1.3 and 1.5 therefore suggests the machine word, the word-pair, the blockette, and the block as natural units. The machine word is irreducible relative to most operations, the word-pair and the blockette are irreducible only relative to certain transfer operations, and the block is irreducible only relative to transfers to and from magnetic tape. Choosing the machine word as an item ensures simplicity of internal operations but requires grouping items into larger units (*packing*, see Sec. 2.7) for purely technical reasons whenever access to magnetic tape is necessary. Choosing the block as an item ensures simplicity of transfers to and from tape but since the block is not irreducible relative to internal operations any internal operation on the whole block must be synthesized from elementary instructions.

On the other hand, whenever operations are to be performed on some subset of an item, it is necessary that the item be reducible relative to these operations. Consider, for example, items representing whole Russian sentences. Should operations on individual Russian words in the sentence be necessary, the sentence item must be reducible relative to operations on Russian words. An item organized as a sequence of machine words arranged as in Fig. 16*b* has the requisite reducibility properties. The representation illustrated in Fig. 16*a* is reducible over an alphabet of English words only on the further assumption that English words themselves are reducible over the ordinary alphabet. Isolating an English word entails executing appropriate unpacking instructions. These instructions will be relatively complex, since machine words are reducible only relative to appropriate shift and extract instructions. In this case the reducibility properties of an item which can be readily factored into machine words are not isomorphic to those of a sentence which requires factoring into English words.

Russian words are first transcribed into machine tokens at the transcription stage, step 3 of Fig. 36, and the problem of item format is first met at this point. The input item format used in our experimental work is shown in Fig. 39. Each item is a group of five machine words, of which the first three, numbered 0, 1, 2, are allocated to Russian words. Positions 1–6 of the fourth word are allocated to the class mark. Unallocated and unused positions

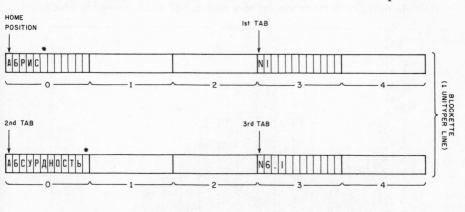

SPACE-FILL (Δ's) IS USED THROUGHOUT

*THE RANKED COMPUTER REPRESENTATION FOR RUSSIAN IS USED HERE.

†CLASS MARKS ARE IN THE NATURAL REPRESENTATION.

Fig. 39. The format of input items on tape (Type 1).

are automatically filled with spaces when the space bar of the Cyrillic Unityper (Sec. 2.8) or tabulation keys set as marked in Fig. 39 are struck.

The choice of the format of Fig. 39 was governed by two major considerations: ease of typing and simplicity of further automatic processing of the transcribed material. Emphasis on ease of typing is extremely important, for reasons both humane and technical. The humane reasons are obvious. As for technical reasons, it has been observed that mistakes in transcription of input data create one of the most serious, recurrent, irritating, and costly problems in automatic information processing. These mistakes can be reduced to a minimum by designing and using equipment in a manner which its operators find comfortable and to which they are already accustomed, or, at least, can easily become accustomed. Had the words to be transcribed occurred in texts it would have been best so far as ease of typing is concerned to choose a mode of transcription as close as possible to normal typing practice. This choice was indeed made with regard to the transcription of texts as described in Sec. 8.2. In typing isolated words a columnar layout is equally satisfactory to a typist, provided that normal use can be made of typewriter tabulation stops; this layout has the advantage of producing items of a standard format well suited to automatic processing.

Because the length of the typewritten line produced by the Unityper corresponds precisely to one blockette of information, a columnar layout is possible only for items made up of a number of machine words which is a

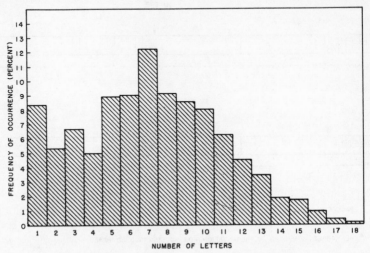

Fig. 40. Distribution of Russian word length
(based on a 6000-word sample of texts).

divisor or a multiple of 10. In general it is also desirable for item size to be a divisor or multiple of 60 to avoid undesirable straddling of block boundaries by items. Higher multiples of 10 are ruled out on the ground that most of the machine words allocated to an item would never serve a useful purpose. Items of one, two, five, or ten machine words may be considered further. One-word items are ruled out, since the data of Fig. 40 show that many Russian words can be expected to be more than 12 characters long. Two-word items might do, if the data of Fig. 40 could be trusted, but there would be no room left available in the item for information other than a Russian word and its class mark.

It seems desirable to allocate three machine words to a Russian word. Strings like "спектрофотоэлектрического" are rare, but their occurrence as exceptions would create a recurrent nuisance if a format based on two machine words per Russian word were used. Only five- and ten-word items remain to be considered. Taking into account later operations on the input items, it seems adequate to have two machine words available for the class mark and for other information it might be desirable to introduce into an item. This suggests a five-word item.

A five-word item meets the restrictions imposed by the typewriter; it is factorable into a three-word representation of the Russian word and two words representing all other information; the representation of the Russian word is unfortunately not irreducible, for instance, relative to the transfer in memory of Russian words, nor is its factorization into three machine words an isomorphic image of any desirable factorization of Russian words; however, ten-word items present identical problems relative to the given machine. The remaining controlling factors seem to be the desirable irreducibility of ten-word items relative to internal transfers (Univac Y and Z instructions) and the desirable compactness of five-word items. The decision, reflected in Fig. 39, was made in favor of five-word items.

In retrospect, this decision seems somewhat unfortunate. The major cost to be paid for using ten-word instead of five-word items is a doubling of the size of certain intermediate files to be stored on reels of magnetic tape. The "waste" of at least half the space on these reels is not significant in itself, since the cost of a reel of magnetic tape is negligible in comparison with other experimental and operating costs. What appears to be significant is the increase in tape travel time ensuing from the need to pass over many unused words. However, this time is significant only if the time required for a computer program to operate on one item is shorter than the time necessary to obtain the next item from tape. A program with this property is called *tape-limited* (Sec. 1.6). It turns out that, with one or two exceptions, most of the operations described in this chapter are not at all tape-limited. In fact, an appreciable portion of the programs for these operations merely effects the transfer of five-word items, either by 5 one-word transfers, or by 2 two-word transfers followed by a one-word transfer. Transferring blockettes by single instructions would have enabled a substantial increase in the over-all speed of operations.

Some writers (Rhodes, 1959; Korolev, 1957; Oettinger, 1953), noting the redundancy (Shannon and Weaver, 1949) inherent in the conventional representations of Russian or English, have urged the use of less redundant representations to save storage space. This expedient seems to have little value for experimental work, because it aggravates packing problems, and also because "compressed" representations are difficult for people to interpret (Sec. 2.6, Table 5), thereby creating a major nuisance on the frequent occasions when peeking at the contents of files or of machine memories is necessary. The value of compression in routine large-scale production remains to be assessed.

The foregoing does not imply that experimental work should not proceed unless the factors affecting item format can be exhaustively analyzed. Indeed, a thorough analysis often requires very acute hindsight. Our description of the criteria of choice has only heuristic value in suggesting problems that should be considered whenever possible in planning experiments, and always when planning for large-scale production. The art or science of information processing has not yet progressed to a stage where surer guides can be provided.

Given a machine capable of dealing more directly with items of varying lengths, the substance but not the spirit of the preceding arguments would be different; likewise, although decisions about the format of other types of items entering into dictionary compilation must be based on other factors, the spirit in which these decisions are made is not sufficiently different to warrant detailed analysis in each case. New item structures will be introduced without comment as the need arises in later sections of this chapter.

> *Exercise 7-1.* Consider a real or hypothetical automatic information-processing machine with facilities for storing items of variable length on magnetic tape. What format would you choose for input items capable of representing the same type of information as those of Fig. 39? It will be no accident if most of your thought about this exercise is given to determining what factors must be taken into account.

7.3. The Structure of Files

Some general characteristics of information-storage media and of the problems of organizing stored information are described in Secs. 1.6 and 1.7. Here our concern will be with one concrete aspect of the organization of magnetic-tape files.

Although it may seem too obvious to warrant stating explicitly, a tape file must have a beginning and an end. Most physical tape mechanisms are equipped with sensing devices that can detect the approach of a physical end of a reel of tape when the tape is traveling in either direction. The tape motion is normally stopped when an end is sensed and an appropriate electric signal is transmitted to the control unit to halt the execution of the program. The end-sensing circuits, which are rarely called into action, often fail when they are needed, as circuits that are seldom used are apt to do. The results may be mildly catastrophic. Moreover, the length of a tape file need not coincide with the length of a reel of tape. It is therefore essential that recorded information be available on tape to signal the beginning and end of the tape file, thus distinguishing between the actual end of the tape file and the accidental interruption of a tape file occupying several reels by the physical end of a tape reel.

Characters indicating the end of a tape file are commonly called *sentinels*. The main requirement imposed on sentinel characters is that they be different

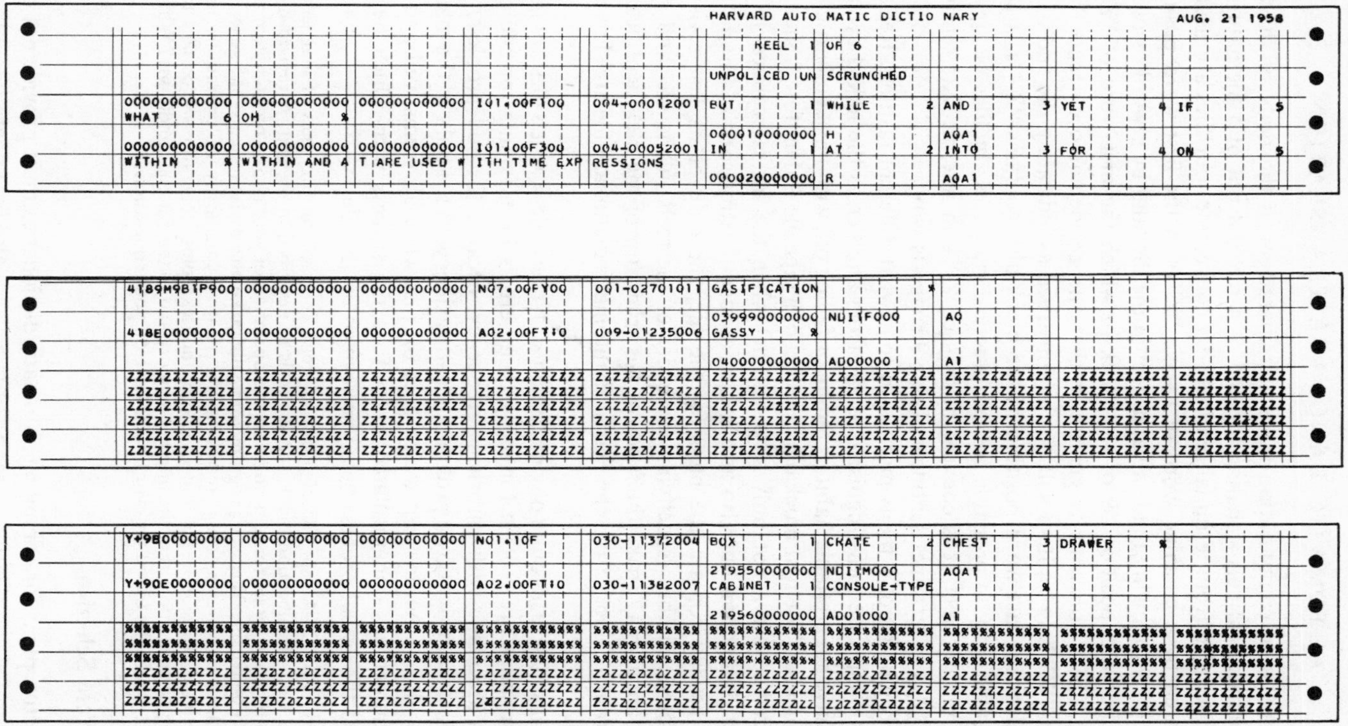

Fig. 41. Identification block and sentinels in the Harvard automatic dictionary file: *(top)* identification block; *(middle)* sentinel block, reel 1; *(bottom)* sentinel block, reel 6.

in nature or in configuration from any characters in the body of the file. Intermediate sentinels signaling the end of a reel but not of a file naturally must differ also from terminal sentinels. By convention, our files are usually terminated by a machine word of twelve "Z" 's in the first word position immediately following the last valid item in the file, and by a word of twelve "Z" 's in word position 59 of the block in which the first word of "Z" 's occurs. These conventions guarantee that the accidental occurrence of a word of "Z" 's in an item will not be construed as indicating the end of the file, unless the last word position of the block also contains "Z" 's, a joint event sufficiently improbable for all practical purposes. In files occupying more than one reel, it is convenient to use "Z" 's in intermediate sentinels and " % " 's in terminal sentinels (Table 9 suggests why).

The beginning of a file is marked in somewhat different fashion. Reels of magnetic tape are indistinguishable one from another to the naked eye and labels have a regrettable habit of fading or falling off, with most embarrassing consequences. The first block of every reel of a tape file is therefore frequently organized to serve not only as a sentinel when the file is scanned in the backward direction but also as a means of identification. Such *identification blocks* usually contain the name of the file, the date of its last revision, consecutive numbers specifying the order of the reels in the file, and such other data as may strike the fancy of the originator of the file. Identification and terminal blocks from the Harvard automatic dictionary file are shown in Fig. 41.

All this may seem quite obvious and it is. Nevertheless, when files are made and programs are written for using them, errors in providing for end conditions are made with surprising frequency, because the end problems are easily lost from sight by a programmer struggling with the bulk of the matter. The result is a booby trap which goes off just when all seems well and the programmer has congratulated himself on the success with which he has handled the actual process being performed on the file.

> *Exercise 7-2.* A tape file of five-word items is stored on two reels of magnetic tape, with identification block and sentinels as described in the text. A revised file is to be made by performing some simple operation of your choice on each item in the original file. Write a program that will make the revised file. What general conclusions can you draw regarding the relative complexity of operations on individual items in a file and of those "housekeeping" operations necessary to bring items in and out of the internal memory and to test for end conditions? What effect would the use of ten-word items have on your conclusions?

7.4. Word Selection

Effective plans for compiling automatic dictionaries must provide not only for the initial compilation of each dictionary, but also for updating to improve the accuracy of existing entries and to add new entries. Because automatic translation is still in an experimental stage, adequate flexibility of procedures must be retained. The need for skilled manual intervention must be minimized, with as much of the routine work as possible to be performed by the machine. Bilingualism and *expertise* in each technical field to which the words in an automatic dictionary are to apply do not necessarily go hand in hand and do so only rarely; it is therefore desirable to separate those tasks requiring linguistic skill from those in which a knowledge of the subject matter of texts is of paramount importance.

The initial selection of entries for a dictionary can be effected by either of two major processes, each having peculiar advantages and disadvantages. The first method may be likened to panning for gold. In this method some number of texts of the type eventually to be translated are scanned, and a glossary of all distinct forms occurring in this sample is compiled. The main advantage of this method is that every form obtained in this way is in current use and therefore a suitable candidate for inclusion in the glossary. If the scanning is to be performed automatically, the text must necessarily be transcribed onto some automatically readable medium, and hence become available for other purposes, including the eventual testing of translation methods. The major disadvantage of the procedure is that it becomes progressively less and less productive. While nearly all of the first few words of the first text scanned are likely to be nuggets, as the work progresses more and more gravel must be handled before another nugget is found. Relatively few word forms account for the vast majority of form occurrences in any text: these forms are found over and over again and must be rejected over and over again. The second disadvantage is that, in pure panning, only those particular inflected forms of any word that actually occur in the sample under study will be entered in the glossary. Until all of the forms constituting the paradigm of a word have occurred in some text, a complete characterization of this word is not available. This precludes the possibility of early systematic treatment of inflectional processes by means more elegant, reliable, and economical than the handling of each inflected form as a unique entity—a method to which many investigators have resorted (Sec. 10.7).

The second approach may be called the "fish-net" method. Available dictionaries are dragged for words useful according to some reasonable criteria. Anything caught in the net is retained. The chief advantage of this procedure is that a large selection of useful words can be obtained very rapidly. The major disadvantage is that the resulting dictionary is only as good as the criteria of selection or as the dictionaries from which the selection was made. Moreover, words without current utility may well be caught in the net.

Although the vocabulary obtained by panning is excellent for the panned texts, neither method is guaranteed to yield a vocabulary precisely suitable for the first arbitrary texts to which the dictionary is applied; but either lends itself to the addition of new words or new forms whenever these appear in a text that is being processed. How rapidly the ratio of text words not in the

Table 39. The 1000 most frequent word forms
in a sample (from Dewey, 1923).

Distinct word forms	Cumulative percent occurrence in 100,000 words
50	46
100	54
150	59
200	62
250	64
300	66
350	67
400	69
450	70
500	71
550	72
600	73
650	74
700	75
750	75
800	76
850	77
900	77
950	78
1000	78

dictionary to those in it will diminish to a satisfactory level is still open to conjecture. For that matter, so is a satisfactory definition of this level.

The optimum size of a dictionary must be determined and this determination poses some vexing questions. The data of Table 39, obtained by Dewey (1923), show that 100 most common word forms make up over 50 percent of a sample of 100,000 words of running English text and that 1000 most common word forms account for well over 75 percent of all words in the 100,000 word sample. This leads to what Bull, Africa, and Teichroew (1955) have called the *80-percent fallacy*.

If Dewey's data are used uncritically, it is easy to jump to the conclusion that a relatively small vocabulary is adequate for automatic translation, particularly if initial experiments are restricted to some well-defined narrow area of knowledge. According to the 80-percent fallacy, a dictionary with 1000 word forms in it will probably account for more than 80 percent of all the words occurring in a text; therefore almost everything one might wish to say can be said with a very small vocabulary. Leaving aside statistical fluctuations which would affect the validity of this estimate for any particular text, we may consider just what constitutes this 80 percent of the words.

Figure 42 is an excerpt from a book (Churchill, 1901) used as a source in

---

The ---- ------, the ------ ------, and the --- ------ the ----- ----- to ----. The ------ ----- --------- in the ------- ----- of -----, and a ---- ---- over the ------------. Not a ----- --------- the ----- of the ---- -------- to the ---------, and the ---- had ------ the ---- of ----- from over it. On the ---- ------- ------ ---- down to ------- -----; men ------ on the ------ and ---------- on the --------- ----, as if this were some -------- ----------. ----- ----- to the other ----- -------- after. -------- was -----, and ------. ---. ------ ------- in her ----- and ----- to ----.

"--------, where are we going?"

-------- did not ----.

"-----!"

She ------. In that ---- she ---------- that great good-------- man, her ------'- -------, and for his ---- ------- ------ had put up with much from his ----'- -------in----. She could ---- over, but never ------- what her ---- had said to her that ---------. ---. ------ had ----- been ----- before, and -------------. But as the ---- thought of the ------, ------- out on the ------, it -------- -------- to her that no ---- would have ------- it. In all her life she had never -------- ---- now that her ---- was not a ----. From that time ----- --------'- -------- ------ her ---- was -------.

She ---------- -------, -------, and -------- ----------, and ---- out ---------- to ---- the ------- and ------- the ----------- of the ----. Not that this -------- much to her. At the ---- of the ------------- ------- to the ------ ---- she ---, of all ------, --. ---------- ------ ------- on the ----, and ---- ------ ------------- on the ---- of the ------house. In ------- ---- -------- would have -------, for at ----- of her he ------------ ------------, ------ his ---- into his -----, and ------- his --- with more ---- than the -------- --------- he ------- -------- to the ---. ------- --------- would not have ------ the ----------.

Fig. 42. Passage containing only Dewey's 200 most frequent word forms.

Dewey's original count. All words whose frequency in that count put them below the 200th most frequent word were omitted and replaced by dashes.

It is quite obvious that the dashes in turn could be replaced by almost anything. All that is left when the least frequent words are taken out and only the 200 most frequent words left in is a grammatical shell useful perhaps for parlor games but hardly informative. The 200 most frequent words are mostly so-called *function* words which provide the relational framework of a sentence but carry very little referential meaning and are of little value in differentiating one text from another. Figure 43 is the same passage, this time with all words belonging to the set of 750 most frequent left in. While the text is somewhat fuller it is still difficult to discern any significance in the passage and considerable freedom in filling in the blanks remains.

The ---- ------, the ------ ------, and the --- turned the ----- ----- to ----. The ------ ----- --------- in the ------- ----- of light, and a ---- ---- over the ------------. Not a ----- --------- the ----- of the ---- -------- to the ---------, and the ---- had ------ the ---- of ----- from over it. On the ---- ------- ------ ---- down to ------- -----; men ------ on the ------ and ---------- on the --------- ----, as if this were some -------- ---------. Women ----- to the other ----- -------- after. -------- was heard, and ------. ---. ------ ------- in her ----- and began to talk.

"--------, where are we going?"

-------- did not ----.

"-----!"

She turned. In that hour she ---------- that great good-------- man, her ------'- -------, and for his ---- ------- ------ had put up with much from his ----'- -------in-law. She could pass over, but never ------- what her ---- had said to her that afternoon. ---. ------ had often been ----- before, and ------------. But as the girl thought of the ------, ------- out on the ------, it -------- -------- to her that no ---- would have ------- it. In all her life she had never -------- till now that her ---- was not a ----. From that time ----- --------'- attitude toward her ---- was -------.

She ---------- -------, however, and -------- something, and went out ---------- to find the ------- and ------- the ----------- of the ----. Not that this -------- much to her. At the ---- of the ------------ ------- to the ------ ---- she saw, of all ------, --. --------- ------ ------- on the ----, and ---- ------ ------------ on the ---- of the ------house. In another ---- -------- would have ------, for at sight of her he ------------ ------------, ------ his ---- into his -----, and ------- his --- with more ---- than the -------- --------- he usually accorded to the ---. ------- --------- would not have ------ the ---------.

Fig. 43. Passage containing only Dewey's 750 most frequent word forms.

The same text is given once more in Fig. 44, where the words absent in Fig. 43 are given, and none of the 750 most frequent words appear. In fact, few of the words in Fig. 44 are sufficiently frequent to appear in Dewey's list at all, since he gives only words that occurred more than ten times in the sample of 100,000 words. We find here that the ingredients of the story become quite clear, although there is little indication of any relations among the elements named by the words given in the text. The text can no longer be mistaken for, say, scientific writing, and a few experiments conducted with volunteer students demonstrated that something of the flavor and style of

---

--- rain ceased, --- clouds parted, --- --- sun ------ ---
muddy river -- gold. --- bluffs shone May-green -- --- western
flood -- -----, --- - haze hung ---- --- bottom-lands. --- - sound
disturbed --- quiet -- --- city receding -- --- northward, ---
--- rain --- washed --- pall -- smoke ---- ---- --. -- --- boat
excited voices died ---- -- natural tones; --- smoked -- ---
guards --- promenaded -- --- hurricane deck, -- -- ---- ---- ----
pleasant excursion. ----- waved -- --- ----- boats flocking
-----. Laughter --- -----, --- joking. Mrs. Colfax stirred -- ---
berth --- ------- -- ----.
    "Virginia, ----- --- -- -----?"
    Virginia --- --- move.
    "Jinny!"
    --- ------. -- ---- ---- --- remembered ---- ----- -----natured
---, --- mother's brother, --- --- --- sake Colonel Carvel ---
--- -- ---- ---- ---- --- wife's sister-------. --- ----- ----
----, --- -----forgive ----- --- aunt --- ---- -- --- ---- --------.
Mrs. Colfax --- ----- ---- cruel ------, --- inconsiderate.
--- -- --- ---- ------- -- --- speech, staring --- -- --- waters,
-- suddenly occurred -- --- ---- -- lady ----- ---- uttered --.
-- --- --- ---- --- --- ----- realized ---- --- ---- --- aunt ---
--- - lady. ---- ---- ---- forth Virginia's -------- ------ ---
aunt --- changed.
    --- controlled herself, -------, --- answered --------, ---
---- --- listlessly -- ----- --- Captain --- inquire --- destination --
--- boat. --- ---- ---- mattered ---- -- ---. -- --- foot -- ---
companionway leading -- --- saloon deck --- ---, -- --- people,
Mr. Eliphalet Hopper leaning -- ---- rail, --- pensively expectorating
-- --- roof -- --- wheel------. -- ------- mood Virginia -----
---- laughed, --- -- ----- -- --- -- straightened convulsively,
thrust --- quid ----- --- cheek, --- removed --- hat ---- ---- zeal
---- --- grudging deference -- usually accorded -- --- sex. Clearly
Eliphalet ----- --- ---- chosen --- situation.

Fig. 44. The words absent in Fig. 43.

the original remain. Indeed, it is possible for some people to date the work accurately.

Several attempts have been made to guess at the size of vocabulary that will be suitable for work in certain fields. Oswald and Lawson (1953) examined 16 German articles on brain surgery, a total of about 200 pages. They concluded that a dictionary with 4328 entries would enable dealing accurately with nearly 90 percent of the running nouns, verbs, and adjectives, and close to 80 percent of the different items (single lexical entries) encountered in samples of the literature of brain surgery. This estimate includes words of common usage as well as technical terms. A Russian-English mathematical glossary prepared for the American Mathematical Society (1953) contains approximately 2800 entries. These include occasional inflected forms (for instance, "докажем" as well as "доказать") and words of common usage, but no words whose meanings become clear upon mere transliteration. A *Russian-English Electronics Dictionary* prepared by the Department of the Army (1956) has about 22,000 entries, but many of these are phrases whose components recur in many entries. A standard Russian-English desk dictionary (Smirnitskij, 1949) has about 50,000 entries. These estimates do not provide a clear guide for deciding how to balance the nuisance of not finding words in a dictionary against the waste of having many entries that are used only rarely, if at all. Such a guide may be obtained in the near future as a result of experimental operation of automatic dictionaries and translators.

Our word-selection procedures represent a combination of panning and fishing. An initial set of over 10,000 words was selected from the general, electronics, and mathematics dictionaries cited in the preceding paragraph. The goal was to obtain both words of common currency and as complete as possible a coverage of technical terms in the two areas. In the process of selecting from the two technical dictionaries (Matejka, 1957), Russian entries with inflected keys were ignored, except in the absence of an equivalent entry with a key in the classical canonical form; in such cases, the inflected key was transformed to the classical canonical form. Distinct individual component words were culled from the phrases. Words of doubtful utility were usually included; it seemed more efficient to carry a few doubtful words through routine compilation procedures with provision for their automatic removal (based on a criterion of frequency of use) after the accumulation of operating experience, than to have valuable personnel spend time on intricate, arbitrary, and often inconclusive selection procedures.

All words encountered in texts submitted for translation that are not in the dictionary are listed as a by-product of dictionary operation (Sec. 8.4). These new words from text sources serve as the raw material for an updating procedure almost identical to the compilation process. For updating, therefore,

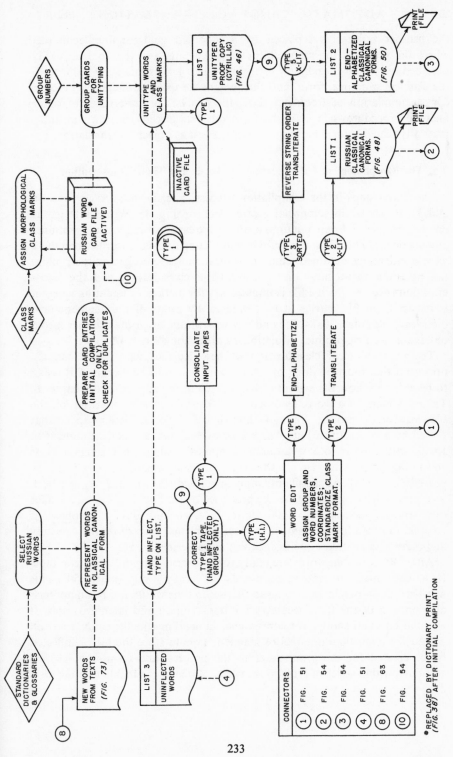

Fig. 45. The compilation process, Part I.

233

the fish-net process is replaced by a modified gold-panning technique, where the gravel is the main object of processing, and the gold an easily obtained by-product; each new form found in a text is manually reduced to the classical canonical form, and then enters the usual compilation process. The accumulation of frequency data, the analysis of contexts, and other analytic procedures can be carried out on texts already recorded in automatically readable form because of their interest as objects of translation.

7.5. The Morphological Classification of Words and Their Transcription

The initial steps in the compilation process, already outlined in Figs. 36 and 37, are shown in somewhat greater detail in Fig. 45. Words selected as described in Sec. 7.4 were put onto ordinary file cards, stored in a conventional alphabetic file. This file was helpful in eliminating duplicate words when these were accidentally obtained from the several sources and generally in maintaining order throughout the morphological classification and the transcription processes. Such a file is unnecessary for dictionary updating where it is replaced, for old entries to be corrected, by prints of the magnetic-tape dictionary file (Figs. 38, 84, 85) and, for new words, by prints or other media obtainable as a by-product of dictionary operation (Sec. 8.4).

The morphological classification of new Russian words is a manual operation that must be done by people with enough knowledge of Russian to be able to follow the synoptic charts given in the appendix to Chapter 6. The class mark of a word, as determined from the charts, is written on the file card for the word. "Enough" knowledge of Russian seems to be that gained by a good student after a year or two of standard college courses in Russian. It is useful to have a qualified linguist available in a supervisory or consulting capacity to help classifiers with unusual situations. A satisfactory procedure is to ask the classifiers simply to lay aside the card for any questionable word. These cards may then be processed by the linguist. A good classifier can handle approximately 1000 words in a working day. This phase of compilation therefore presents no serious problems. Procedures for detecting and correcting mistakes in classification are described in Sec. 7.6.

After classification, words are ready for transcription onto magnetic tape. It is convenient to transcribe words in groups of about 750. This number was chosen large enough so that setup time would not be an appreciable portion of the time necessary for transcription and later processes. It was chosen small enough for convenience in handling the file cards, to avoid a need for more than one reel of magnetic tape to store the file of inflected forms created for the group later in the process, and to present clerical assistants with a series of finite jobs instead of one interminable-looking task.

Transcribing is done with the Cyrillic Unityper described in Sec. 2.8. A group number serving as the name of the group is included in the identification block for each group. The structure of the file item recorded on magnetic tape for each word is described in Sec. 7.2. One page of hard copy produced by the Unityper for proofreading purposes (List 0, Fig. 45) is shown in Fig. 46.

HARVARD COMPUTATION LABORATORY
UNITYPER HARD COPY

Problem _____ Tape No. 019 Page Block Count _____ Total Block Count _____ Page 3 Of ___ Pages

Word	Code	Word	Code
МОЛОТОЧЕК	N1.3	МОЛОТОЧКОВЫЙ	A3
МОМЕНТНЫЙ	A2	МОМЕНТОМЕТР	N1
МОНАУРАЛЬНЫЙ	A2	МОНЕЛЬ	N3
МОНЕТНЫЙ	A2	МОНИТОР	N1
МОНОАТОМНЫЙ	A2	МОНОЛИТНЫЙ	A2
МОНОСКОП	N1	МОНОТРОН	N1
МОНОХРОМАТИЧЕСКИЙ	A6	МОНОЖНИК	N1К1
МОНТАЖНЫЙ	A2	МОНТЕР	N1
МОНТЕРСКИЙ	A6	МОРОЗОСТОЙКИЙ	A7
МОРОХОСТОЙКОСТЬ	N6	МОРСКОЙ	A3
МОСКОВСКИЙ	A6	МОСТ	N1
МОСТИК	N1.1	МОСТОВОЙ	A3
МОТАЛКА	N4.32	МОТОК	N1.3
МОТОР	N	МОТОРГЕНЕРАТОР	N1
МОТОРНЫЙ	A2	МОТОРЧИК	N1.1
МОЩНОСТЬ	N6	МОЩНОСТНЫЙ	A2
МОЩНЫЙ	A2	МУВИТОН	N1
МУЗА	N4	МУЗЫКАЛЬНЫЙ	A2
МУЛЬТИВИБРА...	A2	МУЛЬТИПЛИКАЦИЯ	N7
МУСКОВИТ	N1	МУТАТОР	N1
МУФТА	N4	МУФТОВЫЙ	A3
МО	I1	МОЛЬТИПЛЬНЫЙ	A2
МЯГКИЙ	A6	НАБЕГАНИЕ	N1θ
НАБИВКА	N4.31	НАБОРI	N1
НАБОРНЫЙ	A2	НАБРЫВ	N1.1
НАВАГЛАЙД	N1	НАВАГЛЯБ	N1
НАВАМАНДЕР	N1	НАВАР	N1
НАВАСКАП	N1	НАВАСКРИН	N1
НАВАСПЕКТОР	N1	НАВЕДЕНИЕ	N1θ
НАВИВКА	N4.31	НАВИГАЦИОННЫЙ	A1
НАВИГАЦИЯ	N7	НАВОДИТЬ	V4
НАВОДКА	N4031	НАГЛУКО	I1
НАГНЕТАТЕЛЬ	N3	НАГРЕВ	N1
НАГРЕВАНИЕ	N1θ	НАБЕГАТЬ	V1
МНОГООБЕЩАТЬ	V1	НАБИРАТЬ	V1
НАБРАСЫВАТЬ	V1	МИТИНГОВАТЬ	V3
МИКРОФОНИРОВАТЬ	V3	МОДУЛИРОВАТЬ	V3
МОНТИРОВАТЬ	V3	МИРИТЬ	V4
МИРИТЬСЯ	V4	МИКРОФОНИТЬ	V4
МОСТИТЬ	V4.1	МОЛЧАТЬ	V6
	V4.1		
НАБВОСТИ	V8.2		

Fig. 46. Cyrillic Unityper proof copy (List 0, Fig. 45).

A character is not irretrievably committed to magnetic tape until a whole line (one blockette) has been typed and a special *trip key* has been struck. Striking the trip key also prints a dash in a special column, as shown in Fig. 46, to indicate to the typist that the line has been completed. Until she strikes the trip key the typist is free to backspace or even to erase a whole line in case of mistake. Several places where the typist caught her own mistakes are marked by overprints in Fig. 46. Only the last struck of the overprinted characters is recorded on tape but the overprinting later calls the attention of the proofreader to spots requiring careful checking.

The tape produced by the transcription operation is said to be of *Type 1*. A print of a segment of one such tape is shown in Fig. 47. Note that the ranked computer representation (Table 10) is used for Russian, while the class marks are represented in the natural alphabet. If different portions of a single group have been recorded on separate reels of tape the contents of these reels are automatically consolidated onto a single reel prior to further processing.

7.6. Detecting and Correcting Mistakes in Classification and Transcription

Mistakes are inevitable in any manual operation. Unlike mistakes made by machines, which tend to be gross and systematic and therefore easily detected and corrected, human mistakes are often subtle and unpredictable. Detecting mistakes is tedious and difficult work, itself subject to mistakes. Correcting mistakes is a manual process, capable of doubling the original number of mistakes, hence possibly worse than leaving well enough alone. Once the nobler tasks have been accomplished, once the big picture has been painted with the broad brush, automatic information processing often becomes a wearing manual tug of war between man and his own mistakes, with the machine standing by as a very expensive automatic referee. Because mistakes are often regarded, for obvious reasons, as a subject not fit for public discussion, it is worth emphasizing again that they create one of the most harassing, recurrent, and costly problems of "automatic" information processing.

Detection of mistakes is generally more easily and effectively accomplished by a reviewer than by the one who made them. The typist who has finished transcribing a group is glad to be done with it, and usually unenthusiastic at the prospect of proofreading. Moreover, a mistake once made often looks so natural to the person who made it that it is quite likely to be overlooked on a simple visual check. To improve on visual mistake detection, a frequent practice, especially in punched-card installations, is to do transcription twice, the second time on a *verifier* which compares holes already punched on one card with those being punched on a second one by a different operator. Discrepancies are automatically signaled, and a verifier can be made to lock, preventing further operation until something is done about the mistake.

An alternative procedure is to use an automatic machine to rearrange transcribed information into an array where mistakes stand out; this procedure is well suited to the present work. For this reason, proofreading for classification and transcription mistakes is preceded by the *word-edit* operation shown in Fig. 45.

The word-edit operation rearranges the information in a Type 1 file into a form more suitable for subsequent automatic processing. Class marks are written originally in the format shown on Figs. 46 and 47 to minimize key strokes and mistakes in transcription. They are standardized by machine into the format shown on Fig. 48. The machine also assigns a serial number to each canonical form. This number is compounded of the group identification given on the Type 1 tape and a rank within the group calculated by the machine. Figure 48 shows words 173 through 237 of Group 19. The same serial number is later given to all forms derived from a given canonical form and provides a valuable means of controlling machine operations, especially when giving instructions to machine operators who know no Russian or as a more convenient alternative to deciphering the ranked computer representation. The serial number occupies the first eight character positions of machine word 4 (Fig. 39) in each item.

The list of Fig. 48 was obtained by transliterating the file of *Type 2* produced by the word-edit operation. The ranked computer representation is used in the Type 2 file as it is in all files intended for further automatic processing. Files that are to be printed are transliterated so that people may easily read the prints. As indicated in Sec. 2.8, transliteration could be avoided simply by replacing the natural type slugs on the printer by Cyrillic ones, but this alternative was too costly for experimental purposes.

List 1 (Fig. 48) is used in conjunction with List 0 (Fig. 46) to check for transcription mistakes. Four misspelled words (Ms) detected in Group 19 by scanning these lists are shown in Fig. 48. Note that only one Russian word appears on each line of Fig. 48, as contrasted with two on each line of Figs. 46 and 47, making List 1 easier to scan than the others. This is achieved by embedding each five-machine-word item in the Type 2 file into a ten-word item, as part of the transliteration process. The last five words of each new item always contain spaces. This detail is not profound, but it does illustrate once again how the structure of machine tokens must be taken into account in planning automatic processes. Careful attention to modest details of this kind may not directly or dramatically contribute to the advancement of knowledge but it often makes the difference between producing useful material and creating an ocean of waste paper.

The word-edit operation also determines the *coordinates* of the end of each Russian string, a pair of numbers indicating in which of the three machine words allocated to a Russian string (Fig. 39) and in which character position of that machine word the string ends. The last three positions of machine word 4 in each item are allocated to the coordinates. As an example, "микрофонировать" (019–0229, Fig. 48) is 15 characters long, ends in

```
DEG4GP6HET;              A2        DEG4GP9C9E5I G3T;        A3
DEG4GQ1)KGKE T;          A2        DEG4GQ9)C6EE G)KU        N6
DEG4GQ9)C6EE T;          A1        DEG4GQC6E                N1
DEG4GQC6EET;             A1        DEG4GRB)CUET :           A2
DEG4GRKTI63G ;           A3        DEG4GVC6BKIG 5ET;        A2
DEG4GVC6D6EK ET;         A2        DEG4GVK)7ET;             A2
DEG4GYQ6;BG3 T;          A3        DE676)K36EEG )KU         A6
DEG76)K36EET ;           A1        DEG76)K36EET ;           A1
DEG79DT;                 A3        DEG79K6CU                N3
DEG79KU                  V4.01     DEG79KU)Y                V4.01
DG56CU                   N6        DG56D                    N1
DG5LC9IL6DG) KU          N6        DG5LCGD6KI               N1
DG5LCU                   N3        DG5LCYKGI                N1
DG5LCYKGIET;             A2        DG5LCYP9GEET ;           A1
DG5LCYP9Y                N7        DG819B1                  N4.1
DG8)9QET;                A2        DGBIT;                   A3
DGC                      N1        DGC6BLCYIET;             A2
DGC9256EG3T;             A3        DGC9EY                   N7
DGCE6GK3G5               N1        DGCE96GK3G5              N1
DGCGKGQ6B                N1.3      DGCGKGQBG3T;             A3
DGD6EKET;                A2        DGD6EKGD6KI              N1
DGE1LI1CUET;             A2        DGE6CU                   N3
DGE6KET;                 A2        DGE9KGI                  N1
DGEG)KGDET;              A2        DGEGC9KET;               A2
DGEG)BGH                 N1        DGEGKIGE                 N1
DGEGNIGD1K9Q 6)B9;       A6        DGEK17E9B                N1.1
DGEK)7ET;                A2        DGEK6I                   N1
DGEK6I)B9;               A6        DGIG8G)KG;B9 ;           A7
DGIGNG)KG;BG )KU         N6        DGI)BG;                  A3
DG)BG3)B9;               A6        DG)K                     N1
DG)K9B                   N1.1      DG)KG3GI                 A3
DGK)CB1                  N4.31     DGKGB                    N1.3
DGKGI                    N1        DGKGI46E6)1K GI          N1
DGKGIET;                 A2        DGKGIQ9B                 N1.1
DG+EG)KU                 N6        DG+EG)KET;               A2
DG+ET;                   A2        DL39KGE                  N1
DL81                     N4        DL89B1CUET;              A2
DLCUK9HC6B)E T;          A2        DLCUK9HC9BGP 9Y          N7
DL)BG39K                 N1        DLK1KGI                  N1
DLMK)                    N4        DLMKG3TI                 A3
DX                       I1        DXCUK9HCUET;             A2
DY4B9:                   A6        E12641E96                N10
E1293B1                  N4.31     E12GI                    N1
E12GIET;                 A2        E12IT84                  N1.1
E1314C1;5                N1        E1314CG2                 N1
E131D1E56I               N1        E131I                    N1
E131)B1H                 N1        E131)BI9E                N1
E131)H6BKGI              N1        E13656E96                N10
E1393B1                  N4.31     E13941P9GEET ;           A1
E13941P9Y                N7        E13G59KU                 V4
E1365B1                  N4.31     E14CLNG                  I1
E14E6K1K6CU              N3        E14163                   N1
E14I631E96               N10       E12641KU                 V1
DEG4GG26+1KU             V1        E129I1KU                 V1
E12I))T31KU              V1        D9K9E4G31KU              V3
D9BIGMGE9IG3 )KU         V3        DG5LC9IG31KU             V3
DGEK9IG31KU              V3        D9I9KU                   V4
D9I9KU)Y                 V4        D9BIGMGE9KU              V4
```

Fig. 47. Print of a tape containing transcribed Russian classical canonical forms (Type 1, Fig. 45).

```
MONTERSKIJ            A06.00   019-0173 010
MOROZOSTOJKI J        A07.00   019-0174 101
MOROKOSTOJKO ST,      N06.00   019-0175 103   o Ms
MORSKOJ               A03.00   019-0176 007
MOSKOVSKIJ            A06.00   019-0177 010
MOST                  N01.00   019-0178 004
MOSTIK                N01.10   019-0179 006
MOSTOVOJ              A03.00   019-0180 008
MOTALKA               N04.31   019-0181 007
MOTOK                 N01.30   019-0182 005
MOTOR                 N01.00   019-0183 005
MOTORGENERAT OR       N01.00   019-0184 102
MOTORNYJ              A02.00   019-0185 008
MOTORCHIK             N01.10   019-0186 008
MOSHCHNOST,           N06.00   019-0187 008
MOSHCHNOSTNY J        A02.00   019-0188 010
MOSHCHNYJ             A02.00   019-0189 006
MUVITON               N01.00   019-0190 007
MUZA                  N04.00   019-0191 004
MUZIKAL,NYJ           A02.00   019-0192 011   o Ms
MUL,TIPLEKSN YJ       A02.00   019-0193 102
MUL,TIPLIKOT SIJA     N07.00   019-0194 102   o Ms
MUSKOVIT              N01.00   019-0195 008
MUTATOR               N04.00   019-0196 007
MUFTA                 N04.00   019-0197 005
MUFTOVYJ              A03.00   019-0198 008
MJU                   I01.00   019-0199 002
MJUL,TIPL,NY J        A02.00   019-0200 012
MJAGKIJ               A06.00   019-0201 006
NABEGANIE             N10.00   019-0202 009
NABIVKA               N04.31   019-0203 007
NABOR                 N01.00   019-0204 005
NABORNYJ              A02.00   019-0205 008
NABRYZG               N01.10   019-0206 007
NAVAGLAJD             N01.00   019-0207 009
NAVAGLOB              N01.00   019-0208 008
NAVAHANDER            N01.00   019-0209 010
NAVAR                 N01.00   019-0210 005
NAVASKAP              N01.00   019-0211 008   o Ms
NAVASKRIN             N01.00   019-0212 009
NAVASPEKTOR           N01.00   019-0213 011
NAVEDENIE             N10.00   019-0214 009
NAVIVKA               N04.31   019-0215 007
NAVIGATSIONN YJ       A01.00   019-0216 101
NAVIGATSIJA           N07.00   019-0217 009
NAVODIT,              N01.00   019-0218 006
NAVODKA               N04.31   019-0219 007
NAGLUXO               N03.00   019-0220 007
NAGNETATEL,           N01.00   019-0221 011
NAGREV                N10.00   019-0222 006
NAGREVANIE            V01.00   019-0223 010
NABEGAT,              V01.00   019-0224 008
MNOGOOBESHCH AT,      V01.00   019-0225 012
MABIRAT,              V01.00   019-0226 008
MABRASYVAT,           V03.00   019-0227 011
MITINGOVAT,           V03.00   019-0228 103
MIKROFONIROV AT,      V03.00   019-0230 012
MODULIROVAT,          V03.00   019-0231 011
MONTIROVAT,           V04.00   019-0232 006
MIRIT,                V04.00   019-0233 008
MIRIT,SJA             V04.00   019-0234 011
MIKROFONIT,           V04.10   019-0235 007
MOSTIT,               V06.00   019-0236 007
MOLCHAT,              V08.20   019-0237 008
NABLJUSTI
```

Fig. 48. Edited items: transliterated classical canonical forms (List 1, Fig. 45).

the third character position of machine word 1, and has coordinates "103"; the first digit in "103" denotes the machine word and the last two denote position within the word. Since these coordinates refer to the ranked computer representation, and since a single character in that representation may be replaced by several characters on transliteration, the coordinates on Fig. 48 may seem to be wrong; for example, "moshchnost'" (019–0187) has 11 characters, but "мощность" only 8.

An algorithm for determining the location of the end of a Russian word, and hence for computing its coordinates, could be used each time this information is required by the machine. By computing the coordinates once and incorporating them in the item for each word the necessity for repeated execution of the algorithm is eliminated, but obviously at the price of storage space and of the execution of a table look-up algorithm whenever a process depends on knowledge of where a word ends.

It is a frequently used technique, of great value in appropriate circumstances, to trade the repetitive use of one algorithm for storage space and the repetitive use of another presumably simpler algorithm. For example, once a table of logarithms has been calculated, it is simpler for a person to keep this table on his bookshelf and refer to it whenever he needs a logarithm than it is to compute individual logarithms. In automatic calculations it often proves simpler and more economical to use an algorithm for calculating the logarithm of an arbitrary number than to store and consult an extensive table.

Exercise 7-3. Write a program for computing the coordinates of a Russian word. Hint: the 36 characters in the first three machine words of each item factor into the representation of the Russian word and a string of spaces.

The word-edit operation also produces a file (*Type 3*) of items made up as shown in Fig. 49. These are identical to items in the Type 2 file, except that each string representing a Russian word is in the reverse of the normal sequence, in other words, the last character of the string is in the first character position of word 0 of the item, and so on. This odd procedure is necessary to produce List 2, of which a section is shown in Fig. 50. The key role of this list in detecting mistakes in classification will be described after it has been shown how it is produced.

The reader who compares Fig. 39 with Fig. 49 will note that △-fill is specified in one, and 0-fill in the other. Because shift operations introduce "0"'s (Example 1-2) and shift instructions are useful in producing the items in a Type 3 file, *some* of the original "△"'s are replaced by "0"'s. It is too much trouble to replace the "0"'s by "△"'s in the way illustrated in Example 1-2. A more satisfactory expedient is to treat "△" and "0" as *equivalent* in the ranked computer representation (Table 10). This is done by regarding all characters ≤ "0" (Table 9) as equivalent to "△", which works because "i" and "-" are not among the characters used in the ranked computer representation. (What are the consequences of a mistake whereby an "i" or a "-" *is* introduced into the representation of some word? Why do shift instructions introduce "0" 's?)

The strings in Fig. 50 are in end-alphabetic order, that is, they are ordered by terminal characters, within terminal characters by penultimate characters, and so on, as in a rhyming dictionary. The ordinary alphabetic order is by first characters, within first characters by second characters, and so on. The T and Q instructions (Secs. 1.4 and 2.7), which play a central role in any

NNN: WORD GROUP IDENTIFICATION.

SSSS: WORD RANK WITHIN GROUP.

C_1:   MAJOR WORD COORDINATE (LOCATION OF FINAL CHARACTER, WORD 0, 1, OR 2)

C_2:   MINOR WORD COORDINATE (LOCATION OF FINAL CHARACTER POSITION, 1 THROUGH 12)

'RUSSIAN IN RANKED COMPUTER REPRESENTATION

Fig. 49. Item containing an inverted Russian word.

program for ranking items, naturally enough were designed to weight the character in the first position more heavily than that in the second, and so on, in keeping with characteristics of the conventional representations of words and integers (Sec. 2.6). One can imagine a machine with "turned around" T and Q instructions, in which end-alphabetization could be effected after merely lining up the ends of strings (right-justification) rather than their beginnings (left-justification). A little reflection will show that the normal T and Q instructions will induce the desired end-ordering on words only if they are represented in the manner illustrated in Fig. 49.

A sorting program (Fig. 45) rearranges the items in the Type 3 file into

01 REGULIROVAT,	VO3-00	23-0171	102
MODULIROVAT,	VO3-00	19-0230	012
NE MODULIROVAT,	VO3-00	20-0167	102
NEDO MODULIROVAT,	VO3-00	20-0172	104
PA NORAMIROVAT,	VO3-00	24-0213	102
NEE HKRANIROVAT,	VO3-00	21-0166	102
PA RAFINIROVAT,	VO3-00	24-0214	102
NE REZONIROVAT,	VO3-00	21-0175	102
NE RAZONIROVAT,	VO3-00	21-0177	102
NE OBRONIROVAT,	VO3-00	21-0180	102
MIK ROFONIROVAT,	VO3-00	19-0229	103
NE EVIBRIROVAT,	VO3-00	20-0177	101
NE FOKUSIROVAT,	VO3-00	21-0168	102
NEKO RREKTIROVAT,	VO3-00	20-0168	104
O RIENTIROVAT,	VO3-00	23-0165	:01
MONTIROVAT,	VO3-00	19-0231	011
OSTIROVAT,	VO3-00	23-0173	010
NA PUNSHIROVAT,	VO3-00	20-0174	101
NE SMONTIROVAT,	VO3-00	21-0172	102
N ESOGLASOVAT,	VO1-00	21-0147	101 WCm
OPRESSOVAT,	VO3-00	23-0167	011
PER EPOLJUSOVAT,	VO3-00	24-0218	102
OSVINTSOVAT,	VO3-00	22-0204	008
OBORVAT,	VO5-20	22-0171	012
OBRAZOVYVAT,	VO1-00	23-0158	010
OSNOVYVAT,	VO1-00	23-0138	103
OTF IL1TROVYVAT,	VO1-00	24-0196	012
PEREKIDYVAT,	VO1-00	23-0162	012
OPROKIDYVAT,	VO1-00	22-0172	010
OBREZYVAT,	VO1-00	24-0201	012
PEREDELYVAT,	VO1-00	23-0154	010
OTDELYVAT,	VCI-00	21-0141	010
OBLAMYVAT,	VO1-00	24-0194	011
PEREKRYVAT,	VO1-00	19-0227	011
NABRASYVAT,	VO1-00	23-0143	010
OTSASYVAT,	VO1-00	22-0176	009
OPISYVAT,	VO1-00	24-0184	101
P EREXVATYVAT,	VO1-00	24-0211	010
OXVATYVAT,	VO1-00	20-0143	010
NAMATYVAT,	VO1-00	21-0177	011
OPECHATYVAT,	VO1-00	21-0146	011
NEUCHITYVAT,	VO1-00	19-0224	008
NABEGAT,	VO1-00	21-0143	008
OBZHIGAT,	VO1-00	24-0199	010
PEREZHIGAT,	VO1-00	21-0142	008
OBLADAT,	VO1-00	24-0208	006
FADAT,	VO1-00	24-0107	008
OTPADAT,	VO1-00	21-0149	011
NERAZREDAT,	VO1-00	24-0193	010
OXLAZHDAT,	VO1-00	24-0210	009
N EPOVREZHDAT,	VO1-00	21-0153	012
OSVOBOZHDAT,	VO1-00	23-0161	011
NEOTRAZHAT,	VO1-00	21-0157	010
NADLEZHAT,	VO6-00	20-0011	009
PEREMEZHAT,	VO1-00	24-0193	010
OPEREZHAT,	VO1-00	22-0181	009
N EDREBEZZHAT,	VO4-20	20-0198	012 OWCm
PEREEZZHAT,	VO1-00	24-0200	010
PEREDERZHAT,	VO1-00	24-0231	011
NAGRUZHAT,	VO1-00	20-0155	009
PEREGRUZHAT,	VO6-00	24-0202	011
OKRUZHAT,	VO1-00	22-0182	008
PE REZARJAZHAT,	VO1-00	24-0198	012
OIKAZAT,	VO5-00	23-0192	008
NEZAMERZAT,	VO1-00	23-0150	011
NEDOSVJAZAT,	VO4-00	20-0180	011 OWCm

Fig. 50. End-alphabetized classical canonical forms (List 2, Fig. 45).

end-alphabetic order. The sorted Type 3 file is then subjected to the inverse of the transformation that reversed the sequence of characters; it is worth noting that this particular transformation is its own inverse because it follows that only one program need be written to do both jobs. The characters, now in normal sequence, are right-justified and transliterated to produce a file of *Type 5* which, when printed, produces List 2.

> *Exercise 7-4.* (*a*) Can you prove that any transformation which inverts the order of a string, for instance, which maps "123" into "321", is its own inverse?
> (*b*) Write a program for producing an item in the Type 3 file from one in the Type 2 file. Hint: the coordinates of each word are useful for this purpose.

The value of List 2 (Fig. 50) in catching mistakes arises from the fact, evident upon examining the appendix of Chapter 6, that the morphological classification of words is determined to a high degree by the last few characters in their representations. That the classification cannot be completely determined in this way is the reason why classifying could not be done completely automatically. It was therefore done wholly manually, although in the future it might well be worth doing semiautomatically. In any case, the degree of determination is high enough so that when words are end-alphabetized long runs of members of the same class occur and mavericks are easily spotted and checked. The mistakes in Fig. 50 stand out among the deviations from the pattern when the class-mark column is scanned. One need only compare Fig. 50 with Fig. 48 to see how much easier it is to catch classification mistakes on the former than on the latter. Using an automatic machine to rearrange data into some order where a pattern is likely to emerge is a powerful technique not only for catching mistakes, but also in the study of morphology and syntax, as shown in Secs. 5.9, 8.5, and 10.4.

When using well-designed punched-card files, it is relatively easy to correct a mistake as soon as it is caught. With caution and good luck, it is necessary only to take from the file the card on which the mistake occurs and to replace this card with a correct one. This simple technique can obviously not be used with tape files. The analogous procedure would be to transcribe data from the tape with the mistake onto another until the faulty item is reached. A replacement for this item is then recorded onto the new tape, followed by the remaining items on the original tape. Under certain circumstances, as when the ratio of items with mistakes to total items is high, this procedure is efficient. At the other extreme, if there is only one mistake in a whole reel of tape, the procedure can become quite unpleasant. When mistakes are few, it can be more efficient and less conducive to chaos to let bad items ride along with the good until they can be eliminated as a natural by-product of some necessary operation. For such reasons, once mistakes have been marked on Lists 1 and 2, no further action is taken on them until the stage described in Sec. 7.8 is reached.

7.7. Direct and Inverse Inflection; Condensation

The reasons for producing a complete generated paradigm (Sec. 5.2) for each Russian word to be put in the dictionary file have been presented in Sec. 5.6. The Type 2 file produced by the word-edit operation (Figs. 45 and 48) serves as initial input to a series of programs for direct inflection (Fig. 51) called *inflected-form generators* (Foust, 1958, 1957).

An inflected-form generator executes one or more of the algorithms described under the heading "Generation Rules" in the appendix to Chapter 6. The new file produced by an inflected-form generator contains many more items than the original file, namely one for each member of the generated paradigm of the word. The format of each item is substantially that of items in the Type 2 file. In fact, the item containing the classical canonical form is unchanged except for the insertion of a mark indicating that it has been processed by an inflected-form generator. This mark also appears in character position 9 of machine word 4 of every newly generated item, as may be seen in Fig. 52, where "H" is used as the process mark. The process mark is significant only as a means for keeping track of the operations that have been performed on the items in the file.

For all items other than those containing the original classical canonical forms an appropriate new set of coordinates (Sec. 7.6) is calculated by each inflected-form generator and placed in the standard position. Items containing members of the generated paradigms obtained from those adjectival canonical forms which are themselves generated from verbal forms as described in Sec. 6.1 are subjected to special modifications. The inflected-form generators for verbs produce adjectival items in which the verbal class mark is replaced by an adjectival class mark as directed by the generation rules for verbal paradigms. To distinguish the adjectival classical canonical forms produced in this way from those introduced as described in Secs. 7.4 and 7.5, the serial number of the parent verb is given to these forms after a slight modification. For example, "набегающий", "набегавший", "набегаемый", and "набеганный", all generated from "набегать" (019–0224, Figs. 46 and 52), have serial numbers "019A0224", "019B0224", "019C0224", and "019D0224" respectively (Fig. 52). With this exception all members of a generated paradigm have identical serial numbers.

Exercise 7-5. Write programs for obtaining the complete generated paradigms of words in classes *N1*, *N4.31*, and *V1*.

Several inflected-form generators were written in turn as successive portions of the tables of the appendix to Chapter 6 became available. The reducibility properties of machine tokens for inflected forms do not exactly match those required by the algorithms that generate these forms, as the reader who has done Exercise 7-5 will have noticed. Consequently, some complexity is unavoidable in the inflected-form generators. Because of the experimental nature of the work, effort was spent mainly on getting programs to operate satisfactorily, not on making them as elegant, compact, and fast-operating as possible. Inflected-form generators for all classes of words could not therefore be simultaneously loaded into the Univac's thousand-word memory.

For all these reasons, the generation of inflected forms had to proceed in several *passes* through each group. The file of Type 2 serves as input to the first pass and the output is a file of type 12ISM in which words of only certain classes have been inflected and given a process mark identifying the pass. This intermediate file then serves as input to another pass where another set of classes are inflected. This process is iterated until all classes have been inflected, as indicated by the closed loop through the direct-inflection box of Fig. 51. The order of application of the inflected-form generators was arranged so that the classes with fewest members (Table 30) would be processed first; the inflection of adjectives naturally had to follow that of verbs. With the greater care in programming that is made possible by hind-sight, and on a machine with a larger internal memory than the Univac, the whole process of direct inflection could be done easily and rapidly in a single pass.

Not all words are inflected automatically. Those belonging to classes *N99.99*, *A99.99*, or *V99.99* (Sec. 6.1) obviously cannot be. In several instances of sparsely populated classes with complex generating rules it seemed expedient to postpone writing inflected-form generators and to do manual inflection for the time being. A simple *extract* program was written to test all items in the Type 12ISM file for the presence of a process mark. Items without a process mark, namely those that have not been inflected, are segregated into a new file of Type UF (Fig. 51). (Invariable words, with class mark *I*, are given a process mark at an appropriate stage.) Periodically, tapes containing these files are merged, transliterated, and printed to obtain a list (List 3) of uninflected words. The generated paradigms for these words are written out by hand and the resulting material, treated as a new group, once more enters the compilation process (Fig. 45). Groups of this kind are treated like others, with a few exceptions of which the most significant is the omission of direct-inflection passes.

Once all inflected-form generators have been applied to a group, the strings

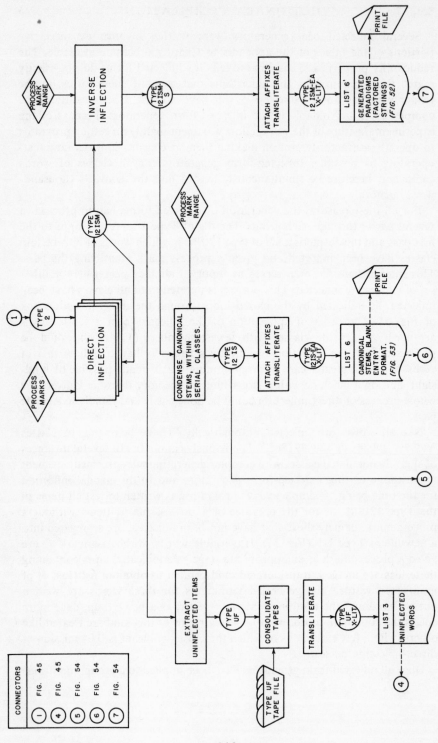

Fig. 51. The compilation process, Part II.

NAGREV-E	N01.00	600	019-0222H007
NAGREV-Y	N01.00	T00	019-0222H007
NAGREV-OV	N01.00	G30	019-0222H008
NAGREV-AM	N01.00	1D0	019-0222H008
NAGREV-AMI	N01.00	1D9	019-0222H009
NAGREV-AX	N01.00	1N0	019-0222H008
NAGREVANI-E	N10.00	600	019-0223H010
NAGREVANI-JA	N10.00	Y00	019-0223H010
NAGREVANI-JU	N10.00	X00	019-0223H010
NAGREVANI-EM	N10.00	6D0	019-0223H011
NAGREVANI-JA X	N10.00	YN0	019-0223H011
NAGREVANI-J	N10.00	;00	019-0223H010
NAGREVANI-JA M	N10.00	YD0	019-0223H011
NAGREVANI-JA MI	N10.00	YD9	019-0223H012
NAGREVANI-I	N10.00	900	019-0223H010
NABEGA-T'	V01.00	KU0	019-0224H008
NABEGA-JU	V01.00	X00	019-0224H007
NABEGA-ESH'	V01.00	6RU	019-0224H009
NABEGA-ET	V01.00	6K0	019-0224H008
NABEGA-EM	V01.00	6D0	019-0224H008
NABEGA-ETE	V01.00	6K6	019-0224H009
NABEGA-JUT	V01.00	XK0	019-0224H008
NABEGAL-	V01.00		019-0224H007
NABEGAL-A	V01.00	100	019-0224H008
NABEGAL-O	V01.00	G00	019-0224H008
NABEGAL-I	V01.00	900	019-0224H008
NABEGA-J	V01.00	;00	019-0224H007
NABEGAJT-E	V01.00	600	019-0224H009
NABEG-AJA	V01.00	1Y0	019-0224H007
NABEGA-V	V01.00	300	019-0224H007
NABEGA-VSHI	V01.00	3R9	019-0224H009
NABEGAJUSHCH -IJ	A04.00	9;0	019A0224H010
NABEGAJUSHCH -EGO	A04.00	64G	019A0224H011
NABEGAJUSHCH -EMU	A04.00	6DL	019A0224H011
NABEGAJUSHCH -IM	A04.00	9D0	019A0224H010
NABEGAJUSHCH -EM	A04.00	6D0	019A0224H010
NABEGAJUSHCH -AJA	A04.00	1Y0	019A0224H010
NABEGAJUSHCH -EJ	A04.00	6;0	019A0224H010
NABEGAJUSHCH -UJU	A04.00	LX0	019A0224H010
NABEGAJUSHCH -EE	A04.00	660	019A0224H010
NABEGAJUSHCH -IE	A04.00	960	019A0224H010
NABEGAJUSHCH -IX	A04.00	9N0	019A0224H010
NABEGAJUSHCH -IMI	A04.00	9D9	019A0224H011
NABEGAJUSHCH -	A04.00		019A0224H008
NABEGAJUSHCH -A	A04.00	100	019A0224H009
NABEGAJUSHCH -O	A04.00	G00	019A0224H009
NABEGAJUSHCH -E	A04.00	600	019A0224H009
NABEGAJUSHCH -I	A04.00	900	019A0224H009
NABEGAVSH-IJ	A04.00	9;0	019B0224H010
NABEGAVSH-EG O	A04.00	64G	019B0224H011
NABEGAVSH-EM U	A04.00	6DL	019B0224H011
NABEGAVSH-IM	A04.00	9D0	019B0224H010
NABEGAVSH-EM	A04.00	6D0	019B0224H010
NABEGAVSH-AJ A	A04.00	1Y0	019B0224H010
NABEGAVSH-EJ	A04.00	6;0	019B0224H010
NABEGAVSH-UJ U	A04.00	LX0	019B0224H010
NABEGAVSH-EE	A04.00	660	019B0224H010
NABEGAVSH-IE	A04.00	960	019B0224H010
NABEGAVSH-IX	A04.00	9N0	019B0224H010
NABEGAVSH-IM I	A04.00	9D9	019B0224H011
NABEGAVSH-	A04.00		019B0224H008
NABEGAVSH-A	A04.00	100	019B0224H009
NABEGAVSH-O	A04.00	G00	019B0224H009
NABEGAVSH-E	A04.00	600	019B0224H009
NABEGA-VSHI	A04.00	3R9	019B0224H009
NABEGAEM-YJ	A03.00	T;0	019C0224H010
NABEGAEM-OGO	A03.00	G4G	019C0224H011
NABEGAEM-OMU	A03.00	GDL	019C0224H011
NABEGAEM-YM	A03.00	TD0	019C0224H010
NABEGAEM-OM	A03.00	GD0	019C0224H010
NABEGAEM-AJA	A03.00	1Y0	019C0224H010
NABEGAEM-OJ	A03.00	G;0	019C0224H010
NABEGAEM-UJU	A03.00	LX0	019C0224H010
NABEGAEM-OE	A03.00	G60	019C0224H010
NABEGAEM-YE	A03.00	T60	019C0224H010
NABEGAEM-YX	A03.00	TN0	019C0224H010
NABEGAEM-YMI	A03.00	TD9	019C0224H011
NABEGAEM-A	A03.00	100	019C0224H009
NABEGAEM-O	A03.00	G00	019C0224H009
NABEGAEM-Y	A03.00	T00	019C0224H009
NABEGA-EM	A03.00	6D0	019C0224H008
NABEGANN-YJ	A01.00	T;0	019D0224H010
NABEGANN-OGO	A01.00	G4G	019D0224H011
NABEGANN-OMU	A01.00	GDL	019D0224H011
NABEGANN-YM	A01.00	TD0	019D0224H010
NABEGANN-OM	A01.00	GD0	019D0224H010
NABEGANN-AJA	A01.00	1Y0	019D0224H010
NABEGANN-OJ	A01.00	G;0	019D0224H010
NABEGANN-UJU	A01.00	LX0	019D0224H010
NABEGANN-OE	A01.00	G60	019D0224H010
NABEGANN-YE	A01.00	T60	019D0224H010
NABEGANN-YX	A01.00	TN0	019D0224H010
NABEGANN-YMI	A01.00	TD9	019D0224H011
NABEGAN-	A01.00		019D0224H007
NABEGANEN-	A01.00		019D0224H009
NABEGANN-A	A01.00	100	019D0224H009
NABEGANN-O	A01.00	G00	019D0224H009
NABEGANN-Y	A01.00	T00	019D0224H009
NABEGAN-A	A01.00	100	019D0224H008
NABEGAN-O	A01.00	G00	019D0224H008
NABEGAN-Y	A01.00	T00	019D0224H008
MNOGOOBESHCH A-T'	V01.00	KU0	019-0225H012
MNOGOOBESHCH A-JU	V01.00	X00	019-0225H011
MNOGOOBESHCH A-ESH'	V01.00	6RU	019-0225H101
MNOGOOBESHCH A-ET	V01.00	6K0	019-0225H012
MNOGOOBESHCH A-EM	V01.00	6D0	019-0225H012

Fig. 52. Generated paradigms (factored strings) (List 6′, Fig. 51).

in the resulting file of Type 12ISM are subjected to inverse inflection (Fig. 51). The basic algorithm for inverse inflection has been described in Chapter 5, particularly in Sec. 5.5. Items in the resulting file of Type 12ISM-S have the now familiar structure. Affixes of order zero, if any, are put in character positions 11 and 12 of machine word 3. Affixes of order one are put in character positions 8–10 of the same word. They may be seen in the ranked computer representation in Fig. 52. Only canonical stems remain in machine words 0–2.

The information in the 12ISM-S file is helpful in the process of writing English correspondents. To produce a list suitable for visual inspection, canonical stems and affixes are recombined into strings where they are separated only by a dash, then transliterated. The printed form of the resulting file is List 6', illustrated in Fig. 52.

The entry keys for the automatic dictionary (Sec. 5.3) are obtained by eliminating all duplicate stems (for certain minor exceptions, see Sec. 9.6) resulting from the inverse inflection of the members of a generated paradigm, an operation called "condensation" (Fig. 51). All items having identical serial numbers are examined during testing for duplicate stems. In condensation, the stem obtained by factoring the classical canonical form is always retained and distinguished from those stems, if any, obtained from other members of the reduced paradigm of the word by a character "F" in position 7 of machine word 3 (Fig. 53). The file of Type 12IS, which is the condensation product, when joined with English correspondents and grammatical marks, forms a piece of the dictionary file.

A medium suitable for the clerical operations necessary to assign English correspondents and grammatical marks is prepared from the 12IS file by procedures similar to those used to obtain List 6'. The resulting List 6 is shown in Fig. 53. In transliteration, each five-machine-word item is embedded in several blockettes of spaces, to produce a format allowing ample space for handwriting. For convenience in associating each stem on List 6 with the word to which it belongs, a typical affix with which it may combine appears next to it but separated by a dash as a reminder that one is now dealing with stems.

7.8. English Correspondents and Grammatical Codes

Once Lists 6 and 6' (Figs. 53 and 52) for a group of words have been made available by the automatic operations described in Sec. 7.7, the group is ready for a major manual process: the correction of mistakes caught earlier and the assignment of English correspondents and of grammatical codes (Fig. 54). The product of these operations is an annotated version of List 6 (Fig. 55).

NABEGA-T.	V01.00FKU0	019-0224H008
NABEGAL-	V01.00	019-0224H007
NABEGAJT-E	V01.00 600	019-0224H009
NABEG-AJA	V01.00 1Y0	019-0224H007
NABEGAJUSHCH -IJ	A04.00F9I0	019A0224H010
NABEGAVSH-IJ	A04.00F9I0	019B0224H010
NABEGA-VSHI	A04.00 3R9	019B0224H009
NABEGAEM-YJ	A03.00FTI0	019C0224H010
NABEGA-EM	A03.00 6D0	019D0224H008
NABEGANN-YJ	A01.00FTI0	019D0224H010
NABEGAN-	A01.00	019D0224H007

Fig. 53. Canonical stems; blank entry format (List 6, Fig. 51).

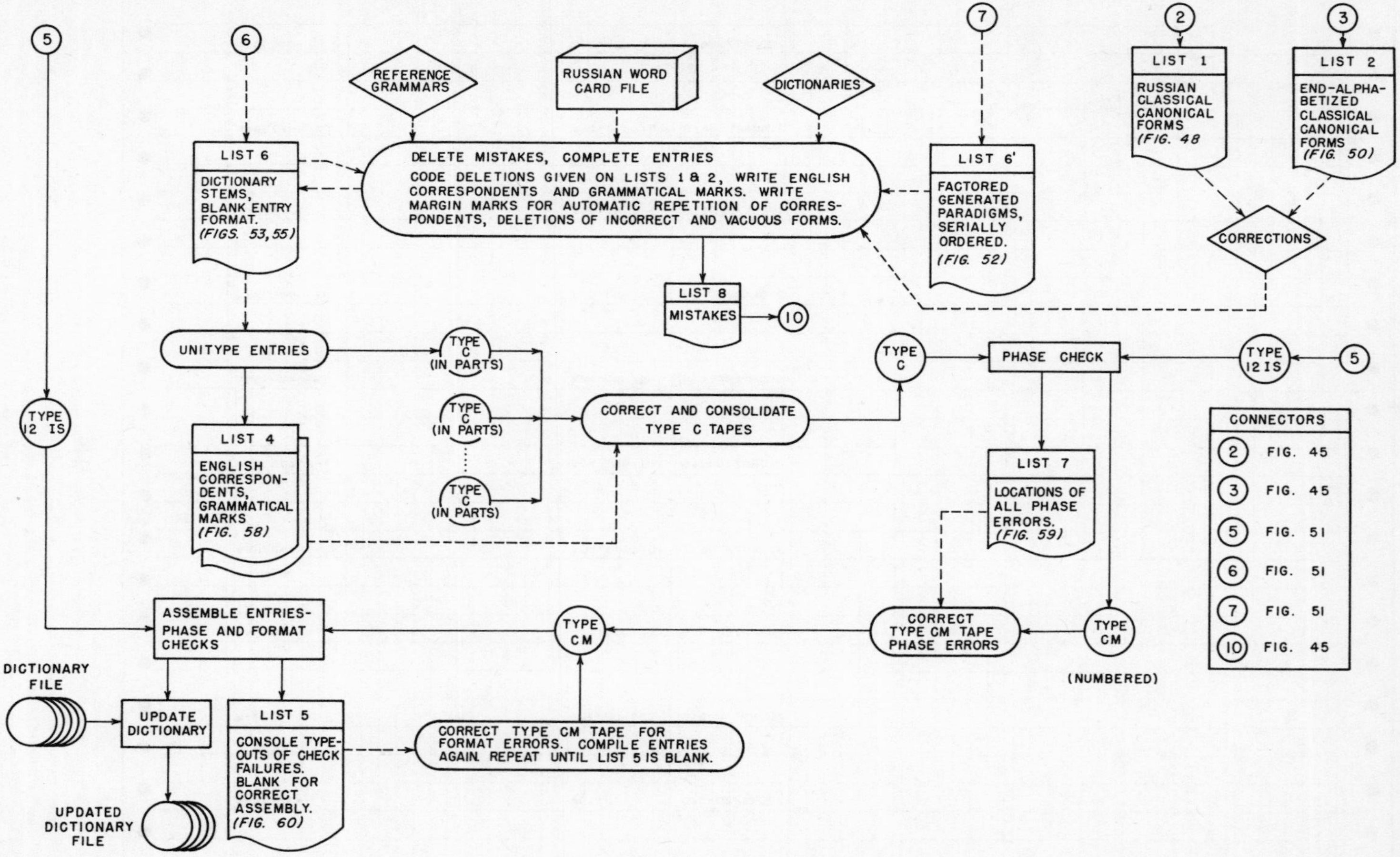

LIST 2 — END-ALPHA-BETIZED CLASSICAL CANONICAL FORMS *(FIG. 50)* ③

LIST 1 — RUSSIAN CLASSICAL CANONICAL FORMS *(FIG. 48)* ②

CORRECTIONS

⑤ TYPE I2 IS

CONNECTORS

②	FIG. 45
③	FIG. 45
⑤	FIG. 51
⑥	FIG. 51
⑦	FIG. 51
⑩	FIG. 45

PHASE CHECK

TYPE CM (NUMBERED)

LIST 7 — LOCATIONS OF ALL PHASE ERRORS. *(FIG. 59)*

CORRECT TYPE CM TAPE PHASE ERRORS

⑦ LIST 6' — FACTORED GENERATED PARADIGMS, SERIALLY ORDERED. *(FIG. 52)*

DICTIONARIES

TYPE C

DELETE MISTAKES, COMPILE ENTRIES CODE DELETIONS GIVEN ON LISTS 1 & 2, WRITE ENGLISH CORRESPONDENTS AND GRAMMATICAL MARKS. WRITE MARGIN MARKS FOR AUTOMATIC REPETITION OF CORRES-PONDENTS, DELETIONS OF INCORRECT AND VACUOUS FORMS.

RUSSIAN WORD CARD FILE

LIST 8 — MISTAKES ⑩

REFERENCE GRAMMARS

CORRECT AND CONSOLIDATE TYPE C TAPES

TYPE CM

CORRECT TYPE CM TAPE FOR FORMAT ERRORS. COMPILE E ENTRIES AGAIN REPEAT UNTIL LIST 5 IS BLANK.

⑥ LIST 6 — DICTIONARY STEMS, BLANK ENTRY FORMAT. *(FIGS. 53, 55)*

UNITYPE ENTRIES

TYPE C (IN PARTS)   TYPE C (IN PARTS) .... TYPE C (IN PARTS)

LIST 4 — ENGLISH CORRESPON-DENTS, GRAMMATICAL MARKS *(FIG. 58)*

ASSEMBLE ENTRIES-PHASE AND FORMAT CHECKS

LIST 5 — CONSOLE TYPE-OUTS OF CHECK FAILURES. BLANK FOR CORRECT ASSEMBLY. *(FIG. 60)*

UPDATE DICTIONARY

DICTIONARY FILE

UPDATED DICTIONARY FILE

⑤ TYPE I2 IS

Fig. 54. The compilation process. Part III.

250

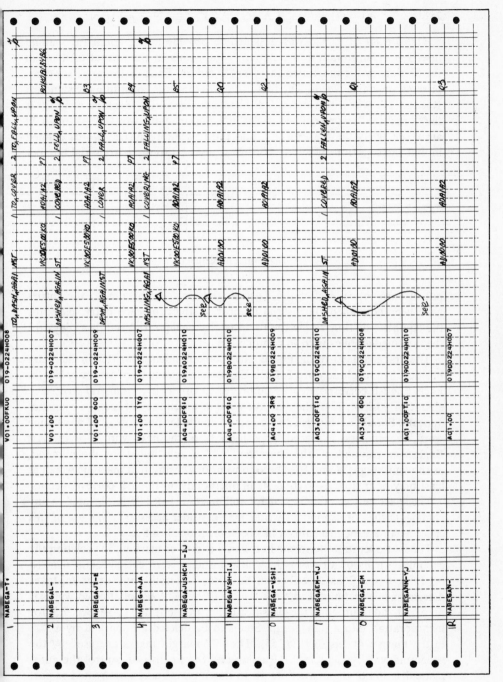

Fig. 55. Annotated version of List 6.

The format of List 6 was designed to provide ample room for manual insertion of information in a pattern conforming to the format of items in the dictionary file illustrated in Figs. 38 and 56. Each dictionary item comprises 30 machine words. The first five of these words constitute the standard Russian five-word item described throughout Sec. 7.7. Machine word 26 is the "organized" word defined in Sec. 6.5, and machine words 27 through 29 are respectively the first, second, and third "semiorganized" words defined in that same section. Machine word 25 is reserved for an entry serial number

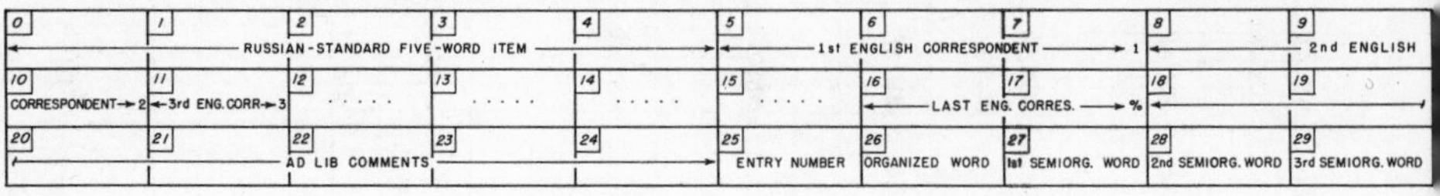

0	1	2	3	4	5	6	7	8	9
←———————— RUSSIAN-STANDARD FIVE-WORD ITEM ————————→					←————— 1st ENGLISH CORRESPONDENT ————→ 1			←———— 2nd ENGLISH	
10	11	12	13	14	15	16	17	18	19
CORRESPONDENT→2←3rd ENG. CORR→3			· · · · ·	· · · · ·	· · · · ·	←————LAST ENG. CORRES. ——→ % ←—			———→
20	21	22	23	24	25	26	27	28	29
←——————————— AD LIB COMMENTS ———————————→					ENTRY NUMBER	ORGANIZED WORD	1st SEMIORG. WORD	2nd SEMIORG. WORD	3rd SEMIORG. WORD

Fig. 56. The format of dictionary items.

put in automatically after the Russian and English portions of the entries have been combined. This serial number, distinct for each entry, plays for the complete dictionary file the role that the group serial number plays for the intermediate files used in compilation.

The first character of the first English correspondent is written in character position 1 of machine word 5. Each English string may be of arbitrary length. Each English correspondent is numbered in order, by a numeral written in character position 12 of the machine word in which the last English character appears. Should an English string end on character position 12 of a machine word, the numeral is placed in position 12 of the next word. For the last English correspondent, the character " % " is substituted for the numeral, as shown in Figs. 38 and 56. These conventions ensure that machine words at least may be treated as irreducible in operations on English correspondents.

The space between the " % " mark and machine word 25 is allocated to any comments the correspondent writer may consider necessary. The use of this space as an escape mechanism has been mentioned in Sec. 6.3.

The first operation performed on List 6 is that of marking for deletion the canonical stems of all words in which mistakes were detected and marked on Lists 1 and 2 as described in Sec. 7.6. Naturally, any other mistakes detected while scanning List 6 are treated likewise. A list of affected words is prepared and returned to the classification or transcription stage (Sec. 7.5) for incorporation into a new group. Stems to be deleted are marked by a "0" in the margin of List 6 (Fig. 55). The "0" margin marks are used by the programs which assemble the Russian and the English to control the automatic deletion of the marked items. In this way, mistakes are treated as part of a far less painful routine than any we contemplated applying at the time they were detected.

The selection of English correspondents is a relatively straightforward process once it has been decided to use existing dictionaries as the basis for the initial compilation of the automatic dictionary file (Sec. 7.4); it remains only to select among the correspondents given in those dictionaries. It is desirable to rank correspondents in the order of decreasing likelihood of their occurrence in technical texts. When a word appears in more than one source dictionary, the correspondents in the technical dictionaries are to be given precedence. With this restriction, the choice and ranking of correspondents can be left to the discretion of the correspondent writers. The only sensible alternative to this procedure would be direct reference to occurrences of the words in actual contexts, something much more readily accomplished as a by-product of actual operation of the automatic dictionary. Plans for updating the dictionary therefore include not only the addition of new words found in texts, but also the alteration of the correspondents of existing entries, both in substance and in ranking, on the basis of experience with occurrences of the words in these entries in texts.

Since the source dictionaries must be consulted to determine the English correspondents for each word, the information necessary to enter the appropriate grammatical coding in the organized and semiorganized machine words is readily available at the same time. The classification tables of Sec. 6.3 through 6.5 are used to determine what characters should be entered in these words. Marks in the first semiorganized word, with first character "A", are used to make a record of the sources consulted in preparing the dictionary entry: $A0 =$ Smirnitskij (1949); $A1 =$ Department of the Army (1956); $A2 =$ Segal (1953); $A3 =$ American Mathematical Society (1953).

As indicated in Sec. 6.1, the generated paradigm of a word may include vacuous strings that do not belong to the reduced paradigm. These strings are of two kinds, called "artificial" and "academic". The strings "дамей" and "комнатей" (Fig. 34) are typical artificial strings, which are alien to the Russian inflectional system. Academic strings, on the other hand, are quite plausible morphologically, but happen not to be used, at least according to the "Sprachgefühl" of native informants or in the opinion of some accepted authority. Figure 57 shows the 87 strings produced for *писать** by the inflected-form generators. Of these, only 55 are considered to belong to the reduced paradigm, the remainder being artificial or academic. Of the twelve stems remaining as representations of *писать** and its derivatives after condensation, five arise from artificial or academic forms exclusively and may be deleted.

The deletion of vacuous stems is not mandatory. If they remain in the

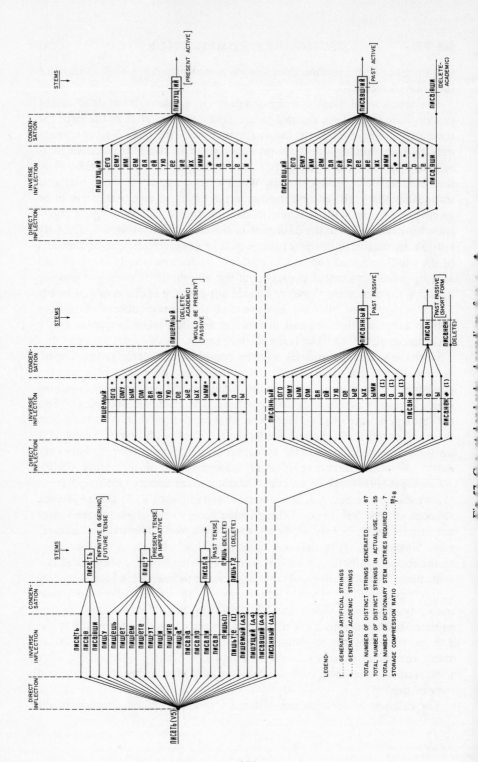

| | DIRECT INFLECTION | INVERSE INFLECTION | CONDENSATION | STEMS |

LEGEND:

I....GENERATED ARTIFICIAL STRINGS

*....GENERATED ACADEMIC STRINGS

TOTAL NUMBER OF DISTINCT STRINGS GENERATED.........87

TOTAL NUMBER OF DISTINCT STRINGS IN ACTUAL USE.....55

TOTAL NUMBER OF DICTIONARY STEM ENTRIES REQUIRED.....7

STORAGE COMPRESSION RATIO .... $\frac{55}{7} = 8$

dictionary and are indeed vacuous they will never be used and can eventually be removed automatically as part of the updating process; the price of retaining them is a mild increase in both the labor of compilation and the volume of the dictionary. If supposedly vacuous stems are deleted but their corresponding strings do in fact occur, this occurrence will be detected during dictionary operation and the stems may then be added to the dictionary again as part of the updating process. Since the problem of deletion is not critical (Sec. 9.8), the decision to delete or not was left to the discretion of the correspondent writers. As in the case of mistakes, deletion is effected by using a "0" margin mark.

As indicated in Chapter 5, and evident in Fig. 55, a given Russian word may be represented in List 6 by more than one stem. The question therefore arises, should all stems for one word be given the same English correspondents? To answer this question, it is necessary to consider briefly the place of the automatic dictionary in the development of automatic translation. Standard dictionaries with classical canonical forms of the domain language as entry keys also give classical canonical forms as correspondents in the range language. The proper inflected form in the range must be generated after look-up, as suggested in Sec. 5.6. In designing an automatic dictionary, two basic possibilities present themselves, corresponding to two modes of dictionary organization and operation. In a dictionary based on stems or other canonical entries (canonical dictionary, Sec. 5.3), it is natural to provide only canonical correspondents and to generate inflected forms with an appropriate algorithm after look-up. In a dictionary based on distinct entries for distinct members of reduced paradigms (paradigm dictionary), it is natural to attempt to provide each entry with appropriately inflected correspondents (Secs. 10.2, 10.3). These inflected correspondents might be generated automatically during compilation by that same algorithm which, with a canonical dictionary, would be used each time a form is found in a text. (This different use of algorithms is another illustration of the general phenomenon mentioned in Sec. 7.6 in connection with the calculation of coordinates.)

Initially, it is convenient to stand between these extremes by taking advantage of the fact that certain stems correspond to restricted subsets of reduced paradigms, and therefore choosing fitting inflected correspondents rather than canonical ones. Pending the development and application of more sophisticated translation algorithms, this hybrid procedure can provide at once word-by-word translations that are slightly more readable than those that could be produced if only canonical correspondents were present in the dictionary. The rules for writing correspondents of verbs therefore provide for assigning inflected correspondents to stems which permit it. In case of any doubt, the canonical correspondent takes precedence.

For nouns and adjectives canonical correspondents are used exclusively.

In deciding whether or not inflected correspondents may be assigned to a given stem, List 6′ (Fig. 52) plays an important role. By consulting this list, the correspondent writer can easily determine which strings among those with a common serial number have a common stem. If these strings clearly admit of a common inflected correspondent, such a correspondent is used; otherwise, for safety, the canonical form is used. Figure 55 illustrates the application of these conventions to the verb набегать*. The detailed conventions are stated in Table 40; they are by no means rigorous, but have been found adequate and fail-safe as a stopgap measure pending the completion of developments described in Secs. 10.1 and 10.2.

Table 40. Conventions for assigning correspondents to verbal forms.

1. Correspondents should be assigned in the infinitive form unless:

    (a) The stem belongs to strings admitting a single common correspondent form other than the infinitive, for example, past indicative; or
    (b) The stem belongs to frequently occurring strings admitting a single common correspondent form, and to strings of rare occurrence, for example, representations of the imperative.

    In doubtful cases, the infinitive should always be used.

2. Characteristics of the strings to which a stem belongs should be marked in the third semiorganized word as follows:

B0—infinitive	B4—imperative
B1—present indicative	B5—present gerund
B2—perfective future	B6—past gerund
B3—past indicative	K0—ambiguity

    The interpretation of the majority of the strings to which a stem belongs is marked in character positions 1 and 2. If the stem belongs to other strings, the first mark is followed by "K0", followed in turn by marks of the interpretations of these other strings, in any order.

3. Verbal paradigms with "-" in their group serial are treated as distinct from participial paradigms with "A", "B", "C", or "D" in the group serial. All members of the reduced paradigms associated with each serial (-, A, B, C, or D) may be found in List 6′ for the word group.

4. When the conditions of 1 permit, the correspondents of gerundial and participial stems are to be inflected as follows:

    Present gerund—English present participle ("ing" form)
    Past gerund—English past participle
    Present participle active—"ing" form
    Present participle passive—English past participle
    Past participle active—"ing" form
    Past participle passive—English past participle

Margin marks "1", "2", "3", and so forth are used to distinguish among stems with the same serial numbers but with distinct correspondents (Fig. 55). When two or more stems with the same serial numbers have identical correspondents and coding, the first is given an ordinary margin mark, for instance, "1", and the others are marked "1R", to indicate that the information marked by "1" under the same serial number applies to them. This convention is simply a labor-saving device designed to avoid writing identical information several times.

Once List 6 has been filled in as shown in Fig. 55, all the handwritten material is transcribed onto magnetic tape. Of the information originally printed on List 6, only the serial numbers are transcribed, to provide a positive means of associating the newly transcribed information with that in the 12IS file from which List 6 was made (Fig. 51). The proof copy obtained on transcribing the information in Fig. 55 is shown in Fig. 58. The files produced in this fashion are labeled "Type C" (Fig. 54).

7.9. Mistake Detection and Correction; Assembly

Obvious mistakes made in preparing the Type C files and caught by proofreading are corrected. Of the remaining mistakes, the most pernicious—which are, unfortunately, fairly frequent and hard to detect—are those "phase" mistakes due to the accidental omission, duplication, or mutilation of whole or partial items during typing or during presumable correction of mistakes detected by proofreading. The presence of mistakes of this type makes a proper merging of the Type 12IS file with the Type C file impossible.

A special program (Fig. 54) compares the Type 12IS and Type C files, and produces a slightly modified Type C file, labeled "Type CM". Whenever the serial numbers on the two input tapes get out of phase, the program moves them ahead in a manner calculated to bring them back in phase if at all possible, and then to detect later phase errors, if any. At the same time mistake-indicating information is recorded in normally unused portions of the items in the neighborhood of the mistake on the Type CM file. A count of the number of blocks recorded on the Type CM tape is inserted in every other 30-word item, to provide points of reference for use in correcting the detected phase mistakes. A print of a section of a Type CM file containing phase mistakes is shown in Fig. 59. The mistake occurred in the vicinity of a stem with serial number 018-1439; the checking program managed to get the input tapes back in phase at serial number 018-1441, as indicated by the mark "PANIC END AT". Mistakes detected in this way are then corrected, and the corrected Type CM file is subjected once more to

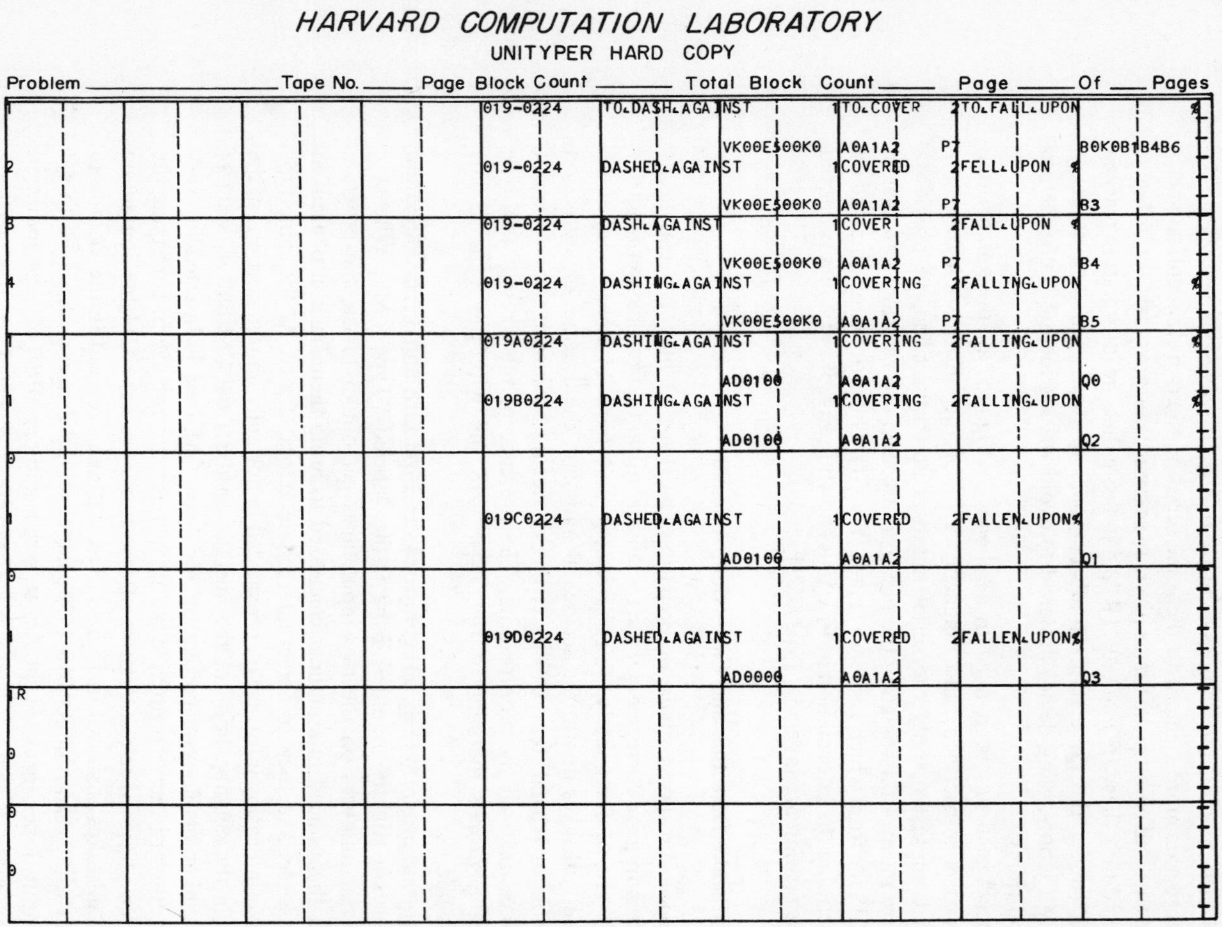

HARVARD COMPUTATION LABORATORY
UNITYPER HARD COPY

Problem _____ Tape No. _____ Page Block Count _____ Total Block Count _____ Page _____ Of _____ Pages

	Page Block Count	Total Block Count		Page	Of
	019-0224	1TO₂DASH₂AGAINST	1TO₂COVER	2TO₂FALL₂UPON	
2	019-0224	VK00E₂500K0 DASHED₂AGAINST	A0A1A2 1COVERED	P7 2FELL₂UPON	B0K0B1B4B6
3	019-0224	VK00E₂500K0 DASH₂AGAINS	A0A1A2 1COVER	P7 2FALL₂UPON	B3
4	019-0224	VK00E₂500K0 DASHING₂AGAINST	A0A1A2 1COVERING	P7 2FALLING₂UPON	B4
	019A0224	VK00E₂500K0 DASHING₂AGAINST	A0A1A2 1COVERING	P7 2FALLING₂UPON	B5
	019B0224	AD0100 DASHING₂AGAINST	A0A1A2 1COVERING	2FALLING₂UPON	Q0
		AD0100	A0A1A2	2FALLING₂UPON	Q2
	019C0224	DASHED₂AGAINST	1COVERED	2FALLEN₂UPON₂	
		AD0100	A0A1A2		Q1
	019D0224	DASHED₂AGAINST	1COVERED	2FALLEN₂UPON₂	
R		AD0000	A0A1A2		Q3

258

Fig. 58. Transcribed English correspondents and grammatical marks (List 4, Fig. 54).

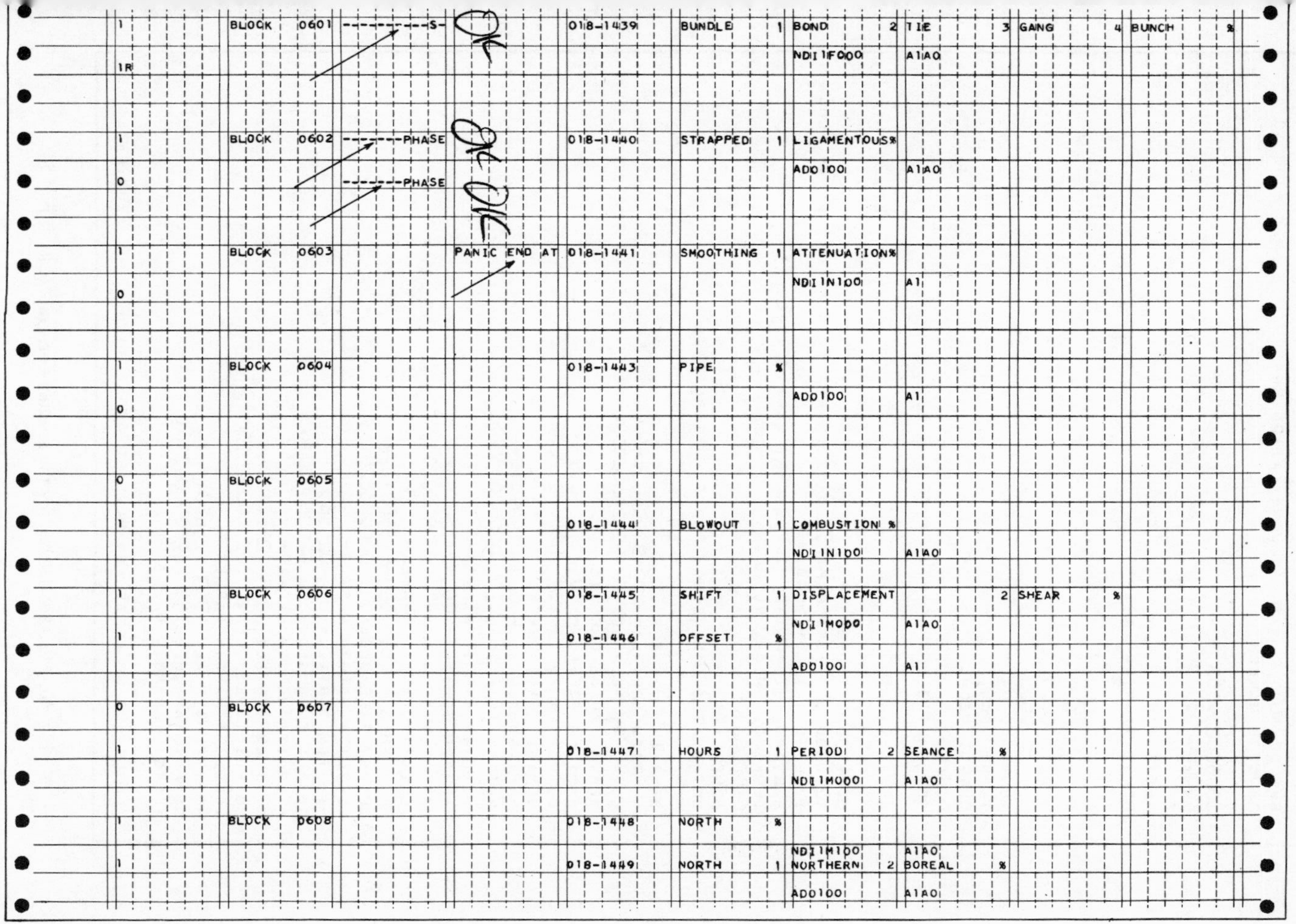

```
1   BLOCK 0601 ----------S-    018-1439 BUNDLE    1 BOND          2 TIE     3 GANG   4 BUNCH  %
                                                    ND11F000        A1A0
1R
1   BLOCK 0602 -------PHASE    018-1440 STRAPPED  1 LIGAMENTOUS%
0                                                   AD0100          A1A0
               -------PHASE
1   BLOCK 0603     PANIC END AT 018-1441 SMOOTHING 1 ATTENUATION%
                                                    ND11N100        A1
0
1   BLOCK 0604                 018-1443 PIPE      %
                                                    AD0100          A1
0
0   BLOCK 0605
1                              018-1444 BLOWOUT   1 COMBUSTION %
                                                    ND11N100        A1A0
1   BLOCK 0606                 018-1445 SHIFT     1 DISPLACEMENT          2 SHEAR   %
                                                    ND11M000        A1A0
1                              018-1446 OFFSET    %
                                                    AD0100          A1
0   BLOCK 0607
1                              018-1447 HOURS     1 PERIOD        2 SEANCE    %
                                                    ND11M000        A1A0
1   BLOCK 0608                 018-1448 NORTH     %
                                                    ND11M100        A1A0
1                              018-1449 NORTH     1 NORTHERN      2 BOREAL    %
                                                    AD0100          A1A0
```

Fig. 59. Detected phase errors (List 7, Fig. 54).

```
RUN OF A09        TRIAL COMPILATION RUN         12 JUNE 58
PROCESS MARKER:

00000000M000
000000008000
•
CLEAR 1;0
REPOSITIONING TAPES•

1       0000
2       0000
4       0000
3       0002
ABOVE FORMAT;
INPUT MECH,
OUTPUT MECH,
OTHERS
SET BKPT 8, FORCE NO TRANSF FOR RR


CLEAR 1;0

DATE TODAY DD-MM-YY
12 JUNE 58  •SET BKPT 8 FORCE TRANSF TO COPY OLD I• D• BLOCK
TYPE IN 4 I• D• WORDS
HARVARD AUTO•MATIC DICTIO•NARY PART 8 •GROUP 3      •
DUMP EDIT MECH, BLOCK;
040000040000
DUMP OVER
SET BKPT 8, FORCE NO TRANSF TO TERMINATE

SERIAL -F- CHECK;
---ITEM SERIAL;--003-01904006--
OUTPUT TAPE;BLOCK;     020000020020
SERIAL PHASE CHECK;
---ITEM SERIAL;--003A01904009--
OUTPUT TAPE;BLOCK;     020000020021
SERIAL PHASE CHECK;
---ITEM SERIAL;--003B01904010--
OUTPUT TAPE;BLOCK;     020000020022
ILLEGIT• MARGIN MARKER-
---ITEM SERIAL;--003-02556103--
OUTPUT TAPE;BLOCK;     020000020159
ILLEGIT• MARGIN MARKER-
---ITEM SERIAL;--003-02556012--
OUTPUT TAPE;BLOCK;     020000020159
ILLEGIT• MARGIN MARKER-
---ITEM SERIAL;--003-02556102--
OUTPUT TAPE;BLOCK;     020000020160
ILLEGIT• MARGIN MARKER-
---ITEM SERIAL;--003-02566102--
OUTPUT TAPE;BLOCK;     020000020160
ILLEGIT• MARGIN MARKER-
---ITEM SERIAL;--003-02566011--
OUTPUT TAPE;BLOCK;     020000020161
ILLEGIT• MARGIN MARKER-

TAPES OUT OF PHASE AND REQUIRE CORRECTION    RUN TERMINATED
```

Fig. 60. Mistakes detected by the assembly program (List 5, Fig. 54).

VYVERJAL			V01.00	003-01868007	ADJUSTED 1	REGULATED %			
						VN00P70000	AO		B3
VYVERJAJT			V01.00 600	003-01868009	ADJUST 1	REGULATE %			
						VN00P70000	AO		B4
VYVER			V01.00 YYO	003-01868007	ADJUSTING 1	REGULATING %			
						VN00P70000	AO		B5
VYVERJAJUSHC	H		A04.00F9:0	003A01868010	ADJUSTING 1	REGULATING %			
						AD0100	AO		Q0
VYVERJAVSH			A04.00F9:0	003B01868010	ADJUSTING 1	REGULATING %			
						AD0100	AO		Q2
VYVERJAEM			A03.00FT:0	003C01868010	ADJUSTED 1	REGULATED %			
						AD0100	AO		Q1
VYVERJANN			A01.00FT:0	003D01868010	ADJUSTED 1	REGULATED %			
						AD0100	AO		Q3
VYVOZH			V04.00 L00	003-01908006	TO REMOVE 1	TO TAKE OUT2	TO EXPORT %		
ENTRY ERROR	ENTRY ERROR	ENTRY ERROR	ENTRY ERROR	ENTRY ERROR	ENTRY ERROR	ENTRY ERROR	ENTRY ERROR	ENTRY ERROR	ENTRY ERROR
ENTRY ERROR	ENTRY ERROR	ENTRY ERROR	ENTRY ERROR	ENTRY ERROR	ENTRY ERROR	ENTRY ERROR	ENTRY ERROR	ENTRY ERROR	ENTRY ERROR
VYVOZ			V04.00 9RU	003-01908008	REMOVE 1	TAKE OUT 2	EXPORT %		
						VN00P30000	AO		B1
VYVOZJASHCH			A04.00F9:0	003AD1908009	REMOVED 1	TODK OUT 2	EXPORTED %		
ENTRY ERROR	ENTRY ERROR	ENTRY ERROR	ENTRY ERROR	ENTRY ERROR	ENTRY ERROR	ENTRY ERROR	ENTRY ERROR	ENTRY ERROR	ENTRY ERROR
ENTRY ERROR	ENTRY ERROR	ENTRY ERROR	ENTRY ERROR	ENTRY ERROR	ENTRY ERROR	ENTRY ERROR	ENTRY ERROR	ENTRY ERROR	ENTRY ERROR

Fig. 61. Assembly in the presence of mistakes.

NAGREV · N01.00 100 · 019-0222P007 HEATING | WARMING · %
N01N100 · AOA1

NAGREVANI · N10.00F600 · 019-0223P010 HEATING | WARMING · %
N01N100 · %

NABEGA · V01.00FKU0 · 019-0224P008 TO DASH AGAI NST · 1 TO COVER · 2 TO FALL UPON · %
AOA1
VKOOE500KO · AOA1A2 · P7 · BOKOB1B4B6

NABEGAL · V01.00 · 019-0224P007 DASHED AGAIN ST · 1 COVERED · 2 FELL UPON · %
VKOOE500KO · AOA1A2 · P7 · B3

NABEGAJT · V01.00 600 · 019-0224P009 DASH AGAINST · 1 COVER · 2 FALL UPON · %
VKOOE500KO · AOA1A2 · P7 · B4

NABEG · V01.00 1Y0 · 019-0224P007 DASHING AGAI NST · 1 COVERING · 2 FALLING UPON · %
VKOOE500KO · AOA1A2 · P7 · B5

NABEGAJUSHCH · A04.00F910 · 019A0224P010 DASHING AGAI NST · 1 COVERING · 2 FALLING UPON · %
ADD100 · AOA1A2 · QO

NABEGAVSH · A04.00F910 · 019B0224P010 DASHING AGAI NST · 1 COVERING · 2 FALLING UPON · %
ADD100 · AOA1A2 · Q2

NABEGAEM · A03.00FT10 · 019C0224P010 DASHED AGAIN ST · 1 COVERED · 2 FALLEN UPON%
ADD100 · AOA1A2 · Q1

NABEGANN · A01.00FT10 · 019D0224P010 DASHED AGAIN ST · 1 COVERED · 2 FALLEN UPON%
ADD000 · AOA1A2 · Q3

NABEGAN · A01.00 · 019D0224P007 DASHED AGAIN ST · 1 COVERED · 2 FALLEN UPON%
ADD000 · AOA1A2 · Q3

MNOGOOBESHCH AJUSHCH · A04.00F910 · 019A0225P102 PROMISING · 1 HOPEFUL · 2 LIKELY · %
ADD100 · AO · QO

NABIRA · V01.00FKU0 · 019-0226P008 TO DIAL · 1 TO SET UP · 2 TO COLLECT · %
VNOOP100KO · AOA1A2 · P7 · BOKOB1B4B6

Fig. 62. An assembled portion of Group 19.

phase checking to ensure that the corrections were properly made. Only when the phase-checking program signals no mistakes is the Type CM file released for assembly.

An assembly program (Fig. 54) combines the Russian information in the Type 12IS file with the English correspondents and grammatical codes in the Type CM file to form complete dictionary entries with the structure illustrated in Fig. 56. Whenever a margin mark "0" is detected in an item in the Type CM file, this item and its mate in the Type 12IS file are deleted, that is, no dictionary entry item is recorded on the output tape.

The assembly program also checks for a variety of mistakes such as the absence of the "F" mark (Sec. 7.7) in an item from the Type 12IS file when the first item with a new serial number appears in the Type CM file or the improper use of margin marks, for instance, the use of "2R" if "2" has not been used. A check is also made for phase mistakes. Whenever the assembly program detects a mistake an appropriate print-out is made on the console typewriter, as shown in Fig. 60. At the same time, the affected entries are deliberately and conspicuously mutilated (Fig. 61) to provide a signal in case correction procedures are accidentally omitted or not properly followed.

Should a phase mistake, indicated by repeated print-outs, be present, assembly cannot be completed. Figure 60 shows a series of local mistakes in Group 3, followed by a phase mistake. Detecting and correcting phase mistakes at this point is horribly inefficient, since it is only after the first such mistake has been corrected and assembly has once more been attempted that other similar mistakes can be caught. This lesson was learned only with Group 3, our first mass-produced group, the first two small groups having been "babied" through all procedures with loving care for check-out purposes. The phase-checking program was then developed for use with later groups.

The only mistakes that can be detected effectively by either the phase-checking or the assembly program are those which violate the conventions established regarding the format of items in the Type 12IS or Type CM files. Mistakes in the spelling of English correspondents or in the grammatical codes that have escaped detection by proofreading pass through into the assembled dictionary; some of these are detected later by procedures described in Sec. 9.10.

An assembled portion of Group 19 is shown in Fig. 62. This portion was assembled from the pieces shown in Figs. 53 and 58. Each newly assembled group dictionary file must then be merged with those for other groups into a single dictionary file. A more detailed description of the structure of the dictionary file is given in Secs. 9.6 and 9.9. At this point it is sufficient to indicate that items in this file are ordered primarily according to the alphabetic order of canonical stems and that the file occupies several reels of magnetic tape. A print of a portion of this file is shown in Fig. 38. At the time of writing, the file contained about 22,000 stem entries, representing approximately 10,000 Russian words.

CHAPTER 8

DICTIONARY OPERATION

8.1. The Function of the Harvard Automatic Dictionary

The dictionary file compiled as described in Chapter 7 has been put to use as part of an automatic system, the Harvard Automatic Dictionary. The basic function of this system is to adjoin to strings occurring in a text the information about these strings that is stored in the dictionary file. The result is a file containing, at least in theory, all the information about a text string that can be obtained by considering the string as an isolated unit, that is, apart from its context. This file, obtained in a way to be described, is called an *augmented text* (Fig. 70); it consists essentially of dictionary entries, in general one for each word in the original text, arranged in the order in which the words occur in this text.

The augmented text is a basic tool for the investigation of the syntactic and semantic properties of Russian and English necessary to create algorithms capable of translating automatically, smoothly, economically, and above all, accurately; it is also the raw material on which such algorithms will operate. A description of some uses of augmented texts in research is given in Chapter 10; applications of augmented texts to fundamental linguistic studies not necessarily related to automatic translation will readily suggest themselves to the reader.

Storing dictionaries on magnetic tape or other media readable by a machine is a novel development. As more is learned about efficient ways of compiling and updating dictionary files automatically, especially about efficient ways of reflecting current usage of words in these files, the periodic publication of conventional dictionaries far more accurate and up-to-date than any made heretofore becomes an attractive possibility. Interesting problems of lexicography that have been ignored or found intractable can be brought to light, with renewed hope for solutions.

When edited into a format more suitable for visual examination than that illustrated in Fig. 70, the entries in an augmented text provide a rough, word-by-word approximation to a translation of the original text. Such approximate translations have been found useful in themselves, in a manner described in Sec. 8.5. The augmented text and other intermediate files participating in the operation of the Harvard Automatic Dictionary are also useful in several

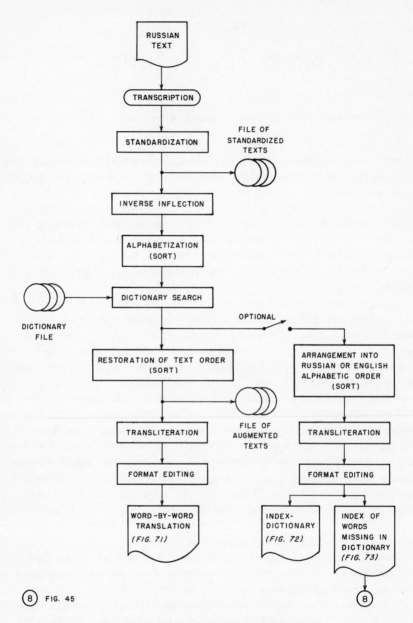

FIG. 45

Fig. 63. The Harvard Automatic Dictionary.

auxiliary functions, of which some are described in this chapter and others in Chapters 9 and 10.

A flow chart showing the major steps in the operation of the automatic dictionary is given in Fig. 63 (Giuliano, 1959a; Jones, 1959; Barnes, 1958). Except for transcription, the whole process described by the flow chart is fully automatic. The boxes in Fig. 63 represent distinct passes through a text. At the end of each pass, the program for the next is automatically loaded into memory and, in general, the output tape of one pass serves as the input tape for the next.

The necessity for several passes is in part an accident: it proved expedient to use a succession of programs written independently for other purposes instead of going through the labor of combining these programs into a single one, which in effect means writing a new program; in part, separate passes are necessary whenever combining several operations into one pass would require a program exceeding the memory capacity of the machine; finally, because some tape sorting is essential in the process, a minimum number of distinct passes is unavoidable.

8.2. The Initial Transcription of Texts; Standardization

The ink-on-paper tokens by which most texts intended for translation are represented are not, as we have seen in Sec. 1.5, suitable for direct automatic handling by contemporary information-processing machines. A manual transcription into machine tokens is therefore unavoidable. It is evident from the descriptions of the transcription operations necessary in compilation given in Chapter 7 that such operations can cause considerable grief.

The instrument used for transcribing texts is the Cyrillic Unityper described in Sec. 2.8. As in the transcription of words to be entered in the dictionary, described in Sec. 7.2, the choice of a format for transcribed material was governed by a desire to maximize both ease of transcription and simplicity of further automatic processing of the transcribed material. In the transcribing of texts it is highly advantageous to adhere as closely as possible to normal typing practice. Unfortunately, the resulting format is awkward for subsequent automatic processing, because no direct relation is established between the reducibility properties of strings of words and those of machine tokens. It is therefore necessary to interpose an automatic format-standardizing operation between transcription and dictionary look-up to provide, so to speak, "impedance matching" between man and machine.

A Russian text to be processed by the automatic dictionary is shown in Fig. 64. Part of the copy produced by the Unityper during transcription is shown in Fig. 65. A few special conventions, already illustrated in Fig. 19

Доклады Академии наук СССР
1956. Том 109, № 5

ФИЗИКА

К. С. ВУЛЬФСОН

НОВЫЙ МЕТОД ИЗМЕРЕНИЯ СКОРОСТИ СВЕТА

(Представлено академиком Г. С. Ландсбергом 14 I 1956)

Развитие электроники позволило значительно усовершенствовать метод Физо измерения скорости света. Применяя остроумный нулевой метод, Бергштранд ([1]) довел точность измерений до 0,25 км/сек, т. е. до $1 \cdot 10^{-6}$. Электромагнитные методы измерения скорости света также доведены до высокой степени точности ([2-5]). Несмотря на это, разброс средних значений скорости света в различных работах превосходит величину вероятной ошибки, что не дает возможности надежно установить значение скорости света. Сопоставление результатов измерений, выполненных за большой промежуток времени, наводит даже на предположение, что скорость света меняется со временем ([6]). Представляется поэтому желательным дальнейшее повышение точности измерения скорости света путем улучшения существующих методов и нахождения новых более совершенных.

В настоящей работе предлагается новый вариант метода Физо, при помощи которого можно надеяться повысить точность измерения скорости света.

Метод Физо, сущность которого хорошо известна, основывается на: 1) измерении расстояния от модулятора до зеркала; 2) измерении частоты модуляции светового луча и 3) установлении минимума интенсивности светового луча, прошедшего после отражения от зеркала второй раз модулирующее устройство. Неточность в установлении минимума приводит к неправильному значению частоты модуляции. Чтобы снизить влияние ошибки в установлении минимума интенсивности, приходится применять большие расстояния, однако методы измерения больших расстояний не обеспечивают той высокой относительной точности, которой удается достигнуть при небольших расстояниях. Примененный Бергштрандом нулевой метод значительно уменьшил ошибку при установлении минимума и позволил ему тем самым достигнуть блестящих результатов.

В предлагаемом ниже методе измерение 3), являющееся главным источником ошибок, исключается благодаря применению автоматически действующей схемы. Кроме того, новый метод не требует применения больших расстояний, и поэтому измерения могут быть выполнены с очень большой относительной точностью.

Сущность метода состоит в следующем. Источник света или модуляционное приспособление посылает кратковременный световой импульс длительностью порядка $10^{-7} - 10^{-8}$ сек. После отражения от зеркала импульс попадает на фотоэлектрическое приемное устройство. Возникающий в последнем электрический импульс вызывает при помощи специальной схемы посылку повторного светового импульса. Таким образом, после посылки первичного импульса система начинает генерировать световые импульсы, частота повторения которых определяется расстоянием от модулятора или источника света до зеркала и от зеркала до фотоприемника, а также задержкой сигнала в аппаратуре. Определяя период или, что то же, частоту повторения световых и электрических импульсов, возникающих в схеме при

929

Fig. 64. Russian text.

HARVARD COMPUTATION LABORATORY

UNITYPER HARD COPY

Problem TexT OOD Tape No. BSB #2 Page Block Count 3 Total Block Count 11 Page 1 Of 4 Pages

```
082658⌞⌞⌞⌞⌞⌞000000000008

$SOURCE:$⌞⌞ДОКЛАДЫ⌞⌞АКАДЕМИИ⌞⌞НАУК⌞⌞СССР⌞⌞$1956$⌞⌞ТОМ⌞⌞$109,⌞NO.⌞5$⌞⌞ФИЗИКА⌞⌞$AUTHOR:$⌞⌞К.С.ВУЛЬФСОН⌞⌞$TITLE:$⌞⌞НОВЫЙ⌞⌞М
                                                                                                                  Р
ЕТОД⌞⌞ИЗМЕРЕНИЯ⌞⌞СКОРОСТИ⌞⌞СВЕТА⌞⌞*(*⌞⌞ПРЕДСТАВЛЕНО⌞⌞АКАДЕМИКОМ⌞⌞Г.С.ЛАНДСБЕРГОМ⌞⌞$14⌞I⌞1956$⌞⌞*)*⌞⌞$TEXT:$⌞⌞РАЗВИТИЕ⌞⌞Э

ЛЕКТРОНИКИ⌞⌞ПОЗВОЛИЛО⌞⌞ЗНАЧИТЕЛЬНО⌞⌞УСОВЕРШЕНСТВОВАТЬ⌞⌞МЕТОД⌞⌞ФИЗО⌞⌞ИЗМЕРЕНИЯ⌞⌞СКОРОСТИ⌞⌞СВЕТА⌞⌞*.*⌞⌞ПРИМЕНЯЯ⌞⌞ОСТРОУМНЫ

Й⌞⌞НУЛЕВОЙ⌞⌞МЕТОД⌞⌞*,*⌞⌞БЕРГШТРАНД⌞⌞$(FOOTNOTE⌞1)$⌞⌞ДОВЕЛ⌞⌞ТОЧНОСТЬ⌞⌞ИЗМЕРЕНИЙ⌞⌞ДО⌞⌞$0,25$⌞⌞КМ⌞⌞СЕК⌞⌞*,*⌞⌞Т.Е.⌞⌞ДО⌞⌞⌞⌞⌞
$1⌞TIMES⌞10⌞TO⌞THE⌞MINUS⌞6$⌞⌞*.*⌞⌞ЭЛЕКТРОМАГНИТНЫЕ⌞МЕТОДЫ⌞ИЗМЕРЕНИЯ⌞СКОРОСТИ⌞
IMES⌞10⌞TO⌞THE⌞MINUS⌞6$⌞⌞*.*⌞⌞ЭЛЕКТРОМАГНИТНЫЕ⌞⌞МЕТОДЫ⌞⌞ИЗМЕРЕНИЯ⌞⌞СКОРОСТИ⌞⌞СВЕТА⌞⌞ТАКЖЕ⌞⌞ДОВЕДЕНЫ⌞⌞ДО⌞⌞ВЫСОКОЙ⌞⌞СТЕПЕН

И⌞⌞ТОЧНОСТИ⌞⌞$(FOOTNOTES⌞2⌞TO⌞5)$⌞⌞*.*⌞⌞НЕСМОТРЯ⌞⌞НА⌞⌞ЭТО⌞⌞*,*⌞⌞РАЗБРОС⌞⌞СРЕДНИХ⌞⌞ЗНАЧЕНИЙ⌞⌞СКОРОСТИ⌞⌞СВЕТА⌞⌞В⌞⌞РАЗЛИЧНЫ

Х⌞⌞РАБОТАХ⌞⌞ПРЕВОСХОДИТ⌞⌞ВЕЛИЧИНУ⌞⌞ВЕРОЯТНОЙ⌞⌞ОШИБКИ⌞⌞*,*⌞⌞ЧТО⌞⌞НЕ⌞⌞ДАЕТ⌞⌞ВОЗМОЖНОСТИ⌞⌞НАДЕЖНО⌞⌞УСТАНОВИТЬ⌞⌞ЗНАЧЕНИЕ⌞⌞СК
                                                                                         ЗА
ОРОСТИ⌞⌞СВЕТА⌞⌞*.*⌞⌞СОПОСТАВЛЕНИЕ⌞⌞РЕЗУЛЬТАТОВ⌞⌞ИЗМЕРЕНИЙ⌞⌞*,*⌞⌞ВЫПОЛНЕННЫХ⌞⌞ЗА⌞⌞БОЛЬШОЙ⌞⌞ПРОМЕЖУТОК⌞⌞ВРЕМЕНИ⌞⌞*,*⌞⌞НАВО

ДИТ⌞⌞ДАЖЕ⌞⌞НА⌞⌞ПРЕДПОЛОЖЕНИЕ⌞⌞*,*⌞⌞ЧТО⌞⌞СКОРОСТЬ⌞⌞СВЕТА⌞⌞МЕНЯЕТСЯ⌞⌞СО⌞⌞ВРЕМЕНЕМ⌞⌞$(FOOTNOTE⌞6)$⌞⌞*.*⌞⌞ПРЕДСТАВЛЯЕТСЯ⌞⌞ПО

ЭТОМУ⌞⌞ЖЕЛАТЕЛЬНЫМ⌞⌞ДАЛЬНЕЙШЕЕ⌞⌞ПОВЫШЕНИЕ⌞⌞ТОЧНОСТИ⌞⌞ИЗМЕРЕНИЯ⌞⌞СКОРОСТИ⌞⌞СВЕТА⌞⌞ПУТЕМ⌞⌞УЛУЧШЕНИЯ⌞⌞СУЩЕСТВУЮЩИХ⌞⌞МЕТОДОВ

⌞⌞И⌞⌞НАХОЖДЕНИЯ⌞⌞НОВЫХ⌞⌞БОЛЕЕ⌞⌞СОВЕРШЕННЫХ⌞⌞*.*⌞⌞$NEW⌞PARAGRAPH$⌞⌞В⌞⌞НАСТОЯЩЕЙ⌞⌞РАБОТЕ⌞⌞ПРЕДЛАГАЕТСЯ⌞⌞НОВЫЙ⌞⌞ВАРИАНТ⌞⌞МЕ

ТОДА⌞⌞ФИЗО⌞⌞*,*⌞⌞ПРИ⌞⌞ПОМОЩИ⌞⌞КОТОРОГО⌞⌞МОЖНО⌞⌞НАДЕЯТЬСЯ⌞⌞ПОВЫСИТЬ⌞⌞ТОЧНОСТЬ⌞⌞ИЗМЕРЕНИЯ⌞⌞СКОРОСТИ⌞⌞СВЕТА⌞⌞*.*⌞⌞$NEW⌞PARA
```

Fig. 65. Transcribed Russian text.

and evident in Fig. 65, are superimposed on ordinary typing practice. These conventions are the following:

1. *Distinct words must be separated by at least one space.* This, of course, is standard practice; it is emphasized to indicate that any larger number of spaces will do. Allowing this degree of freedom complicates the standardizing program somewhat, but restricting typists to using exactly one space would not be realistic. In fact, typists are encouraged to use two spaces whenever they think of it, since this facilitates correcting such mistakes as the omission of a letter or of a word.

2. *Punctuation marks must be bracketed by asterisks. The string formed in this way must be preceded and followed by at least one space.* This convention, which complicates the job of the typist somewhat, is not absolutely essential, but it appreciably simplifies the problem of standardization. For example, periods serving as punctuation marks are distinguished in this way from periods marking abbreviations. The asterisk later serves as a unique signal marking items for which dictionary look-up is not necessary.

3. *Editorial comments, numerals, equations, and so on must be bracketed by dollar signs. The string formed in this way must be preceded and followed by at least one space.* This convention is essential in handling texts containing characters or configurations not available on the typewriter keyboard. By typing strings such as "△△$EQUATION△1$△△" or "△△$B△SUB△1$△△", the typist can represent formulas and the like, although she has the option of simply marking their place with five dots "△△.....△△". Editorial comments, such as "△△$NEW△PARAGRAPH$△△", are made in the same way. In fact, the typist is allowed to write anything she wishes within dollar-sign brackets, *except Cyrillic characters.* The reason for this restriction is that items marked by "$", like those marked by "*", are not affected by operations such as transliteration. Cyrillic characters in brackets would therefore remain in the ranked computer representation, with obviously awkward consequences.

The special use of "$" and "*" as brackets naturally precludes using these characters in their conventional roles. This restriction is not of any serious consequence. The only other conventions concern the proper insertion of identification and sentinel blocks at the beginning and end of the text. These present no unusual problems.

It is worth noting that the typist who transcribes Russian texts does not need to know any Russian. Any good typist can be trained quite rapidly to recognize the elements of the Cyrillic alphabet and to transcribe efficiently by touch typing. In fact, the typists without knowledge of Russian, once trained, consistently seem to be both faster and more accurate than those who know some Russian. An experienced technical typist will have less difficulty than the average typist in handling equations and other unusual technical material, but this is true also of typing in English.

Following proofreading and correction, the tape containing a transcribed text is used as input to a standardizing program, which produces five-machine-word items of the standard type described in Chapter 7, as illustrated in Fig. 66. Naturally, there are no class marks in these items, since such marks do not accompany words that occur in texts. This program essentially performs an unpacking function, as described in Sec. 2.7. Items containing information originally within asterisk or dollar-sign brackets are marked in the first character position of word 0 by "*" or "$", respectively. It is possible that more than 35 characters will be used within "$" brackets. In this case, more than one five-word item is produced, and each item is marked with "$". The items are also given a text serial number, whose role is analogous to that of the group number and rank of Chapter 7.

> *Exercise 8-1.* (a) Write a program for converting files of the format illustrated in Fig. 65 to files of the format illustrated in Fig. 66.
> (b) What provisions can you include in program (a) to detect such transcription mistakes as the omission of opening or closing dollar signs and asterisks?

8.3. Look-up; the Augmented Text

The standardized file of text is next subjected to the operation of inverse inflection (Fig. 63), which factors each string into a stem for matching against dictionary entry keys and an affix defining the string as a particular member of a reduced paradigm. The inverse-inflection program is that described in Chapter 5 and used also for dictionary compilation in the manner outlined in Sec. 7.7. As pointed out in Sec. 5.6, complete consistency in the operation of an automatic dictionary can be guaranteed, whatever algorithm may be used for inverse inflection, if this algorithm is used both in compilation to generate the stems to be used as entry keys and in operation to factor strings occurring in texts.

Since the dictionary file can be regarded as arranged in increasing alphabetic order of stems (Secs. 7.9, 9.6) and since it is stored on magnetic tape, the use of matched-order interrogation of the file is indicated (Sec. 1.6; Giuliano, 1957a). The text file is therefore arranged into the same order as the dictionary file by a sorting pass. The sorting program used here is equivalent to that used to sort entries that are being added to the dictionary. An identical ordering of the two files is thereby guaranteed, in spite of the fact, mentioned in Sec. 7.9 and elaborated in Sec. 9.6, that the order is not based on the ranking of the stems exclusively.

Both the dictionary file and the ordered text file are used as inputs to the actual look-up pass. Whenever a text stem identically matches a stem in a

```
IZMENENIEM                        00A-0427
DETAL,NOSTI                       00A-0428
REGISTRATSII                      00A-0429
SIGNALA                           00A-0430
RASSMATRIVAETSJA                  00A-0431
NIZHE                             00A-0432
BOLEE                             00A-0433
PODROBNO                          00A-0434
*.                                00A-0435
PREDMETOM                         00A-0436
NASTOJASHCHEGO                    00A-0437
SOOBSHCHENIJA                     00A-0438
JAVLJAETSJA                       00A-0439
ANALIZ                            00A-0440
VOZMOZHNOSTEJ                     00A-0441
ULUCHSHENIJA                      00A-0442
OTNOSHENIJA                       00A-0443
SIGNAL                            00A-0444
*/                                00A-0445
SHUM                              00A-0446
PUTEM                             00A-0447
USREDNENIJA                       00A-0448
PERIODCHESKOGO                    00A-0449
SIGNALA                           00A-0450
I                                 00A-0451
RASSMOTRENIE                      00A-0452
KONKRETNOJ                        00A-0453
SXEMY                             00A-0454
PRIBORA                           00A-0455
*,                                00A-0456
V                                 00A-0457
KOTOROM                           00A-0458
USREDNENIE                        00A-0459
SIGNALA                           00A-0460
OSUSHCHESTVLJAETSJA               00A-0461
SREDSTVAMI                        00A-0462
IMPUL,SNOJ                        00A-0463
RADIOTEXNIKI                      00A-0464
*.                                00A-0465
$PART 2                           00A-0466
METODIKA                          00A-0467
I                                 00A-0468
EE                                00A-0469
VOZMOZHNOSTI                      00A-0470
$TEXT:                            00A-0471
USREDNENIE                        00A-0472
SIGNALA                           00A-0473
OSUSHCHESTVLJAETSJA               00A-0474
```

Fig. 66. Standardized text.

272

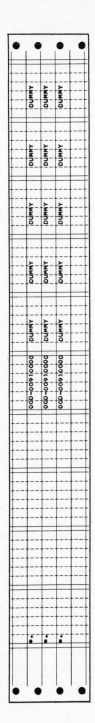

Fig. 67. (*Upper*) dummy "*" or "$" item; (*lower*) dummy item for stem not in dictionary.

dictionary entry, a 30-word item is recorded on the output tape. This item is constructed by replacing some information in the five-word Russian item contained in the 30-word dictionary item (Fig. 56) with information obtained from the text file item. Specifically, the affixes in character positions 8–12 of machine word 3, and the text serial number in character positions 1–8 of

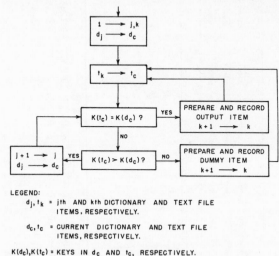

LEGEND:

d_j, t_k = jth AND kth DICTIONARY AND TEXT FILE ITEMS, RESPECTIVELY.

d_c, t_c = CURRENT DICTIONARY AND TEXT FILE ITEMS, RESPECTIVELY.

$K(d_c), K(t_c)$ = KEYS IN d_c AND t_c, RESPECTIVELY.

Fig. 68. Dictionary look-up: skeletal flow chart.

machine word 4 of the text item are transferred into the same positions in the 30-word item. In all other respects, the 30-word output item is identical with the dictionary entry.

Any text item with "*" or "$" in character position 1 of machine word 0 is embedded in a dummy 30-word item (Fig. 67) which is recorded on the output tape. If a text stem matches no dictionary entry key the text item again is embedded in a dummy 30-word item and recorded. A skeletal flow chart for the dictionary look-up process (see Exercise 1-3) is given in Fig. 68. Tests for sentinels in the input files, and some other details, are not shown in this flow chart.

Exercise 8-2. Write a complete program for dictionary look-up based on the flow chart of Fig. 68. Assume that the sentinel conventions for the files are those outlined in Sec. 7.3.

The output of the look-up pass is a file called the "alphabetic subdictionary". A portion of the alphabetic subdictionary for text OEE is shown in Fig. 69. The stem "задач" occurred three times with the affix "и" (OEE–136, OEE–363, OEE–862) and once with the affix "у" The corresponding dictionary entry, with appropriate modifications in the first five machine words, is therefore represented four times in the alphabetic subdictionary. Occurrences of the stems "закон" and "замен", with various affixes, are also shown in Fig. 69. The affixes are given in the ranked representation (Table 10).

The next pass produces the augmented text (Sec. 8.1), by using the text serial numbers as sorting keys. In the same pass, stems and affixes, separated by a dash, are recombined into strings for easier visual inspection. A transliterated portion of the augmented text OOD is shown in Fig. 70. The untransliterated augmented texts are saved for use in experimental processes, as suggested in Sec. 8.1.

8.4. Word-by-Word Translations; Index-Dictionaries and Other By-Products

An editing program selects a portion of the information in the augmented text and arranges it in a format more readily suitable for visual inspection than that illustrated in Fig. 70. This program produces a file which, when printed, looks like Fig. 71. Successive text words are arranged in Fig. 71 in the normal layout of a printed page, that is, words are read from left to right.

The English correspondents for each Russian word are displayed in a vertical array underneath the transliterated Russian. The maximum number of English correspondents to be displayed may be set to any value between 1 and 5. Naturally, if fewer than five correspondents are listed in the dictionary file, only these appear. This layout is reminiscent of the "trots" or "ponies" used by schoolboys, and is intended to serve the same purpose. This product of the automatic dictionary may therefore be regarded as a rough word-by-word translation. Applications of such translations are discussed in Sec. 8.5.

Successive "lines" of text are separated by rows of dashes to aid the eye. It is obvious that the density of information per page in prints like that of Fig. 71 is low or, to put it in other words, that much paper is wasted. One reason for choosing this format was the desire to keep the experimental editing program simple. By allocating a width corresponding to two machine words to each Russian word and its correspondents, the need for calculating the length of the longest of these was avoided.

To assist in cross-reference the serial number of the rightmost word in every second line is printed to the right of the line. Russian words for which no entry was found in the dictionary are represented in transliterated form with the mark "###" added to call attention to this fact. Punctuation marks are represented in the usual way. Information enclosed in dollar-sign brackets

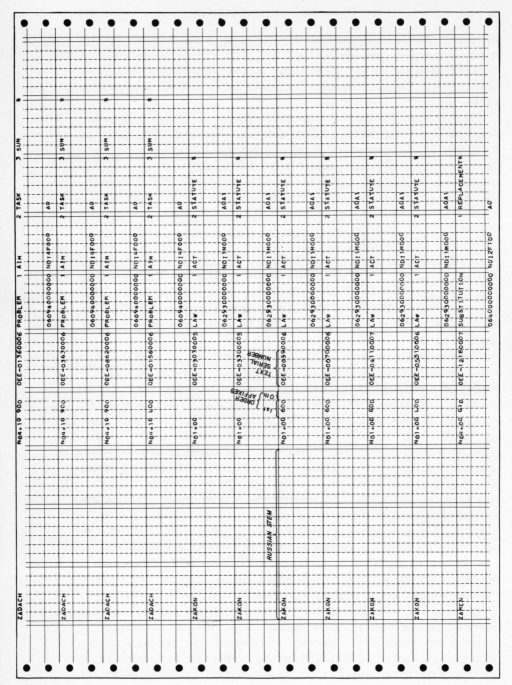

Fig. 69. Alphabetic subdictionary.

276

at the time of transcription is simply reproduced in precisely the form in which it was transcribed. The mark "***" is added to call attention to the fact that such information was not altered by any of the operations of the automatic dictionary.

It is possible for a given stem to match more than one dictionary entry key. This necessarily occurs when two or more homographic words are represented in the dictionary file by distinct entries with identical keys. In such situations, all entries with the given key appear in both the augmented text and the word-by-word translation. The second and subsequent entries are marked "+++++(HOMOGRAPH OF PREV)" to call attention to this unpleasant phenomenon. The use of stems as entry keys adds to the problem. For the sake of simplicity in exposition, questions of homography were deliberately ignored throughout Chapter 7 and most of this chapter; they are treated in Chapter 9 with the care that befits their importance.

Unless an option to terminate the run is exercised after the editing program has been executed, two indexes are prepared from the alphabetic sub-dictionary, as indicated on Fig. 63. The index-dictionary may be prepared with entries arranged either in the Russian alphabetic order of their stems or in the English alphabetic order of the first correspondent; both types of arrangement may be prepared if desired. The format of the Russian-ordered index-dictionary is illustrated in Fig. 72. The stem (A) appears on the first line of each entry. The English correspondents, numbered in the order in which they are stored in the dictionary file, are displayed in a columnar layout (B). The number of occurrences of the stem in the text is also shown (C), together with the serial numbers identifying the location of each occurrence in the text (D). The affix associated with the stem at each occurrence is given (E), although, if the number of occurrences (C) is greater than 40, only the serial numbers (D) and the affixes (E) of the first 40 are shown. Blank numbered spaces (G) are provided to enable writing in of new correspondents when desirable. Finally, the grammatical coding in the dictionary entry is displayed (F).

The index sort (Fig. 63) is so designed that all stems for which no entries were found in the dictionary file are brought together at the head of the index-dictionary file, in Russian alphabetic order. A portion of the index of new words for text OOC is shown in Fig. 73. Ample space is provided for writing in English correspondents determined, in part, by examining the contexts in which the words occur. Indexes for several texts may be combined periodically into a single master index, for use as input to the compilation and updating process (Fig. 36).

Several products and by-products of the operation of the automatic dictionary are indicated in Fig. 74. Illustrative instances of these products

```
DLITEL.NOST- .          N06.00 400  00D-05910012 DURATION    1 TIME       2 PERIOD      %
                                                 ENTRY 005205 NDI1FO        AOA1
EDINICHN-OGO            A02.00 G4G  00D-05920010 SINGLE      1 INDIVIDUAL 2 UNIT        %
                                                 ENTRY 005629 AD0100        AOA1
IZMERENI-JA             N10.00 Y00  00D-05930009 MEASUREMENT1 METERING    2 SOUNDING    3 DIMENSION  %
                                                 ENTRY 007652 NDI1N000      AOA1
OGRANICHEN-A            A01.00 100  00D-05940010 LIMITED     1 RESTRICTED %
                                                 ENTRY 012416 AD0000        A1AOA2                   03
PERIOD-OM               N01.00 GDO  00D-05950008 PERIOD      1 SPELL      %
                                                 ENTRY 014220 NDI1MO        AOA1
IZMERITEL.N- OGO        A02.00 G4G  00D-05960102 MEASURING   1 TEST       2 METER       %
                                                 ENTRY 007659 AD0100        AOA1
PRIBOR-A                N01.00 100  00D-05970007 DEVICE      1 APPARATUS  2 GEAR        3 INSTRUMENT 4 SET       %
                                                 ENTRY 015774 NDI1MO        AOA1
*.                                  00D-05980000  DUMMY         DUMMY        DUMMY        DUMMY        DUMMY
*.                                  00D-05980000  DUMMY         DUMMY        DUMMY        DUMMY        DUMMY
*.                                  00D-05980000  DUMMY         DUMMY        DUMMY        DUMMY        DUMMY
$NEW PARAGRA PH                     00D-05990000  DUMMY         DUMMY        DUMMY        DUMMY        DUMMY
$NEW PARAGRA PH                     00D-05990000  CUMMY         DUMMY        DUMMY        DUMMY        DUMMY
$NEW PARAGRA PH                     00D-05990000  DUMMY         DUMMY        DUMMY        DUMMY        DUMMY
SRAVNENI-E              N10.00 600  00D-06000009 COMPARISON $
                                                 ENTRY 019069 NDI1N000      AO
CHASTOT-                N04.00      00D-06010006 FREQUENCY  $
                                                 ENTRY 021296 NDI2F000      AOA1
```

Fig. 70. Augmented text.

Fig. 71. Interlineated word-by-word translation (trot).

```
                    NEW PARAGRAPH        -V            GAISH-        BYL-O
                    NEW PARAGRAPH        IN            GAISH-        WAS
                                         AT                          EXISTED
                                         INTO                        HAPPENED
                                         FOR
                                         ON
                    ***                                GAISH-
                                                       ###                    OOG-0395

ISPOL,ZOVAN-O       FOTOSOPROTIVLENI-E   TIP-A         FS-           -A1
UTILIZED            PHOTORESISTOR        TYPE          FS-           -A1
TOOK ADVANTAGE (OF) PHOTOCONDUCTIVE CELL PATTERN
                                         SPECIES
                                         CLASS         ###           ***

                    IMEJUSHCH-EE         IMEJUSHCH-EE  MAKSIMUM-     CHUVSTVITEL,NOST-I
                    HAVING               HAVING        MAXIMUM       SENSITIVITY
                                         THERE BEING   THE MOST      SENSITIVENESS

                                         IMEJUSHCH-EE
                                         ++++(HOMOGRAPH OF PREV)                  OOG-0404

-V                  OBLAST-I             -I            -I            KRASN-UJU
                    REGION               2.5 MU        AND           RED
IN                  FIELD                2.5 MU        AND THEN      RUDDY
AT                  DOMAIN                              AND SO        PRETTY
INTO                ***                  ***           TOO
FOR                                                    EITHER
ON

GRANITS-U           (1,2)                OKOL-O        3.5 MU        (
BORDER              (1,2)                OKOL-O        3.5 MU        (
BOUNDARY            ***                  ###           ***

                                                                                 OOG-0414
```

279

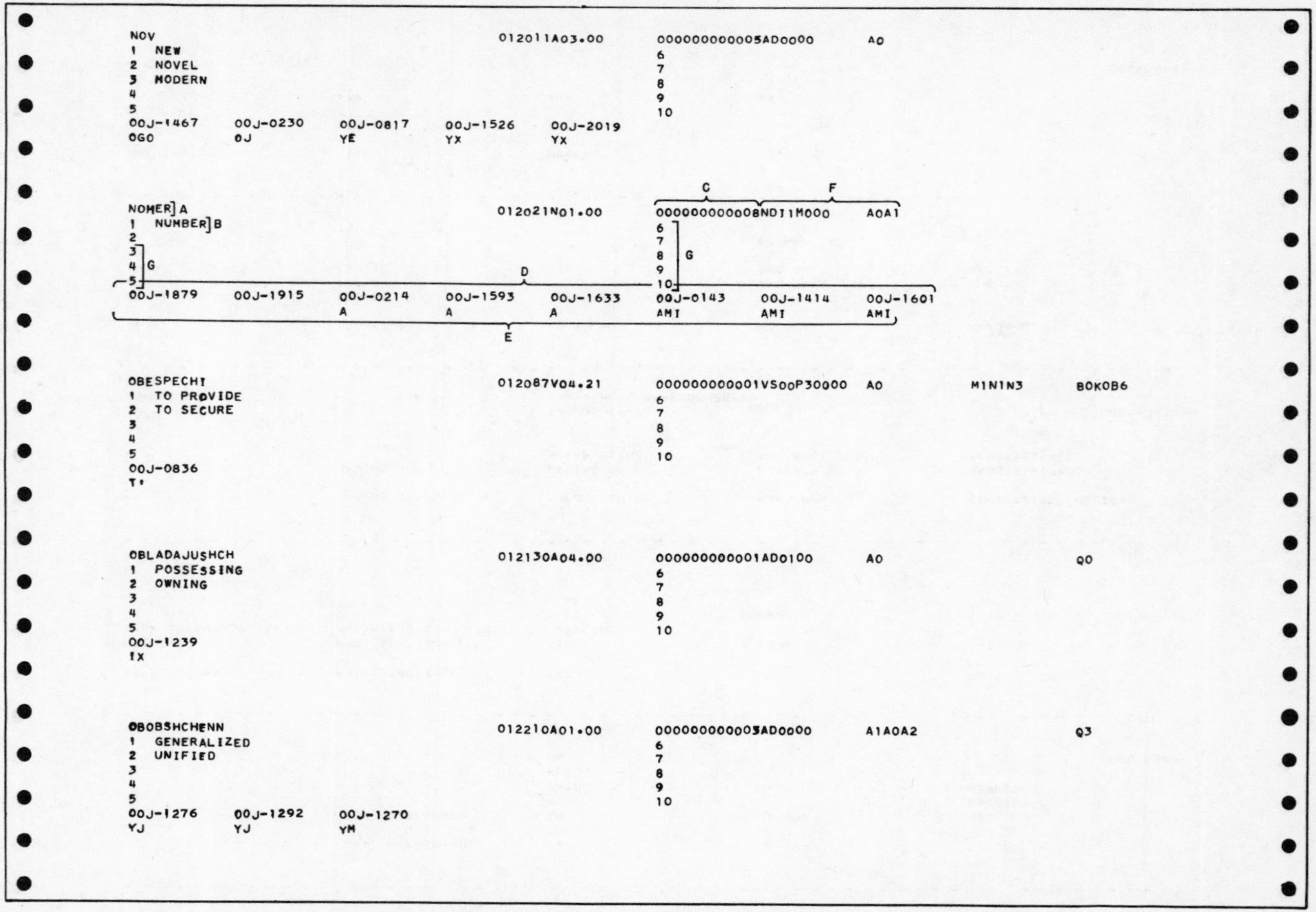

Fig. 72. Index-dictionary (Russian alphabetic order).

```
SCHITYVANI                                    ((((((      000000000008  X-LIT.     X-LIT.     X-LIT.     X-LIT.
1                                                         6
2                                                         7
3                                                         8
4                                                         9
5                                                         10
00C-0278   00C-0069   00C-0284   00C-0295   00C-0362   00C-0449   00C-0501   00C-0521
E          JA         JA         JA         JA         JA         JA         JA

SERDECHNIK                                    ((((((      000000000015  X-LIT.     X-LIT.     X-LIT.     X-LIT.
1                                                         6
2                                                         7
3                                                         8
4                                                         9
5                                                         10
00C-0153   00C-0453   00C-0496   00C-0512   00C-0036   00C-0119   00C-0258   00C-0267   00C-0317   00C-0093
A          A          A          A          AX         AX         AX         AX         E          I

00C-0571   00C-0394   00C-0437   00C-0045   00C-0529
I          I          I          OV         OV
SIGNAL                                        ((((((      000000000001  X-LIT.     X-LIT.     X-LIT.     X-LIT.
1                                                         6
2                                                         7
3                                                         8
4                                                         9
5                                                         10
00C-0156
Y

SKOROST                                       ((((((      000000000001  X-LIT.     X-LIT.     X-LIT.     X-LIT.
1                                                         6
2                                                         7
3                                                         8
4                                                         9
5                                                         10
00C-0283
'

SOZDANI                                       ((((((      000000000001  X-LIT.     X-LIT.     X-LIT.     X-LIT.
1                                                         6
2                                                         7
3                                                         8
4                                                         9
5                                                         10
00C-0341
E
```

Fig. 73. Index: missing words.

and of their application are given in the sections and figures mentioned in Fig. 74.

A table of the frequencies of digrams in a file containing about 18,000 words of technical texts by several authors is shown in Figs. 75 and 76 (Northcutt and Kilbane, 1958). Ordinary punctuation marks are not included in these figures; however, periods occurring in abbreviations and hyphens are

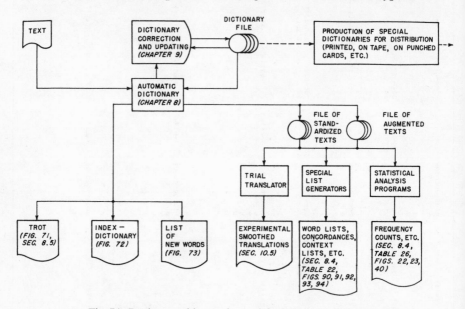

Fig. 74. Products and by-products of the automatic dictionary.

included. Single-letter frequencies derived from the data of Figs. 75 and 76 are tabulated in Fig. 77 and plotted in Fig. 23. Fragments of tables showing the distinct strings occurring in the same texts together with their frequencies are given in Table 41. Figure 78 shows a graph, based on the same sample, of string frequency against rank in the manner of Zipf (1935, 1949) and Apostel, Mandelbrot, and Morf (1957).

8.5. Immediate Applications of Word-by-Word Translations

Although trots of the kind illustrated in Fig. 71 are far from being smooth, idiomatic translations, they may nevertheless perform a number of useful functions in a fail-safe manner (Oettinger, 1954; Giuliano, 1959a). They may prove useful to students equipped with the fundamentals of Russian grammar by enabling them to work on material of immediate interest to them without

Fig. 75. Digram frequencies.

	A	B	V	G	D	E	ZH	Z	I	•	•	J	K	L	M	N	O	P
A	1	209	501	63	207	329	113	433	29	19		31	524	476	384	713	4	298
B	61	1	13		1	73	1	1	36	6			5	101	19	21	492	18
V	631	1		80	68	545		13	389	24			29	357		278	683	
G	110			3	1	115			100	34			9	53		80	730	
D	547	2	109			637			374	10			98	227	2	332	341	11
E	44	64	123	207	569	155	145		11	27			445	717	918	2091	86	105
ZH	67	5	110	8	104	496			135	2			5	81	156	133	203	4
Z	513	14	322	194	46	32			75	1			34	213	683	188	153	73
I	109	167	10	3	106	734	59	146	321	10		263	304	8	7	471	1	11
•	5	2			10	27			6				10			5		
•																		
J										1						35		
K	728		29	17	8	53		1	434	18			21	121	1	23	1051	
L	361	3	1	5		1015	33	213	741	2			21	78	2	94	653	
M	603	5	2	3		951	32	683	363	19			33		103	94	605	
N	1291			12		767		496	1929	15			54	13	5	582	1639	242
O	7	466	1223	707	918	246	289		105	14		245	219	959	818	417	58	374
P	148	8	41	22		544	1		143	11			14	115	12	7	1249	13
R	1600	28	146			1170	61		1102	33			558	535	115	152	1136	5
S	202	2	441			157	1	2	426	53		1	69	10	51	107	605	245
T	783				77	877			692						31	235	1768	13
U	14	10	32	95	918	78	54	169	10	2			91	285	102	55	1	119
F	26				30	43			123	6			2	15	10	2	147	
X	74				15	215		4	24	3				7	2	35	240	
Ts	30		2		18	131		2	336				52	9		205	3	
CH	375					729			311			1						
SH	22	18	168	2	124	193	107		107	1			6	4	222	14	20	39
SHCH	40					288		56	240	3					20	3		3
#						20				4								
Y						237						581						
•			1			11												
EH		17	83	1	26			9	2			160	12	37	19	2		
JU		1	5	1	35			51					89	145	51	429		2
JA				2	97			4	1				43	10		15		
-						140		42					9	3		128		
SPACE	375	365	1628	211	891	196	74	471	1259			1	896	289	789	1218	1228	2441

283

Fig. 76. Digram frequencies.

	R	S	T	U	F	X	TS	CH	SH	SHCH	#	Y	'	EH	JU	JA	-	SPACE
A	260	471	937	12	49	148	118	188	25	48	26	165			208	202		1767
B	182	19	100	97		36	1	2	2	37		334	1			3		21
V	147	111		77		30	23	4	16		2	83	8			41		1129
G	177			37			2	1	3							45		16
D	113	78	7	173						1					20			143
E	1081	745	920	4	25	148	14	73	42	59		10				8		1947
ZH		2	1	16				3	1									7
Z	32	578	586	108	45	350	91	461	43	1		104	10	3	70	10		161
I	141					1		1						28		740		2029
.	2	9	4	1	1				3					1		3		183
J	142	102	6	2	1							67			137	475		1080
K		105	335	167	2		56	15	1			235						335
L	1	10	1	211									855			28	2	108
M		16		188											1			1018
N		102	286	103	6	44		14	1			887	65		2	103	2	115
O	970	1158	942	9	14	15	88	245	46	35				37	14	183	11	1780
P	1459	2	3	192	7							48	1			3		55
R	6	10	45	230	4		6	66	17	2		226	21		13	247		183
S	114	258	1718	103	9	10			7			58	84			522		311
T	645	431	10	169	1	263	21	2		2		179	505		1	23		834
U		46	132	2	6	23		201	7	77				1	174	8		337
F			4	34		48						2						21
X	1	10	2	5	48							2						973
TS				6								12						33
CH			159	15									4					13
SH	3		13			448	2	43	9				30					79
SHCH													20					
#	23	51	103		28	47	2		30	2					62	8		
Y	1	99	105			84			115	1						11		758
.												1				5		608
EH	1	7	176					1	3	280					1	7		4
JU	22	36	154					77		55				1	92			216
JA	3	1	171				3	30										1643
-			3															1
SPACE	705	1956	871	497	288		82	464	79					362	3	108	1	

LETTER	COLUMN SUM	ROW SUM
A	8767	8767
B	1388	1388
V	4990	4990
G	1636	1636
D	3354	3354
E	11202	11202
ZH	971	971
Z	1900	1900
I	9823	9823
'	0	0
.	318	318
J	1283	1283
K	3652	3652
L	4868	4868
M	4522	4522
N	8166	8166
O	13102	13102
P	4016	4016
R	6474	6474
S	6658	6658
T	7794	7794
U	2514	2514
F	534	534
X	1647	1647
TS	561	561
CH	1891	1891
SH	462	462
SHCH	601	601
#	28	28
Y	2400	2400
'	1606	1606
EH	432	432
JU	798	798
JA	2783	2783
-	18	18
SPACE	17832	17832
TOTAL	138991	138991

Fig. 77. Letter frequencies.

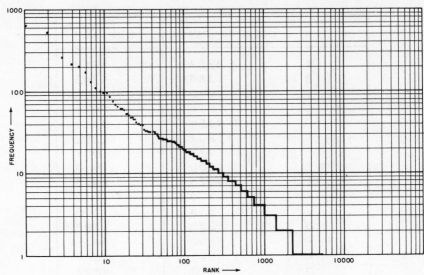

Fig. 78. Rank-frequency relation for Russian strings.

Table 41. Sample word-frequency tables.

Alphabetic		End-alphabetic		Frequency-ordered	
а	55	четных	1	в	659
абсолютная	1	нечетных	1	и	518
абсолютной	2	расчетных	1	на	270
абсолютную	1	электромагнитных	1	с	219
абсолютные	1	трансцендентных	1	при	209
абцисс	1	частотных	9	для	176
автогенераторах	2	высокочастотных	2	что	133
автогенераторов	1	частных	1	от	116
автомат	11	совместных	1	не	112
автомата	18	емкостных	1	по	100
автоматизации	1	плоскостных	4	к	100
автоматизацию	1	высокоскоростных	1	из	88
автоматическая	1	равновероятных	1	времени	78
автоматически	4	полнозначных	1	схемы	70
автоматическим	1	аналогичных	1	или	66
автоматических	2	различных	14	может	64
автоматического	1	симметричных	1	как	64
автоматов	7	несимметричных	1	быть	60
автоматы	5	беспорядочных	1	схема	55
авторов	3	остаточных	16	а	55
авторы	1	восточных	1	то	52
аддитивны	1	промежуточных	1	напряжения	49
адреса	1	обычных	1	его	49
азот	5	тысячных	3	если	47
азота	2	идеальных	1	до	47

tedious repeated reference to dictionaries. They may serve in screening texts prior to translating them in the conventional manner. Examining the trot for a whole text, an abstract, or a portion of a text may be sufficient to indicate whether, on the one hand, the text is irrelevant or unimportant and may be ignored or, on the other hand, it is pertinent and important, and deserves careful translation either in the conventional manner or by human postediting of the trot. A scientist with little or no knowledge of Russian may be able to use a trot directly for his own purposes, and a technical editor may find it useful in checking work done by a bilingual but not technically qualified translator.

The appendix to this chapter contains several pages of an edited trot for the text shown in Fig. 64. Editing operations were deliberately restricted to those it is planned to make automatic in the next stage of progress toward automatic translation. Such edited trots may therefore be used in the trial translating and formula-finding processes described in Secs. 10.5 and 10.6. For other purposes, more radical editing may of course be used. For example, word-order changes can readily be indicated by the lines and arrows. A transcript of the edited trot, made by a typist following the arrows, is also shown in the appendix, together with a page of a translation of the same text prepared by a conventional translator for the American Institute of Physics (1957).

Experiments intended to evaluate the immediate usefulness of trots are still under way. Definitive results are not available at the time of writing, but some preliminary conclusions are warranted. An individual with a scientific or technical background, a rudimentary knowledge of Russian, and, above all, a desire to read technical material in his own field, can understand and translate texts much more rapidly with a trot than without it. Similar individuals without any knowledge of Russian can produce passable translations. They require more time, and some sentences are usually left translated only partially or not at all. Significantly, the chief complaint of such individuals was that 5 or 10 percent of words in the text were not yet in the dictionary file and had to be located in a standard dictionary. Otherwise, they found trots quite helpful. Criticism of the large number of alternative English correspondents given for a single Russian word in some instances was much more frequent in an earlier experiment (Oettinger, 1954), possibly because the correspondents were then displayed in a format more awkward than the present one.

Trots were found only marginally useful by individuals with a good knowledge of literary Russian, but with little or no technical background. In such cases, however, a technical editor without knowledge of Russian could with the aid of the trot improve on the work of the original posteditor.

Expert technical translators seem to be hampered rather than assisted by the trot in its present format. Such individuals, who are unfortunately scarce, can usually rapidly write or dictate a fairly good translation while reading the Russian original. Individuals without technical qualifications, or incapable of expressing themselves fluently in English, seem unable to produce good translations, with or without trots.

More precise evaluations will require not only the accumulation of more data than are now available, but also the development of sound techniques of objective evaluation. As Miller and Beebe-Center (1956) have pointed out, such techniques are difficult to devise.

Edited Trot

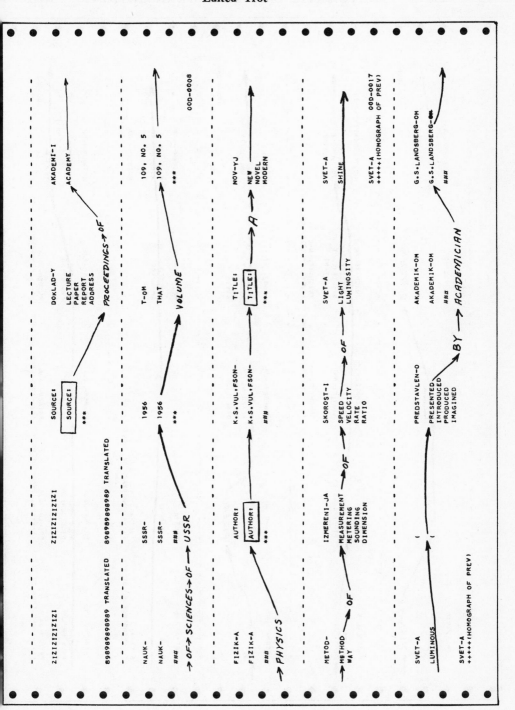

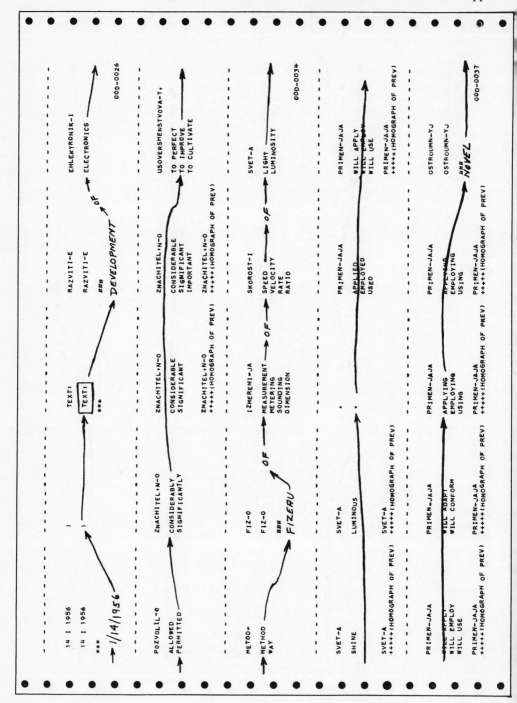

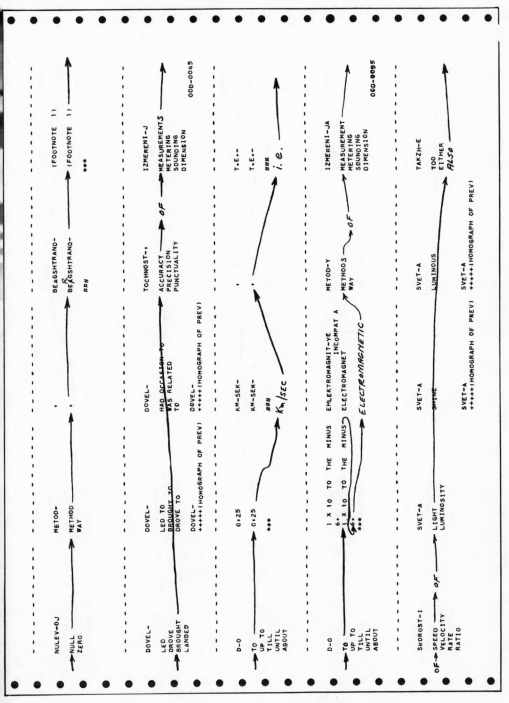

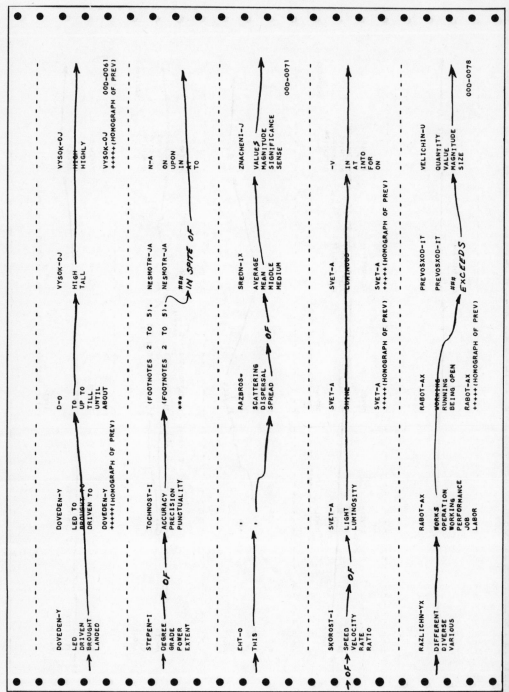

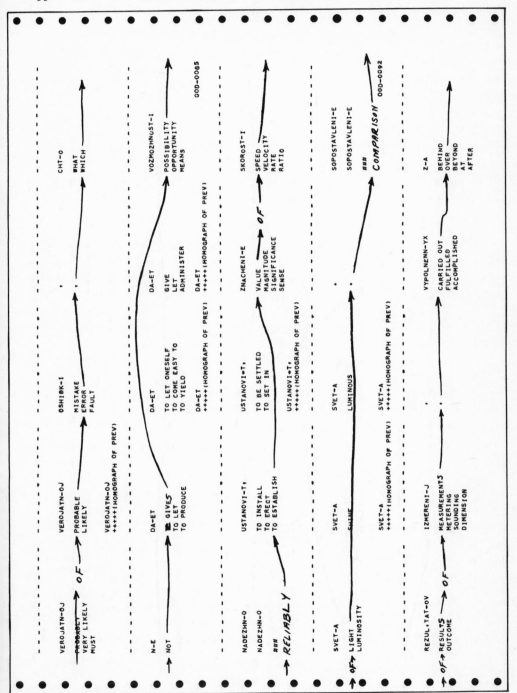

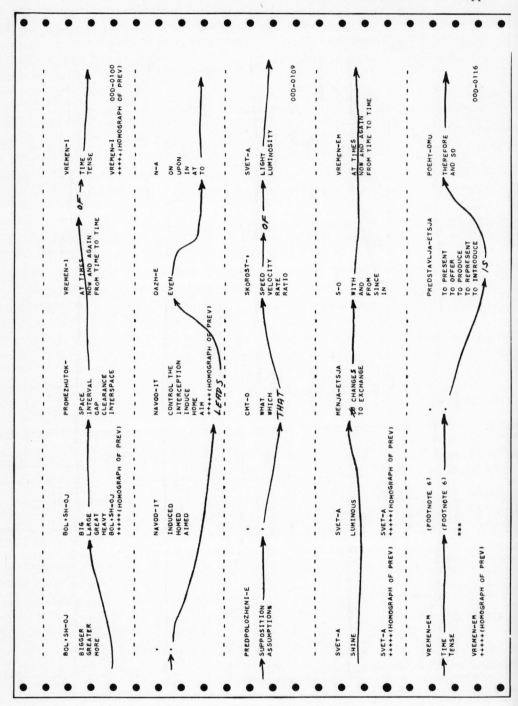

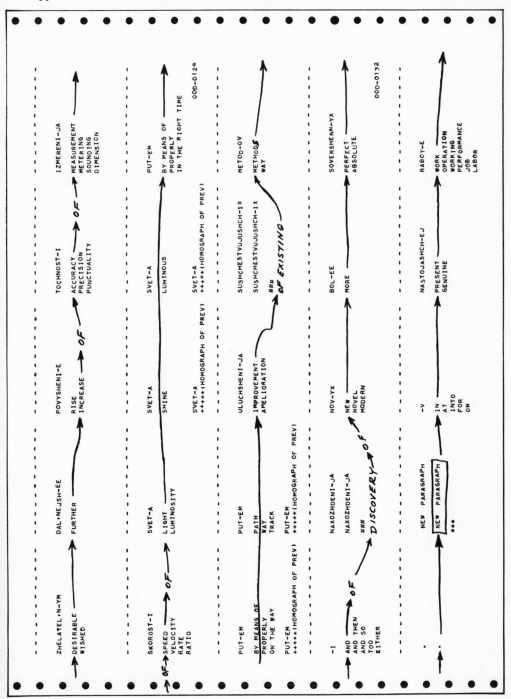

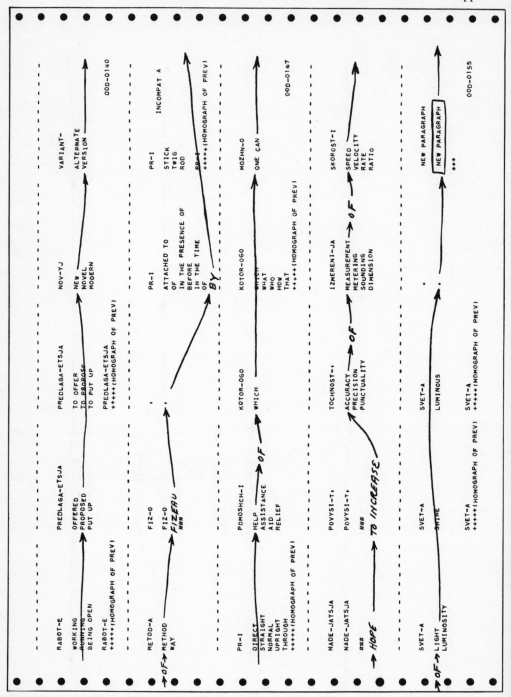

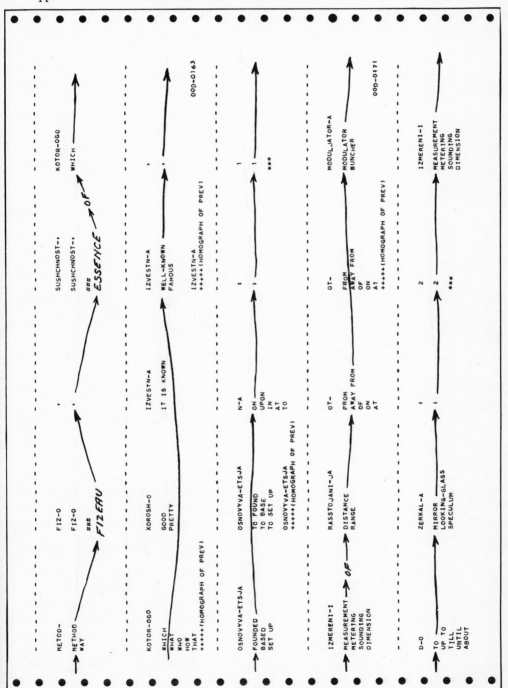

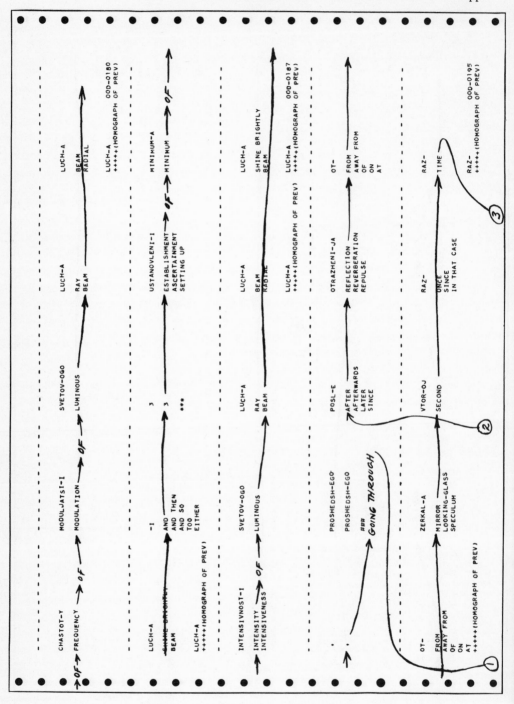

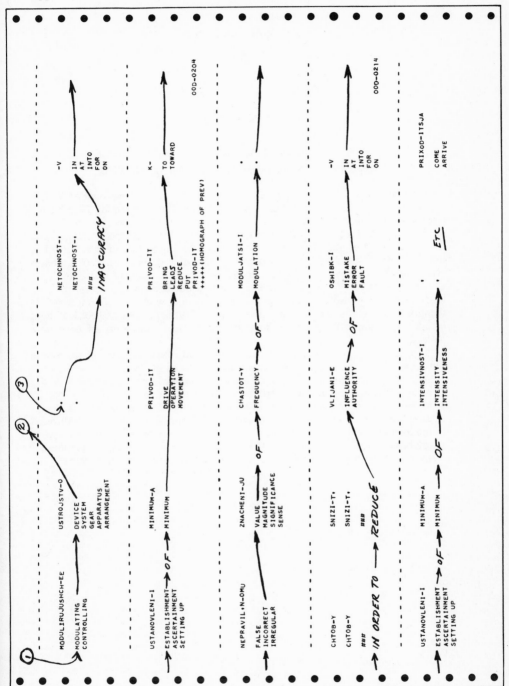

Transcription of Edited Trot

Proceedings of Academy of Sciences USSR
Volume 109, No. 5, Physics, 1956

K. S. Vul'fson

A New Method of Measurement of Speed of Light

(presented by Academician G. S. Landsberg 1/14/1956)

Development of electronics permitted considerably to improve method Fizeau of measurement of speed of light. Applying novel null method, Bergshtrand (footnote 1) brought accuracy of measurements to 0.25 km/sec, i.e. to 1×10 to the minus 6. Electromagnetic methods of measurement of speed of light also brought to high degree of accuracy (footnotes 2 to 5). In spite of this, spread of mean values of speed of light in different works exceeds magnitude of probable error, which not gives possibility reliably to establish value of speed of light. Comparison of results of measurements, carried out over large interval of time, leads even to supposition, that speed of light changes with time (footnote 6). Is therefore desirable further increase of accuracy of measurement of speed of light by means of improvement of existing methods and of discovery of new more perfect.

In present work proposed new version of method Fizeau, by help of which one can hope to increase accuracy of measurement of speed of light.

Method Fizeau, essence of which well-known, founded on:

1. measurement of distance away from modulator to mirror;
2. measurement of frequency of modulation of luminous beam and
3. establishment of minimum of intensity of luminous beam, going through modulating device after reflection from mirror second time. Inaccuracy in establishment of minimum leads to incorrect value of frequency of modulation. In order to reduce influence of error in establishment of minimum of intensity . . .

Conventional Translation

A NEW METHOD OF MEASURING THE VELOCITY OF LIGHT

K. S. Vul'fson

(Presented by Academician G. S. Landsberg January 14, 1956)

The development of electronics has brought about a considerable improvement of Fizeau's method of measuring the velocity of light. Using an ingenious null method Bergstrand [1] increased the precision of the measurement to 0.25 km/sec, i.e. to $1 \cdot 10^{-6}$. Electromagnetic methods of measuring the velocity of light have also reached a high degree of accuracy [2-5]. Nevertheless, the spread of the mean values in various articles exceeds the probable error, so that it is impossible to give a reliable value for the velocity of light. A comparison of measurements made over a long period even suggests that the velocity of light changes in time [6]. It is thus desirable to further increase the precision of light velocity measurements by perfecting existing methods and finding new and more improved methods.

The present article proposes a new variant of Fizeau's method, by which it is hoped to increase the accuracy with which the velocity of light is measured.

Fizeau's method, the essentials of which are well known, is based on: 1) the measurement of the distance from a modulator to a mirror; 2) the measurement of the modulation frequency of a light beam; 3) the fixing of the intensity minimum of the light beam during its second passage through the modulator after reflection from the mirror. Inaccuracy in ascertaining the minimum leads to an incorrect value of the modulation frequency. In order to reduce the effect of the error concerning the minimum intensity long distances must be used, but the methods of measuring long distances do not insure the same high relative precision which can be attained for short distances. Bergstrand's null method considerably reduced the error in locating of the minimum and enabled him to obtain amazing results.

In the method proposed below item 3), which is the principal source of error, is eliminated by the use of an automatic arrangement. In addition, the new method does not require long distances, so that the measurements can be made with very great relative accuracy.

The method is essentially as follows: The light source or the modulating device sends out a short light pulse whose duration is of the order 10^{-7} to 10^{-8} sec. After reflection from the mirror the beam enters a photoelectric receiver. The electrical pulse which is generated in the latter triggers the sending of a second light pulse, through the use of a special circuit. Thus, after emission of the first pulse the system begins to generate light pulses whose frequency of repetition is determined by the distance from the modulator or light source to the mirror and from the mirror to the light receiver as well as by the signal delay in the apparatus. By determining the period, or what amounts to the same, the repetition rate of the light and electrical pulses which are generated for a few easily distinguishable positions of the mirror, it is possible to establish both the velocity of the light and the signal delay in the apparatus.

Since the repetition rate of the light pulses can be determined very precisely the shift of the mirror does not have to be very large, but of the order of a few meters; therefore it can be measured with extreme accuracy by an interferometer. The mirror can be placed in a vacuum chamber in order to eliminate a correction for the refractive index of air. It is expedient to use mirror displacements of such magnitude that the repetition rate of the light pulses will coincide with the harmonics of the standard quartz oscillator which is used in the frequency measurements. The accuracy of the frequency determination will then be especially high.

The repetition rates of the light flashes can be compared with a standard generator by the use of Lissajous figures. The accuracy of such a method of comparison is limited by the average stability of both the standard

CHAPTER 9

PROBLEMS IN DICTIONARY COMPILATION AND OPERATION

9.1. Paradigm Homography

The problem of homography, although introduced in Chapter 5, was ignored in Chapters 7 and 8 to avoid repeated digressions from the outline of techniques for dictionary compilation and operation. In this chapter the problem is considered in some detail, and certain refinements are made on the descriptions given in Chapters 7 and 8.

It is convenient to distinguish two aspects of homography, namely, internal homography within a given paradigm and homography among several paradigms. The existence of internal homography is the motive for introducing, as was done in Sec. 5.2, a distinction between paradigms and reduced paradigms. The use of the representation "студента" for both $студента_{gs}$* and $студента_{as}$* exemplifies internal homography, that is, the homographic use of one representation for more than one inflectional variant of a single word.

The use of the reduced paradigm as the basis for dictionary compilation effectively removes the problem of resolving internal homography from lexicography to syntax, where it belongs. Since people seem capable of determining which of $студента_{gs}$* and $студента_{as}$* the string "студента" is meant to represent in a particular context, the distinction between paradigm and reduced paradigm is meaningful, and there is hope of finding algorithms for the automatic resolution of internal homography. This profession of faith is the best one can do at this time, because no reliable set of algorithms is available as yet. Some comments on the direction of research on the problem are made in Sec. 10.4.

Homography among several paradigms is a more serious problem, complicated by its intimate relation to the problem of polysemanticism. It should be noted that we must speak of homography among paradigms, specifically among reduced paradigms, rather than among words. To say that a word is homographic to another is, strictly speaking, contradictory, for *two* words are distinct and that's that; if such a statement is interpreted metaphorically as implying that the representations of the words are identical,

302

we are really speaking about a property of reduced paradigms. Finally, in speaking of words, we are likely to use their classical canonical representation, and that is misleading. For example, *радиотехник** (radio technician) and *радиотехника** (radio engineering) have distinct classical canonical forms, but the genitive and accusative singular of the first, and the nominative

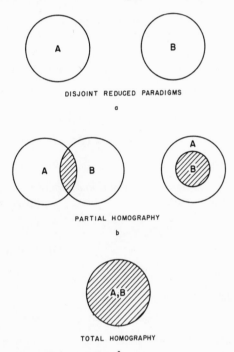

DISJOINT REDUCED PARADIGMS

a

PARTIAL HOMOGRAPHY

b

TOTAL HOMOGRAPHY

c

Fig. 79. Types of homography: (*a*) disjoint reduced paradigms; (*b*) partial homography; (*c*) total homography.

singular of the second share the representation "радиотехника". In fact, Table 42 shows that the reduced paradigms of these two words have eight forms in common.

Every pair of reduced paradigms will be related in one of the three ways illustrated in Fig. 79. First, two reduced paradigms may be disjoint, that is, have no members in common. When this is the case, a problem of homography obviously does not exist. Second, the two reduced paradigms may have some but not all members in common. This situation may be described as *partial homography*, and is illustrated in Table 42. The intersection of the reduced paradigms of *радиотехник** and *радиотехника** has eight members, and only two members of one paradigm and one of the other are not shared.

Finally, two reduced paradigms may have precisely the same membership, a case of *total homography*.

The shaded areas in Fig. 79, representing the intersections of reduced paradigms, also represent the cases where the problem of homography versus polysemanticism arises. The question is one of identification of

Table 42. Partial homography.

Reduced paradigm of радиотехник*

(a)	(1)	"радиотехник"	(e)	(6)	"радиотехники"
(b)	(2)	"радиотехника"		(7)	"радиотехников"
(c)	(3)	"радиотехнику"	(f)	(8)	"радиотехникам"
	(4)	"радиотехником"	(g)	(9)	"радиотехниками"
(d)	(5)	"радиотехнике"	(h)	(10)	"радиотехниках"

Reduced paradigm of радиотехника*

(b)	(1)	"радиотехника"	(a)	(6)	"радиотехник"
(e)	(2)	"радиотехники"	(f)	(7)	"радиотехникам"
(d)	(3)	"радиотехнике"	(g)	(8)	"радиотехниками"
(c)	(4)	"радиотехнику"	(h)	(9)	"радиотехниках"
	(5)	"радиотехникой"			

indiscernibles (Sec. 2.4). Total homography is a nonsensical concept if reduced paradigms alone are considered. There is really no point in distinguishing between two reduced paradigms with "identical" membership. If, however, the properties of the *words* associated with reduced paradigms are invoked, and are not themselves "identical" relative to some technique of resolution, a case can be made for speaking of homography rather than polysemanticism.

Total homography subsumes the theoretically trivial case of duplication; this case has some practical significance because duplicate entries can and do survive the best compilation process; it can be disposed of by stipulating that duplicates be purged as soon as detected. A less trivial, but quite rare instance of total homography may arise if, say, a verb V^* and a noun N^* have identical reduced paradigms. There is hope for distinguishing "V" in "the boy V to the market" as belonging to V^* and "V" in "V are healthful" as belonging to N^*, by some algorithm operating on the contexts of "V".

There is good reason, consequently, for having two dictionary entries with key "V": one with verbal classification marks and correspondents, the other with the same key but with nominal classification marks and correspondents. Relative to the techniques of resolution that may conceivably be based on these data, two distinct words V^* and N^* may therefore be attributed to the "identical" reduced paradigms to which "V" has been assumed to belong.

On the other hand, two nouns N_1^* and N_2^* may be considered distinct and listed separately in a conventional dictionary, on the ground that linguistic history demonstrates that these nouns once had distinct reduced paradigms, which became identical through some accident of phonological change or of spelling reform. The meanings of N_1^* and N_2^* may be quite distinct, but techniques for selecting the appropriate one are much more difficult to envisage than techniques for telling a verb from a noun. Because this problem of semantic resolution also arises on a much larger scale with words having merely different "shades" of meaning, N_1^* and N_2^* are perhaps best regarded as indiscernible, therefore identical, and consequently best encompassed by a single polysemantic entry in the automatic dictionary file.

With partial homography, one course of action is to supply a distinct entry for every word in the set of homographs. For example, in a paradigm dictionary (Sec. 5.3) the entry with key "радиотехника" for *радиотехник** could be supplied with the mark "GS;AS", indicating that the key string represents either the genitive singular or the accusative singular. The entry with key "радиотехника" for *радиотехника**, on the other hand, would be supplied with the mark "NS", indicating that the key string represents the nominative singular. For a given textual instance of the key string, there is a good likelihood that an appropriate algorithm will enable distinguishing whether a nominative on the one hand, or a genitive or accusative on the other hand, is consistent with other determinate properties of the context. On the other hand, there is no fundamental reason why a single entry with key "радиотехника" containing information describing both words should not be usable. Which alternative is chosen depends on criteria of conceptual elegance and practical advantage.

9.2. Stem Homography

In a canonical dictionary (Sec. 5.3) with stems or any other canonical entry keys, the question of homography is superficially complicated by certain operating problems. Once these problems are resolved, the description given in Sec. 9.1 applies.

Clearly, if two reduced paradigms are partially or totally homographic, the sets of distinct stems obtained from these paradigms by the techniques of Sec. 7.7 or similar ones will also overlap. The degree of overlap is determined in part by the elements in the intersection of the reduced paradigms, since distinct canonical stems obtained from these elements will naturally be in the intersection of the sets of distinct canonical stems.

Often, however, the degree of homography of canonical stems will be greater than that of the reduced paradigms. Consider for example two hypothetical paradigms built by adjoining distinct sets of affixes to a single set of canonical stems. The reduced paradigms will be completely disjoint, but the sets of stems will be totally homographic. The factoring of elements of both reduced paradigms of Table 42 produces, in every case, a stem "радиотехник", which is the unique member of the set of canonical stems for each paradigm. In this case, therefore, the reduced paradigms are partially homographic, but the sets of stems are totally homographic.

The homography considered in Sec. 9.1 is called *paradigm homography**. The overlapping of the canonical stems produced by inverse inflection creates a condition that is called *stem homography**. Stem homography in the presence of paradigm homography may be described as *essential*, since it is a property of the language, and stem homography in the absence of paradigm homography as *nonessential*, since it is an artifact which can readily be eliminated. Obviously, no question of stem homography arises when a paradigm dictionary is used. In a canonical dictionary, the question arises because stems or other canonical forms are used as entry keys without immediate regard for the distinctive affixes which can always be used to eliminate all nonessential homography; once means are provided for using affixes as distinctive marks, the question usually disappears in this case also (Sec. 10.1).

In the compilation of the dictionary file for the Harvard Automatic Dictionary, instances of both essential and nonessential stem homography are reflected in the provision of multiple entries with identical stem keys. For example, *радиотехник** and *радиотехника** are entered into the compilation process as distinct words. Consequently, distinct serial numbers are assigned to the classical canonical strings of these words by the *word-edit* operation (Sec. 7.6). After direct and inverse inflection, all members of the reduced paradigms of both words will contain the string "радиотехник". The condensation operation (Sec. 7.7) is designed to eliminate duplicate stems only within sets whose members are identified by the same serial number. Two instances of the stem "радиотехник" will therefore be retained, one with the serial number of *радиотехник**, the other with the serial number of *радиотехника**. From these serial numbers, and Lists 6 and 6′ (Figs. 52 and 53), it is easy to determine what English correspondents and grammatical marks must be attached to each of these homographic canonical stems to produce appropriate entries for the dictionary file.

At the present stage of development of the Harvard Automatic Dictionary, no general provisions are made for the resolution of either paradigm or stem homography. Paradigm homography can be resolved only by advanced techniques for whose development the existence of an operating automatic dictionary is a highly convenient prerequisite. Stem homography is most conveniently resolved in conjunction with the more general problem of using the information supplied by the affixes with which stems occur in texts; this problem is discussed briefly in Sec. 10.1. For the present, as shown in Fig. 71 and the appendix of Chapter 8, whenever two or more entries with the same key are present in the dictionary file, all are transferred into the alphabetic subdictionary and retained throughout the processes leading to the preparation of the augmented text and of the edited word-by-word translation, where their presence is emphasized by special marks; this phenomenon should be regarded as a temporary nuisance, not as a fundamental problem.

From the classical view, nonessential stem homography is relatively rare. There are few ordinary examples where identical stems are hidden in the elements of disjoint reduced paradigms. One major source of such homographs is the class of verbs whose representations are terminated by the affixes of order zero "ся" and "сь" (Sec. 5.4), but this is a special case. The situation, illustrated by "радиотехник", where partial paradigm homography becomes total essential stem homography is much more frequent. By far the most significant source of stem homographs is created by certain characteristics of the inverse-inflection algorithm defined in Chapter 5. These characteristics are analyzed in some detail in the following section.

9.3. The Definition of Affixes

The definition of Russian inflectional affixes given in Sec. 5.4 was concluded with the statement that the affixes as defined in Table 24 closely coincide with the conventional desinences described in textbooks, although some exceptions are evident.

Coincidence with the rather nebulous and often inconsistent conventional definitions is, of course, only one criterion for evaluating any system of inverse inflection. Complete coincidence cannot be expected, especially since the conventional definitions often rely on phonematic criteria not applicable in a context where only graphic criteria are admissible. The desire for some degree of coincidence stems partly from the belief that many results of school linguistics retain their validity, or at least their heuristic value, even from a nominalistic point of view, in spite of their having been obtained by techniques other than those advocated by pure structuralists. Moreover, the continuing prevalence of the Platonic view and of its terminology insures that results consistent with this view will have at least the merit of familiarity, which is of tremendous help in the development of an automatic system, particularly whenever manual intervention in the operation of such a system is necessary for "debugging" purposes.

One reason for using a canonical dictionary is the expected reduction of the size of the file for such a dictionary over the size of the file for a paradigm dictionary; one criterion for evaluating the definition of affixes is therefore the ratio of the number of stems representing a set of words to the number of all members of the reduced paradigms of all these words. In addition, since the factoring of members of a reduced paradigm by design maps these distinct members onto as small as possible a set of stems, the affixes must be used to distinguish members of a reduced paradigm one from another, and the interpretation of the affixes is expected also to yield significant information about the syntactic role of strings occurring in texts; another criterion for evaluating the definition of affixes is therefore the ease, elegance, and accuracy with which this distinction and interpretation can be accomplished.

A. *Invariables*. The treatment of the representations of invariable words is one instance of lack of coincidence between desinences and the affixes defined in Chapter 5. The inverse-inflection algorithm is necessarily applied prior to dictionary look-up. Invariable forms, although distinguished by the class mark *I* in the dictionary file, are therefore treated by the inverse-inflection algorithm precisely like any other forms. Whenever one of the combinations of terminal characters listed in Table 24 occurs in a text string, the string is factored into a stem and one or two affixes.

Table 43 shows how some invariable words are factored. For example, "здесь" is factored into a stem "зд", a first-order affix "е" and a zeroth-order affix "сь". The one-character strings "a", "в", "и", and "o" are each factored into a null stem and a one-character affix. The lack of coincidence with classical definitions is not serious in itself, once persons encountering these forms in the course of their work have become accustomed to the mode of operation of the inverse-inflection algorithm.

The appearance of such factored forms in word-by-word translations (for instance, "в" and "и" in Fig. 71) is somewhat disconcerting. In the future, a slightly more sophisticated editing program can easily join the factored stems and affixes of invariables together again, since the class mark *I* is available in the augmented text. This phenomenon is therefore merely a temporary nuisance. The interpretation of the affixes of invariables presents no serious problem: any interpreting algorithm must be executed *after* dictionary look-up and can therefore be designed to take no action on an affix occurring with a stem whose entry has the class mark *I* beyond checking that it indeed belongs to the representation of the invariable word (Sec. 10.1).

Table 43. Factored invariables.

(1)	"-a"	(9)	"мн-ого"
(2)	"-в"	(10)	"н-а"
(3)	"-и"	(11)	"н-е"
(4)	"-о"	(12)	"откуд-а"
(5)	"благодар-я"	(13)	"очен-ь"
(6)	"зд-е-сь"	(14)	"поэт-ому"
(7)	"л-ишь"	(15)	"пр-и"
(8)	"межд-у"	(16)	"ради-о"

The factoring of invariables also contributes to the creation of stem homographs. For example, "a", "в", "и", and "о", although distinct, have identical stems "#", and "на" and "не" both have the stem "н". This problem is analyzed further in part C of this section, and in Sec. 9.6.

B. *Multiple Stems*. The definitions of affixes given in Sec. 5.4 were designed to minimize the number of instances where more than one canonical stem is required to represent all members of a reduced paradigm. It was shown that for certain words, such as *окно**, a multiplicity of stems is unavoidable because of morphological variation within the classical stem, unless the variation were incorporated into the affixes, in which case both the number of distinct affixes and the problem of stem homography (part C) would grow beyond reasonable bounds. An additional number of multiple stems is created by virtue of the rigid dependence of the inverse-inflection algorithm on the morphological characteristics of strings. Table 44 shows several strings factored by the inverse-inflection algorithm in a way that not only differs from the classical way, but also creates more than one stem to represent the reduced paradigms to which these strings belong.

For example, all members of the reduced paradigm of *аппарат** are factored into "аппарат" plus an affix, except for "аппарат", which is factored as shown in Table 44. Likewise, *расчет** yields the two stems "расч" and "расчет". In most instances, multiple stems are created when a string of characters that serves as a desinence for one class of words occurs fortuitously in the representations of words of some other class. Thus, "ат", "ут", "ть", and "ет", which are primarily verbal desinences and affixes happen to occur at the end of some representations of the nouns *аппарат**, *минута**, *область**, and *расчет**.

Table 44. Multiple stems.

(1)	"аппар-ат"	(5)	"полаг-ая"
	"аппарат-а"		"полага-ю"
(2)	"мин-ут"	(6)	"при-ем"
	"минут-а"		"прием-а"
(3)	"нан-о-ся"	(7)	"радиопри-ему"
	"нанос-ят"		"радиоприем-а"
(4)	"облас-ть"	(8)	"расч-ет"
	"област-и"		"расчет-а"

In some instances, a slight modification of the definition of affixes could eliminate the offending cases. For example, when "ат" occurs in a verbal string, it is usually preceded by "ж", "ч", "ш", or "щ". The appropriate modification of Definition 33 in Table 24*b* would eliminate undesirable factoring in the representations of such words as *аппарат**, *кандидат**, *солдат**, and so forth. Suitable modifications cannot be made in every case, so that the generation of some multiple stems is inevitable.

In nouns and adjectives the generation of multiple stems is sufficiently rare to give little cause for concern, since it causes no substantial increase in the size of the dictionary file. The worst offenders in this respect are verbs. Figure 57 shows that the verb *писать** generates seven distinct stems and an equal number of entries in the dictionary file. Of these, four account for derived participles, and three for the other verbal forms.

Other verbs, without morphological change in their classical stems, will have one or two stems less. It is conceivable that a drastic redefinition of affixes could enable still more substantial reductions. For example, a single stem "пи" would suffice for all strings generated from *писать**, although it is not clear that referring all forms to a single entry would be desirable. It is also obvious that such a definition would have two major corollaries:

first, a vast increase in the incidence of stem homography, and second, a substantial increase in the number and variety of affixes. In the limit, all strings could be factored into a null stem plus an affix consisting of all characters in the string. In this absurd situation, the stem dictionary file would be reduced to nothing, but the methods used for distinguishing homographs and interpreting affixes would be tantamount to using a paradigm dictionary containing "affixes" as keys.

Some reduction in the number of verb stems might be obtained by factoring not only affixes but also such strings as "л", "ла", "нный", "нного", and other infixes or even prefixes. It is not likely, however, that this reduction would be substantial enough to offset the resulting increase in stem homography and in the problems of affix interpretation, especially since many verb forms present so many special problems that it is desirable to have them referred to separate entries. The extension of the set of affixes may be valuable as a technique for obtaining memory addresses of entries in certain look-up systems (Rhodes, 1959; Oettinger, 1953).

The production of multiple stems is, as illustrated in Table 44, associated with the factoring of affixes that are not desinences, and presents problems of interpretation akin to those associated with the factoring of invariables. These problems are, however, chiefly technical. One method of solution (Sherry, 1959), described in slightly greater detail in Sec. 10.1, may be sketched here. Consider, for example, the affix "ат". Given an entry marked "V" in the organized word (Sec. 6.5), the affix "ат" is equivalent to the desinence "ат" which normally indicates the third person plural. If, on the other hand, the first character of the organized word is "N", then the affix "ат" may be treated as equivalent to the null desinence "#" and given the corresponding interpretation. In some cases, the class (for instance, N1) of the word must be taken into account to enable a precise interpretation. In some pernicious but fortunately rare cases (such as "д-ам", "дам-ам", from *дама**) a special mark must be put in the offending entry (with key "д") to signal the need for a special interpretation.

C. *Stem Homography*. As indicated at the end of Sec. 9.2, certain characteristics of the definition of affixes are a source of stem homographs. These characteristics are those leading to the factoring of invariables and the creation of multiple stems. Consider, for example, the words (1) *при**, (2) *прут**, (3) *прямая**, (4) *прямой**, of which the first is a preposition, the second a masculine noun, the third a feminine noun, and the fourth an adjective. The relations of paradigm homography among these four words are represented in Fig. 80a, which shows that *прямая** and *прямой** are partial homographs, but that otherwise all words are disjoint.

Unfortunately, *при**, *прут**, and *прямой** share the stem "пр". This is a consequence of the accidental factoring of an invariable (*при**, Table 43), of the creation of the multiple stems "пр" and "прут" for *прут**, and of the existence of the predicative form "прям" of *прямой**, with the consequent multiple stems "пр" from "прям", and "прям" from all other members of the reduced paradigm. Since *прямая** has no representation "прям" it does not lead to the homographic stem "пр".

Stem homographs created in this peculiar way may be resolved just as the stem homographs of the "радиотехник" variety are. For example, the

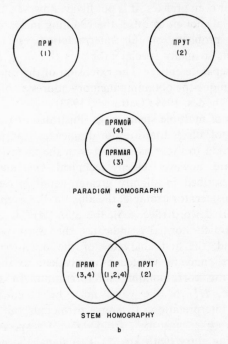

Fig. 80. Homography relations among *при**, *прут**, *прямая**, and *прямой**: (*a*) paradigm homography; (*b*) stem homography.

algorithm responsible for the interpretation of affixes can readily be designed so that, when "пр-и" occurs, the affix "и" will be declared incompatible with all entries with key "пр" except that for *при**. Once this is done, the whole problem of stem homography can be hidden in the inner recesses of the translation algorithm, and, like the spurious factoring of invariables, need never impinge on the consciousness of casual users of a translating machine.

9.4. The Detection of Homographs; Missing Homographs

Once a dictionary file has been compiled and stored on magnetic tape or some equivalent medium, a list of the homographs present in the file can easily be produced (Bossert, 1958). A program capable of scanning the dictionary file and of recording all members of every group of entries found to have the same stem key is all that is required. A portion of the output produced by such a program is shown in Fig. 81.

The list of homographs may be scanned for identical entries resulting from the accidental inclusion of the same word in more than one group of original or updated entries. Once duplicates have been removed, the list displays all sets of homographs present in the dictionary file. As might be expected, many of these homographs belong to certain well-defined classes. Several of the most significant classes of homographs are discussed in Sec. 9.7.

It is clear that the list of homographs displays only those homographs actually present in the dictionary. It is possible, and quite probable, that many words in a given dictionary file have homographs that are used in the language but happen not to have been included in the file. Moreover, the coining of new words may also result in the creation of words homographic to certain dictionary entries. Words that are homographic to dictionary entries but are not themselves included in the dictionary file are called *missing homographs*.

The occurrence of a missing homograph in a text may lead to a nonsafe failure in the operation of the dictionary. The stem factored from a representation of a missing homograph will be referred to an existing entry. A warning that the English correspondents and grammatical marks in the existing entry do not apply to the textual form is usually not possible. It may happen, of course, that the affix occurring in the text will be incompatible with the stem given in the dictionary, which may therefore be rejected; the text string may then be treated as if no correspondent had been found (Sec. 10.1). The correspondents in the dictionary entry may have a meaning so remote from that intended in the context that the error will be obvious either to the ultimate reader of the translation or to a sophisticated program for semantic analysis, if such is possible. Occasions will remain, however, on which the failure cannot be readily detected.

There is no way, during compilation or use of either conventional or automatic dictionary files, of obtaining systematic foreknowledge about missing homographs. The best that can be done is to record all failures that are detected, and to speed the offending words into the dictionary file.

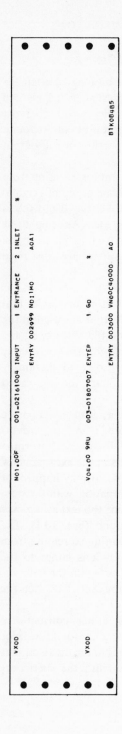

Fig. 81. A pair of stem homographs.

Fig. 82. Set of homographic entries and condensed entry.

9.5. Homograph Condensation

The long-term solution of the problems raised by stem homographs and especially by paradigm homographs must be based on extensive studies as yet only in initial stages. As an intermediate expedient, it is possible to regard every instance of homography as one of polysemanticism and lump all English correspondents in a set of homographs into a single entry.

Writing a program capable of automatically condensing every set of homographs into a single entry presents no serious difficulties. Naturally, no significant grammatical marks can be included in such an entry, as illustrated in Fig. 82, which shows a set of homographs and the resultant composite entry. This expedient is of some value only while nothing better than crude word-by-word translations like that illustrated in Fig. 71 can be produced.

9.6. "Short" Words

It has been pointed out, in Sec. 9.3, part B, that the number of entries in a stem canonical dictionary can be reduced directly as the length and variety of affixes is increased. Any advantages of such a reduction must be balanced against the disadvantage of the increase in the number of stem homographs which also ensues.

We have seen that, even with the relatively small number of short affixes defined in Chapter 5, a fairly serious problem of stem homography arises. While the decision was made to consider this problem as a part of the more general problem of affix interpretation, it proved to be so acute in certain instances that an interim solution had to be found to cover these instances.

As might be expected, words represented by short strings ("short words") present the most pressing problems of stem homography, since little or even nothing is left of their representations when affixes are removed, and since the likelihood of finding identical stems increases as the length of the stems decreases. For example, Table 43 ((1)-(4)) shows how several highly frequent words, each represented by one character, each yield a null stem to which the character is an affix. These words happen to be invariable, but the same problem arises for variable words.

To prevent the creation of an excessive number of stem homographs by short words, certain minor modifications were made to the condensation, sort, and look-up programs described in Chapters 7 and 8. In the condensation operation (Sec. 7.7) no item having a canonical stem of one character or less is eliminated. In the sorting of new groups to be included in the dictionary (Sec. 7.9) or of stems obtained from texts prior to look-up (Sec. 8.3), the stem is used as a major key, but items with identical canonical stems are arranged in alphabetic order of their affixes. In the look-up pass (Sec. 8.3), an identity match between stems is accepted as final only if the stems are longer than one character. Otherwise, an identity match is required between affixes also.

In this fashion, stem homography due to short words is avoided without appreciable increase in the size of the dictionary file. The operation of this expedient is illustrated by the piece of index-dictionary shown in Fig. 83. Stems having two or more characters are sufficiently distinct not to warrant this special treatment and, concomitantly, their number grows so rapidly with length that an undue increase in the size of the dictionary file would result.

The treatment of short words represents a concession toward a paradigm dictionary. As pointed out in Sec. 5.3, many compromises between pure canonical and pure paradigm dictionaries are possible, of which this is one instance.

9.7. Important Classes of Homographs

Certain classes of homographs stand out because of their large membership, or because their resolution requires the solution of difficult problems. Cases of total homography whose resolution requires semantic analysis will not be considered here, not because they are unimportant, but because, at the time of writing, extremely little can be said about effective automatic methods of resolution.

A. *Verbs in "ся" or "сь"*. Verbs whose representations are terminated by "ся" or "сь" create an important class of stem homographs. These verbs may be described in terms of the following three categories:

(a) There exists a verb with representations having the same stems of order zero as the given verb, but terminating in the null affix ("#") of order zero. The meanings of these two verbs are similar, but the verb with "ся" or "сь" may be (1) reflexive (for example, *одеваться**), (2) reciprocal (for example, *встречаться**), (3) passive (for example, *использоваться**).

(b) There exists a verb with representations having the same stems of order zero as the given verb, but terminating in the null affix ("#") of order zero. The meanings of these two verbs are significantly different (for example, *оказываться**).

(c) There exists no verb with representations having the same stems of order zero as the given verb (for example, *бояться**).

For verbs of category (c), there is no problem of homography. Verbs of

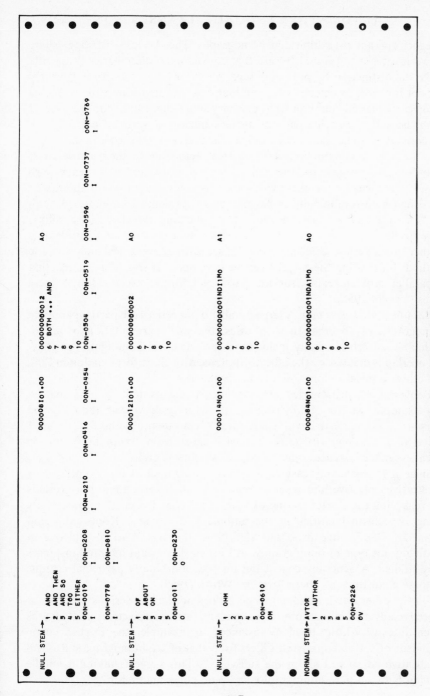

Fig. 83. Short words in the index-dictionary.

317

categories (a) and (b) exhibit stem homography. The reasons for distinguishing between categories (a) and (b) are that conventional dictionaries frequently make the distinction by providing separate entries for verbs in category (b) but not for those in category (a), and that different algorithms might well be designed to handle verbs in each category most efficiently. For example, it might be best to provide pairs of distinct entries for verbs of category (b), the correct one to be chosen according to the presence or absence in the text of "ся" or "сь" in conjunction with the stem belonging to these entries. For verbs of category (a), on the other hand, a single entry for each verb may be sufficient, together with an algorithm capable of modifying the English correspondents as directed by the presence or absence of "ся" or "сь" and by a mark in the entry indicating whether these affixes have reflexive, reciprocal, or passive interpretations. Considerable further investigation seems necessary before satisfactory lexical and syntactic methods for treating these verbs can be developed. It should be noted that a parallel problem exists for the participial forms associated with these verbs (Lynch, 1955).

B. *Adjectival Forms in o.* A large number of adjectives are partial paradigm homographs of adverbs. In such cases, the sole representation of an indeclinable adverb is wholly included in the adjectival paradigm. Typically, the overlap is created by the adjectival predicative short form ending in "o", but other forms also contribute to the problem.

Conventional dictionaries are thoroughly inconsistent in this matter. For example, we may find distinct entries for *бесконечно** (adverb) and *бесконечный** (adjective with short form "бесконечно") and for *хорошо** (adverb) and *хороший** (adjective with short form "хорошо"), but in countless other instances only an adjectival entry is given.

Since it is reasonably easy to determine when a short form represents a predicative adjective and when it represents an adverb, there are grounds for regarding the matter as one of homography, and hence for consistently using two distinct entries in the automatic dictionary. Koutsoudas and Humecky (1957) have proposed algorithms they claim to be effective in resolving this type of homography, at least in the several hundred instances they studied. A generally cogent but on occasion blindly parochial critique of these algorithms has been given by Worth (1959).

On the other hand, although English does not have a serious problem of adjective/adverb homography, the representations of adjectives and adverbs often differ very little. In many instances, for example, the English correspondent of a Russian adverb differs from that of the homographic Russian predicative adjective only in the suffix "ly". This suggests taking a middle ground between pure homography and pure polysemanticism.

A single entry may be provided for each stem that represents both an adjective and an adverb. The English correspondents may be given in the adjectival form. It should not prove too difficult to develop *derivational* algorithms for English, analogous to the inflectional algorithms discussed in Sec. 5.6. On this assumption, once an algorithm for contextual analysis has established that a Russian text string indeed represents an adverb, the derivational algorithm may be called into play to transform the adjectival English correspondents into adverbial forms.

Other compromises are possible, but these are of a technical nature, depending, as we have often seen, on striking a proper balance between table look-up and other types of algorithms.

C. *Adjectives and Participles.* Morphologically, the four participial forms (present active, present passive, past active, past passive) of Russian verbs behave like adjectives, and are classified as such under the system described in Sec. 6.1.

The function of these forms is somewhat more complicated. In English there is a tendency to use morphologically nominal forms in adjectival functions, as in "atom bomb" or "machine word". In Russian there is an inverse tendency to use morphologically adjectival forms in nominal functions, for instance, "в будущем" or "трудящийся".

A given form might therefore function in a verbal, adjectival, or nominal way, with the resultant problems of paradigm homography and poly-semanticism. In some instances, such as "воображаемый", standard dictionaries have a separate entry for the adjectival function of a given string, in others they do not.

In the present automatic dictionary file, only a single entry is provided for the forms serving both adjectival and verbal functions. These forms are distinguished from ordinary adjectival forms by the Q marks described in Sec. 6.5. In certain instances where the functions are verbal and nominal, as in those instances where the functions are adjectival and nominal, two distinct entries are provided, of which one has the standard nominal marks in the organized word (Secs. 6.3 and 6.4).

The present state of this problem is still unsatisfactory, and considerable further investigation seems necessary to improve it.

9.8. The Existence of Strings

Certain strings (vacuous strings, Sec. 6.1) generated by the direct-inflection algorithms defined in the appendix of Chapter 6 do not in fact exist in Russian. In Sec. 7.8 we distinguished between artificial vacuous strings, which are morphologically unacceptable, and academic vacuous strings, which are morphologically acceptable but happen not to be used in certain paradigms.

When the canonical stems obtained from these strings coincide with stems obtained from normal strings, the question of the existence of the artificial or academic forms does not arise for nouns or adjectives. It is only in verbs, where possible functions of stems are explicitly coded (Table 40), that it is necessary in such cases to decide whether or not a particular stem could function as part of, say, a present gerund. This decision must be based on the same criteria that are used for deciding whether or not a stem arising only from artificial or academic forms should be retained.

In general, a stem recognized as arising from an artificial form only may be safely rejected. Such forms are artifacts produced by the inflected-form generators for reasons of convenience and, in any case, the phonetic constraints underlying the concept of artificiality are sufficiently stable to guarantee a small likelihood that such forms will enter the language.

Academic forms are slightly more problematic. Rejecting these forms has two consequences: first, the form may in fact be used, various authorities to the contrary notwithstanding. When an occurrence is detected the form must be added to the dictionary. Second, until an occurrence is actually detected, there is the possibility that earlier occurrences may have been wrongly translated as missing homographs (Sec. 9.4).

Accepting the forms possibly entails some unnecessary labor. Moreover, there is a possibility that a stem for a "nonexistent" form will in fact be looked up, either because it happens to be homographic to some word not in the dictionary, or because of some spelling mistake in transcribing a text. On the other hand, the form may actually be in use, or come into being, in which case it is available in the dictionary.

It is unlikely that the errors due to either acceptance or rejection will be more than an insignificant proportion of all mistakes due to misspellings or errors due to missing homographs. The differences in labor or file size due to acceptance or rejection seem also insignificant. On the whole, therefore, the problem of existence of vacuous strings seems trivial. As indicated in Sec. 7.8, its solution in particular instances is left to the discretion of those who assign correspondents and grammatical marks.

The only reason for dwelling on the problem is that it has seemed to cause inordinate discomfort to those who do not have a professional eye for distinguishing between inconsequential and catastrophic "errors". Until more is known about the incidence of spelling mistakes and of missing homographs and about their effect on the quality of automatically produced translations (or, for that matter, on conventional ones), there seems to be little reason for worrying about a small proportion of such errors.

9.9. The Structure of the Dictionary File

Throughout the description of the compilation and operation of the Harvard Automatic Dictionary given in Chapters 7 and 8, it was assumed that the dictionary file is to be stored on magnetic tape. In addition, the processes of compilation and look-up were based on further assumptions regarding the structure of entries in the file. In this section, these assumptions are examined and refined in the light of the properties of information-storage media described in Secs. 1.6 and 1.7.

The major factor governing the choice of a storage medium for a dictionary file is the size of this file. The 22,000 stem entries at present in the Harvard automatic dictionary file occupy approximately 50,000,000 bits (Sec. 1.3) of storage capacity. To provide flexibility for experimental purposes, this storage capacity is not used at maximum efficiency. However, allowing for increase in both efficiency of storage and file size, the storage requirements for canonical stem dictionaries may be placed in the range of 10^7 to 10^9 bits, with a significant upward adjustment of this range for paradigm dictionaries.

Except for magnetic tape, none of the storage media described in Chapter 1 and in the bibliographic notes of Sec. 1.8 is capable, at present, of providing sufficiently high storage capacity at a sufficiently low cost for experimental purposes. A somewhat wider range of devices may be considered for production purposes, especially as time goes on and some of the devices under development become available. In any case, magnetic tape was the only high-capacity storage medium that could be considered for use with the Univac on which the experimental work described in this book was done. It is important to remember, however, that decisions based on the characteristics of the Univac, or of magnetic tape in general, must periodically be reviewed as technology progresses.

The decision to use 30-machine-word items for dictionary entries was governed by considerations similar to those described in Sec. 7.2 leading to the choice of five-word items for dictionary compilation and for the phases of dictionary operation prior to look-up. Fewer than ten machine words would be inadequate to store all necessary information. Multiples of ten words suggest themselves because of the ease of ten-word transfers, and divisors and multiples of 60 avoid undesirable straddling of block boundaries. The 30-word item organized as shown in Fig. 56 meets these requirements and provides an ample but not extravagant amount of space.

The ordering of items in the file remains to be determined. One factor affecting the ordering of items is the proposed method of look-up. Direct look-up of words as they appear in a text is ruled out by the characteristics of magnetic tape. Magnetic tape is essentially a serial-access medium, which means, in terms of the definitions of Sec. 1.6, that the time required for random-order interrogation is much greater than that for matched-order interrogation. For example, for a thousand-word text, the average access ratio calculated with the formula of Sec. 1.6 is in the neighborhood of 200.

The scanning fraction on which this formula is based, is based in turn on the assumption of equiprobable requests for any items in the file. It is well known from the work of Zipf (1949, 1935) and Mandelbrot (Apostel, Mandelbrot, and Morf, 1957) that the strings occurring in texts are far from being rectangularly distributed (Fig. 78). It is therefore conceivable that by taking advantage of the knowledge of the frequency distribution of words to determine some optimal ordering of entries in the dictionary file the access ratio could be diminished. Matched-order interrogation requires sorting the input data prior to file consultation, and sorting is itself a time-consuming operation. Therefore, if the time for random-order interrogation could be brought below the total time for sorting of input data and for matched-order interrogation, direct look-up of words by random-order interrogation could be attractive. Investigations by Giuliano (1957a) of several file arrangements based on a knowledge of word frequency distribution showed that the access ratio could not be decreased sufficiently. Direct look-up was therefore ruled out.

Matched-order interrogation requires that the text items be sorted into dictionary order prior to look-up. Hence the arrangement of items in the dictionary file cannot depend on anything but stems and affixes or (Rhodes, 1959; Oettinger, 1953) morphologically defined images of these. For example, providing separate file sections for nouns, verbs, and adjectives is ruled out. This does not rule out the possibility that, given sufficient random-access internal memory, it may be worth while to store in it a small subdictionary containing only words of very high frequency, such as prepositions, conjunctions, and the like. Each text string could be looked up in this subdictionary prior to the sort into dictionary order. Only words not found in the subdictionary would then require sorting before look-up. Inasmuch as relatively few distinct forms account for a very large number of running words (Sec. 7.4), such a compromise between matched-order interrogation and random-order interrogation through very fast search or coordinate access in internal memory may prove useful. In view of the well-known inverse relation between word length and word frequency, no time need be wasted in looking up long words in the subdictionary, provided that the length of a word can be determined in less time than would be necessary to look it up.

The order of items in the dictionary file is primarily determined by the alphabetic order of canonical stems. Items with identical stems are further ordered in the alphabetic order of their affixes. For most words this secondary ordering is of no consequence but it is important as a means for ensuring that a proper match can be obtained for "short" words, as described in Sec. 9.6.

Homographic words will be represented in the dictionary file by a set of entries all having the same stem key. In the arrangement described in the preceding paragraph, these entries occur consecutively in the file. In dictionary look-up, whenever a word occurs whose stem is identical with the keys of a set of consecutive homographic entries, the whole set of entries is entered in the augmented-text file, after each member has been modified in the manner described in Sec. 8.3. The editing program (Sec. 8.4) is designed to recognize the presence of homographic entries. The information in every entry of the set is reproduced in the edited file in the normal manner, except that the second and subsequent homographs are marked by "$+++++$(HOMO-GRAPH OF PREV)" to call the attention of the reader to the presence of the homographic set (Fig. 71). A method for the subsequent elimination of certain members of such sets (Sec. 9.3) is outlined in Sec. 10.1.

In other respects, the organization of the dictionary file follows the conventions described in Sec. 7.3; the identification and sentinel blocks are shown in Fig. 41.

9.10. Detecting and Correcting Mistakes in Dictionary Compilation

Much attention has been paid in Sec. 7.6 to the problem of detecting and correcting mistakes in compilation due to misclassification or to poor transcription. In Sec. 7.9 it was pointed out that some gross mistakes in assigning English correspondents and grammatical marks are caught by proofreading. Mistakes more subtle, but likely to prevent the successful merging of the Type 12IS file with the Type C file to form an addition to the dictionary file (Fig. 54), are caught automatically by a program designed for the purpose. There is no practical way, however, to detect subtle mistakes in assigning English correspondents or grammatical marks, prior to the complete compilation of a substantial segment of dictionary file (Coppinger *et al.*, 1959).

Once the dictionary file is available on magnetic tape, it is possible to use the valuable technique, already discussed in Sec. 7.6, of rearranging the dictionary file into an order where a pattern emerges and where deviations from the pattern point to possible mistakes. Mistakes in grammatical marks are detected by one pattern, those in English correspondents by another.

Rearranging the whole dictionary file for purposes of mistake detection is uneconomical. A special program was written (Frink, 1959) to select from each dictionary entry only certain relevant information and place it within ten-word items for ease in processing. Items in the file produced for detecting mistakes in grammatical marks include primarily the morphological class mark and the organized and semiorganized words. For ease in identification and cross reference with files obtained during earlier phases of compilation, the Russian stem, the first English correspondent, the group number, and the dictionary entry number are also included. In addition, the total number of English correspondents in the entry is given.

The resulting condensed file is then automatically arranged in an order determined by a series of keys. The morphological class mark was chosen as the major key, since significant variations in grammatical properties are more likely to emerge if morphologically similar forms are brought together than otherwise. The most significant grammatical properties of Russian words are reflected in the organized machine word, which is therefore chosen as the secondary key. For nouns, the second and third semiorganized words are not used, and the first semiorganized word contains only the names of the dictionaries consulted in compilation. Subsidiary keys were therefore chosen with the properties of adjectives and verbs in mind.

The third semiorganized word, which contains marks identifying participial stems (Sec. 6.4) or describing the possible functions of verbal stems (Table 40), is used as the third key. The fourth key is the second semiorganized word, in which the absence of certain participial forms is marked for stems arising from infinitives of verbs. The first semiorganized word is used as a fifth key, but it is of little interest in this context and has little effect on the order. The most significant information with greatest immediate importance is in the organized words. Much of the information in the semiorganized words is unreliable, and is destined to be a starting point for research, rather than a subject for immediate application.

A portion of the dictionary file, rearranged as described to facilitate the detection of mistakes in grammatical marks, is shown in Fig. 84. Several mistakes appear at the transition between two morphological classes. It is of course a consequence of the rearrangement of dictionary entries that mistakes very likely will be found at all points where there is a transition from one pattern to another. Not all mistakes are displayed so conveniently, not every mistake is brought near a transition, and not all mistakes are caught, but the phenomenon is quite helpful in detecting many mistakes.

It should be noted that a mistake is defined as a deviation from the rules of classification given in Chapters 6 and 7. The reader who examines Fig. 84 doubtless will find marks inconsistent with his own view of Russian morphology or syntax, or with the views of some authority. For example, there is the question of the existence of certain participial forms for particular verbs, or the problem of the prepositions and cases governed by verbs.

					English			
V05.00	VS00P70000	BOKOB6	M1N1N3	AOA2	OKAZA-T.	TO SHOW	03	045-0149 12585
V05.00	VS00P70000	BOKOB6	M1N1N3	AOA2	VXYLOPOTA-T.	TO PROCURE	02	045-0184 03861
V05.00	VS00P70000	B2KOB4		AOA2	VXYLOPOCH-U	WILL PROCURE	03	045-0184 03866
V05.00	VS00P70000	B3		AOA2	OKAZAL-	SHOWED	02	045-0149 12589
V05.00	VS00P70400	B4		AOA2	VXYLOPOTAL-	PROCURED	03	045-0184 03863
V05.00	VS00P70400	BOKOB6	M1N1N3	AOA1A2	VREZA-T.	TO CUT IN	03	045-0169 02684
V05.00	VS00P70400	B2KOB4		AOA1A2	VREZH-U	WILL CUT IN	03	045-0169 02676
V05.00	VS00P70400	B5		AOA1A2	VREZAL-	CUT IN	03	045-0169 02695
V05.00	VS00P70400	B4		AOA1A2	VREZH+T-E	CUT IN	03	045-0169 02678
V05.00	VSROC40000	BOKOB6	M1N1N3N4	AOA1A2	VREZA-T+SJA	TO CUT INTO	03	045-0174 02686
V05.00	VSROC40000	B2KOB4		AOA2	VREZH-US+	WILL CUT INTO	03	045-0174 02677
V05.00	VSROC40000	B4		AOA2	VREZAL-SJA	CUT INTO	03	045-0174 02697
V05.40	VSROC40000	B4		AOA2	VREZH+T-ES+	CUT INTO	03	045-0174 02679
V05.*0	VNOOP30000	B1KOB4B5	N3N4	AOA1A2	BRA-T.	TO TAKE	01	045-0197 00937
V05.40	VNOOP30000	B3		AOA1A2	BRAL-	TAKE	01	045-0197 00692
V05.40	VNROD20000	BOKOB6	N3N4	AOA2	BRA-T+SJA	TOOK	03	045-0201 00938
V05.40	VNROD20000	B1KOB4B5		AOA2	BER-US+	TO UNDERTAKE	03	045-0201 00693
V05.40	VNROD20000	B3		AOA2	BRAL-SJA	UNDERTAKE	03	045-0201 00955
V05.40	VSR0000000	BOKOB6	M1N1N3N4	AOA2	VYBRA-T+SJA	UNDERTOOK	02	045-0205 03043
V05.40	VSR0000000	B2KOB4		AOA2	VYBER-US+	TO GET OUT	02	045-0205 03025
V05.40	VSR0000000	B5		AOA2	VYBRAL-SJA	WILL GET OUT	02	045-0205 03045
V05.41	VS00P30000	BOKOB6	M1N1N3	AOA1A2	VYZVA-T.	GOT OUT	03	045-0209 03249
V05.41	VS00P30000	B2KOB4		AOA1A2	VYZOV-U	TO CALL	03	045-0209 03263
V05.41	VS00P30000	B5		AOA1A2	VYZVAL-	WILL CALL	03	045-0209 03251
V06.00	VN0000000	BOKOB6	N3N4	AOA2	LEZHA-T.	CALLED	03	045-0119 09622
V06.00	VN0000000	B1KOB4B5		AOA2	LEZH-U	TO LIE	03	045-0119 09921
V06.00	VN0000000	B3	N4	AOA2	LAY-	LAY	03	045-0119 09924
V06.00	VNOOP30000	BOKOB6		AOA2	SODERZHA-T.	TO CONTAIN	03	045-0124 18738
V06.00	VNOOP30000	B1KOB4B5		AOA2	SODERZH-U	CONTAIN	03	045-0124 18737
V06.00	VNOOP30000	B3		AOA2	SODERZHAL-	CONTAINED	03	045-0124 18740
V06.10	VNOOP100K0	BOKOB6	N3N4	AOA2	BOJA-T+SJA	TO FEAR	03	045-0132 00933
V06.10	VNOOP100K0	B3		AOA2 Pø	BO-JUS+	FEAR	03	045-0132 00892
V06.10	VNOOP100K0	B4		AOA2 Pø	BOJAL-SJA	FEARED	03	045-0132 00935
V06.10	VNOOP100K0	B4		AOA2 Pø	B-OJSJA	FEAR	03	045-0132 00495
V06.10	VS03000000	BOKOB6	M1N1N3N4	AOA1A2	BOUT-ES+	FEAR	03	045-0132 00901
V06.10	VS03000000	B2		AOA1A2	SOSTOJA-T+SJA	TO TAKE PLACE	03	045-0129 18919
V06.10	VS03000000	B3		AOA1A2	SOSTO-JUS+	WILL TAKE PLACE	03	045-0129 18918
				AOA1A2	SOSTOJAL-SJA	TOOK PLACE	03	045-0129 18921
V06.20	VKOOP2L900	BO	N3	AOA2	VELE-T.	TO ORDER	02	045-0058 01296
V06.20	VKOOP2L900	B1KOB2B4B5B6		AOA2	VELE-JU	ORDER	04	045-0058 01295
V06.20	VKOOP2L900	B3		AOA2	VELEL-	ORDERED	02	045-0068 01298
V06.20	VKOOP30000	B3		AOA2	VYGLJADE-T.	TO LOOK	02	045-0097 03125
V06.20	VKOOP30000	B1KOB2		AOA2	VYGLJAZH-U	LOOK	02	045-0097 03129
V06.20	VKOOP30000	B1KOB2B4B5B6		AOA2	VYGLJAD-ISH.	LOOK	02	045-0097 03124
V06.20	VKOOP30000	B3		AOA2	VYGLJADEL-	LOOKED	02	045-0097 03127
V06.20	VKOOP30000	BO	N3	AOA2	VERTE-T.	TO TURN	05	045-0073 01359
V06.20	VN0000000	BO	N3N4	AOA2	VISE-T.	TO HANG	03	045-0091 01698
V06.20	VN0000000	B1		AOA2	VERCH-U	TURN	03	045-0073 01381
V06.20	VN0000000	B1KOB4B5B6		AOA2	VISH-U	HANG	03	045-0091 01721
V06.20	VN0000000	B1KOB4B6		AOA2	VERT-ISH.	TURN	03	045-0073 01357
V06.20	VN0000000	B3		AOA2	VIs-ISH.	HANG	03	045-0091 01697
V06.20	VN0000000	B3		AOA2	VISEL-	HANG	03	045-0091 01700
V06.20	VN0000000	B3		AOA2	VERTEL-	TURNED	03	045-0073 01363
V06.20	VN0000000	B5	N3N4	AOA2	V-ISJA	HANGING	03	045-0091 01109
V06.20	VNNODE50000	BO		AOA2	GLJADE-T.	TO LOOK	03	045-0102 04243
V06.20	VNNODE50000	B1		AOA2	GLJAZH-U	LOOK	03	045-0102 04247
V06.20	VNNODE50000	B1KOB4B5B6		AOA2	GLJAD-ISH.	LOOK	03	045-0102 04242
V06.20	VNNODE50000	B3	N3N4	AOA2	GLJADEL-ISH.	LOOKED	03	045-0102 04245
V06.20	VNOOF40000	BO	N4	AOA2	ZAVISE-T.	TO DEPEND	01	045-0057 05864

Fig. 84. Dictionary rearranged for detecting mistakes in grammatical marks.

English	Code	Russian	Ref	Code 1	Code 2	Code 3	Version	Number
TO OMIT	%03	OPUSKA-T.	023-0164	VNoQP30000	A1AOA2	BOKOB18uB6	VO1.00	12638
TO OMIT	405	PROPUSKA-T.	004-0407	VNoQP30000	AOA2	BOKOB18uB6	VO1.00	16327
TO OMIT	405	PROPUSTI-T.	004-0408	VSoQP30000	AOA1	BOKOB6	VO4.00	16341
TO OPEN	102	RAZOMKNU-T.	018-1063	VSoOP70000	A1AOA2	BOKOB6	VO2.00	17173
TO OPEN	204	OTPIRA-T.	004-0273	VNoOP70000	AOA1	BOKOB18uB6	VO1.00	13312
TO OPEN	106	VSKRYVA-T.	003-0164	VNoOP70000	AO	BOKOB18uB6	VO1.00	02777
TO OPERATE	304	PEREVODI-T.	024-0225	VNoOP30000	A1AOA2	BOKOB6	VO4.00	13777
TO OPERATE	203	DEJSTVOVA-T.	010-0163	VNoOQ00000	AOA1	BOKOB6	VO3.00	04815
TO OPPOSE	103	PROTIVODEJSTVUVA-T.	018-0565	VNoQP20000	AOA1	BOKOB66	VO3.00	16435
TO ORDER	102	VELE-T.	045-0068	VXoQP2L900	A000	BO	VO6.20	01296
TO ORGANIZE	%03	OBRAZOVYVA-T.	022-0171	VNoQP30000	AOA2	BOKOB18uB6	VO1.00	12275
TO ORGANIZE	203	OBRAZOVA-T.	022-0192	VKoQP30000	AOA2	BOKOB6	VO3.00	12264
TO ORGANIZE	%03	KONSTRUIROVA-T.	006-0086	VKoOP70000	AO	BOKOB66	VO3.00	09196
TO ORIENT	103	ORIENTIROVA-T.	023-0165	VKoQP70000	A1	BOKOB66	VO3.00	12862
TO ORIGINATE	103	ISKOOI-T.	014-0178	VKoQ00000000	AOA1A2	BOKOB6	VO4.00	06432
TO OSCILLATE	203	VIBRIROVA-T.	007-0005	VNo3000000	ACA1	BOKOB6	VO3.00	01583
TO OVERBURDEN	203	PEREGRUZHA-T.	024-0202	VNoQP30000	A1	BOKOB16uB6	VO1.00	13840
TO OVERBURDEN	%02	PEPEGRUZI-T.	024-0223	VSoOP30000	A1	BOKOB6	VO4.00	13852
TO OVERBURDEN	203	ZAGRUZHA-T.	011-0326	VNoOP70000	AOA2	BOKOB45uB1	VO1.00	06049
TO OVERCHARGE	%03	PEREGRUZHA-T.	024-0202	VNoOP30000	A1	BOKOB16uB6	VO1.00	13840
TO OVERCHARGE	102	PEREZARJAZHA-T.	024-0198	VNoOP30000	AOA1A2	BOKOB16uB6	VO1.00	13924
TO OVERCHARGE	203	ZAPRASHIVA-T.	018-0118	VNoOP30000	AOA2	BOKOB16uB6	VO1.00	06695
TO OVERPAY	%02	PREVYSHA-T.	024-0339	VNoOP30000	AO	BOKOB16uB6	VO1.00	15450
TO OVEREXPOSE	102	PEREEMKSPONIROVA-T.	018-0361	VKoQP70000	A1	BOKOB16uB6	VO3.00	14206
TO OVERHEAT	102	PEREGREVA-T.	024-0204	VNoOP70000	AOA1A2	BOKOB18uB6	VO1.00	13824
TO OVERHEAT	102	PEREGRE-T.	024-0203	VSoOP70000	A1	BOKOB2B5	VO1.00	13820
TO OVERLAP	103	PEREKALIVA-T.	024-0197	VNoOP70000	A1AOA2	BOKOB16uB6	VO1.00	13935
TO OVERLOAD	103	PEREKRYVA-T.	024-0194	VNoOP30000	A1AOA2	BOKOB16uB6	VO1.00	13997
TO OVERLOAD	102	PEREGRUZI-T.	024-0202	VNoOP30000	A1	BOKOB16uB6	VO1.00	13840
TO OVERLOAD	%03	ZAVALIVA-T.	011-0255	VSoOP30000	A1	BOKOB6	VO4.00	13852
TO OVERRATE	%03	ZAPRASHIVA-T.	018-0118	VNoOP30000	AOA2	BOKOB16uB6	VO1.00	05792
TO OVERRUN	%02	PEREGREVA-T.	024-0204	VNoOP30000	AOA1A2	BOKOB16uB6	VO1.00	06695
TO OVERRUN	%03	PEREGRE-T.	024-0203	VSoOP70000	A1	BOKOB2B5	VO1.00	13824
TO OVERSHADOW	203	PEPEKALIVA-T.	024-0197	VNoOP70000	AO	AOKOB16uB6	VO1.00	13820
TO OVERTAKE	102	ZATMEVA-T.	013-0023	VNoQ2P30000	A1AOA2	BOKOB16uB6	VO1.00	13035
TO OVERTURN	203	DOGONJA-T.	011-0047	VNoOP30000	AO	BOKOB16uB6	VO1.00	06949
TO OVERWORK-SELF	102	OPROKIDYVA-T.	023-0162	VNoOP30000	A1AOA2	BOKOB16uB6	VO1.00	05317
TO OWE	102	DORABATYVA-T.+SJA	011-0116	VNoO00000000	AOA2	BOKOB16uB6	VO1.00	12809
TO OWN	%02	ZADOLZHA-T.	012-0005	VSoO00000000	AOA1	BOKOB52uB6	VO1.00	05477
TO OWN	%02	OBLADA-T.	021-0142	VNoOP40000	AO	BOKOB16uB6	VO1.00	12123
TO OXIDATE	103	VLADE-T.	003-0086	VNoOP40000	AO	BOKOB185	VO1.00	01797
TO OXIDIZE	%02	OKSIDIROVA-T.	022-0190	VKoOP70000	AOA1A2	BOKOB6	VO3.00	12654
TO OXIDIZE	102	OKSIDIROVA-T.	022-0190	VKoOP70000	AOA1A2	BOKOB6	VO3.00	12654
TO PACK UP	103	OKISLJA-T.	022-0184	VNoOP70000	AOA1A2	BOKOB18uB6	VO1.00	12611
TO PAINT	%03	UKLADYVA-T.	030-0383	VNoOP30OK0	AOA1A2	C4BOKOB18uB6	VO1.00	20318
TO PAINT	203	KRASI-T.	016-0030	VNoOP70000	AOA1A2	BOKOB6	VO4.01	09584
TO PAINT IN	102	VYKRASHIVA-T.	003-0212	VNoOP70000	AO	BOKOB18uB6	VO1.00	03522
TO PAN	%02	RASTSVETI-T.	018-1239	VSoOP70000	A1AOA2	BOKOB6	VO4.00	17484
TO PARADE	%01	PANORAMIROVA-T.	024-0213	VSoOP70000	A1	BOKOB6	VO3.00	13626
TO PARAFFINE	203	VYSTROI-T.	003-0259	VSoOP70KO	AOA1A2	P3BOKOB6	VO4.02	03786
TO PARRY	%01	PARAFINIROVA-T.	024-0214	VNoOP70000	A1	BOKOB6	VO3.00	13663
TO PART	304	OTRAZHA-T.	009-0337	VNoOP30000	AOA1	BOKOB18uB6	VO1.00	13344
TO PARTITION	102	PREDELYVA-T.	024-0201	VNoOP30000	AOA1A2	BOKOB18uB6	VO1.00	13804
TO PASS	%01	PEREGORODI-T.	024-0224	VSoO00000000	A1	BOKOB6	VO4.00	13809
TO PASS	102	PEREDAVA-T.	024-0232	VNoOP30000	A1AOA2	BOKOB4uB6	V12.00	13866
TO PASS	%03	PROVEST-T.	018-0529	VSoOP70KO	AOA1A2	P3BO	VO8.20	16112
TO PASS	203	PROXODI-T.	007-0240	VSoO20000000	AO	BOKOB6	VO8-20	16887
TO PASS	%01	MINU-T.	003-0101	VNoOP30000	AO	BOKOB6	VO2.00	10781
TO PASS	%02	VODI-T.		VNoQ00000000	AO	BOKOB6	VO4.00	01988

Fig. 85. Dictionary rearranged for detecting mistakes in English correspondents.

These problems are not caused by mistakes due to mere carelessness, but reflect an imperfect analysis of the Russian language. Once careless mistakes have been eliminated, lists like that of Fig. 84 may be used in conjunction with dictionary products like those described in Secs. 8.4 and 10.4 to analyze the properties of Russian words with a thoroughness and exactness not heretofore possible.

The file created to detect mistakes in English correspondents is similar in content, but not in ordering, to that created for detecting mistakes in grammatical marks. By arranging the condensed entries in the alphabetic order of English correspondents, patterns suitable for detecting misspellings are created. A portion of a file arranged in this way is shown in Fig. 85. Although flagrant cases of inappropriate assignment of English correspondents may be detected in such lists, these are not so much mistakes as the product of an inadequate analysis of the correspondence between Russian and English. Improving an automatic dictionary file by modifying existing correspondents in accordance with observed usage and by adding new entries is a long-term project. Here again some of the dictionary products described in Sec. 8.4 will play an important role.

The alphabetic list of English correspondents illustrated in Fig. 85 promises to be an important research tool in other respects. It may be regarded as a crude English-Russian dictionary which, with certain modifications, could be used as a basis for translating from English to Russian. It shows quite clearly how certain English words are used as correspondents for more than one Russian word, and therefore lends itself admirably to the study of English homography and polysemanticism, at least relative to Russian. For the study of certain problems of semantics, including those inherent in the construction of thesauri, joint use of the Russian-English and English-Russian files provides a potentially valuable tool (Glantz, 1959).

CHAPTER 10

FROM AUTOMATIC DICTIONARY TO AUTOMATIC TRANSLATOR

10.1. The Interpretation of Affixes

Throughout Chapter 5 and subsequent chapters inverse inflection has been regarded primarily as a process producing canonical stems for use as keys to entries in a stem dictionary. The affixes of orders zero and one, separated from a stem when a string occurring in a text is factored by the inverse-inflection algorithm (Secs. 5.4 and 5.5), also supply valuable information about the syntactic role of the string. During look-up these affixes are transferred into the items constituting the augmented-text file (Sec. 8.3) and thereafter they remain available for interpretation.

The questions involved in the interpretation of zeroth-order affixes are outlined in Sec. 9.7. Here, we are concerned exclusively with first-order affixes (Matejka, 1958; Sherry, 1958, 1959). Broadly speaking, interpreting an affix consists in identifying the text string of which it is a factor as a particular member of a reduced paradigm. It is convenient to extend the idea of interpretation to cover explicit identification of the inflected form or forms which the member of the reduced paradigm represents. For example, the co-occurrence of the affix "ов" and the stem "студент" identifies an occurrence of the member "студентов" (Table 17) of the reduced paradigm of *студент**, and "студентов" in turn is in correspondence with both *студентов$_{gp}$** and *студентов$_{ap}$** (Table 15). Since the stem alone completely identifies the word in question, at least in this case, the interpretation of "ов" might be given simply by "GPAP", standing for "this string represents either the genitive plural or the accusative plural of the word identified by its stem". The internal homography (Sec. 9.1) due to the one-to-many correspondence between the reduced paradigm and the paradigm of *студент** is thus reflected in the notation "GPAP". Such homography can be resolved, if at all, only by the analysis of contexts, as indicated in Sec. 9.1.

When the stem of a string belongs to a set of stem homographs, the affix may also play a significant role in identifying the word in question. How much can be accomplished depends on whether the stem homography is essential or nonessential (Sec. 9.2) and, if the stem homography is essential, on whether partial or total paradigm homography (Sec. 9.1) is involved.

If the stem homography is nonessential, the affix determines a unique word. If the stem homography is essential, and paradigm homography is total, no resolution is possible. The most frequent situation is that in which the stem homography is essential, but the paradigm homography is only partial. In this case if the original text string happens to lie in the intersection of two or more reduced paradigms (Fig. 79*b*, Fig. 80) then again no resolution is possible. If, however, the string happens to lie outside the intersection but within one of the paradigms then a unique word is determined. In this case, as in nonessential homography, several members of the set of homographic entries included in the augmented text during dictionary look-up (Sec. 9.9) may be marked for elimination from further consideration.

For example, the affix "ов" occurring with the stem "радиотехник" (Table 42) identifies a string belonging exclusively to the word *радио-техник**. On the other hand, "ов" has no valid interpretation with the entry for *радиотехника**, which may be marked for elimination by "incompatible" or some equivalent mark. The interpretation of "ов" with the entry for *радиотехник** is "GPAP". Given the occurrence of "радио-техникой" in a text, the interpretation of "ой" is "incompatible" with the entry for *радиотехник**, but it is "IS" (instrumental singular) with that for *радиотехника**. In this case, there remains no unresolved stem, paradigm, or internal homography. On the other hand, "а" with the same stem should be interpreted as "GSAS" (genitive singular, accusative singular) with the entry for *радиотехник**, and as "NS" (nominative singular) with that for *радиотехника**. In this case, since "радиотехника" lies in the intersection of the two paradigms, unresolved paradigm homography remains, and, in one case, unresolved internal homography as well. In addition, the failure of "а" alone to characterize a unique inflectional class, evidenced by the alternative interpretations "GSAS" and "NS" means that our description of the process of identification is not adequate, but must be refined. Much of the remainder of this section will be devoted to this problem.

In cases where the text string belongs to a missing homograph, one of two things will happen. If the affix is compatible with one or more of the set of homographic entries, a failure of the kind described in Sec. 9.4 will occur. If the affix is compatible with none of the entries, it is possible to replace the whole set by a single dummy entry like that illustrated in Fig. 67*b*, except that it may be marked as arising from a homographic set. The word may thereafter be treated like any other not found in the dictionary and, when it appears in the index of missing words (Fig. 73), it will signal its own occurrence. In this case, the failure of the dictionary to contain a member of a homographic set is safe.

The development of algorithms for automatic affix interpretation is not a trivial task. Even if problems of stem or paradigm homography and other theoretical problems were absent, the structure of Russian affixes would not lend itself to the simple kind of algorithm illustrated by Example 2-7, nor even to the somewhat more complex kind required in Exercise 2-3.

A major source of difficulty that has already been noted is the failure of certain affixes to characterize unique inflectional classes. This phenomenon may be called *affix homography**. In some instances affix homography is due to an underlying *desinence homography* to be defined; in others it is an artifact arising from certain peculiarities of affixes described in Sec. 9.3 and leading to a many-to-many correspondence between the set of classical desinences and that of affixes. Affix homography in the presence of underlying desinence homography will be called *essential**, and otherwise *nonessential**.

Consider the desinence "a". In nouns, this desinence may denote the nominative singular, genitive singular, accusative singular, nominative plural, or accusative plural. In verbs, it occurs in the representations of feminine pasts and of present gerunds. In adjectives, it occurs in feminine predicative forms. This failure of the desinence "a" to characterize unique inflectional classes, or, in other words, its performance of multiple functions (Sec. 6.3), is an illustration of desinence homography. The affix "a", which, in general, corresponds uniquely to the desinence "a", therefore exhibits essential affix homography.

It is possible to classify Russian words in such a way that desinence homography can be resolved once it is known to what class a stem associated with a given desinence belongs. The refinement of the morphological classification of Sec. 6.1 by the functional classification of Sec. 6.3 was designed so as to resolve desinence homography in nouns, as shown by the table in the appendix to this chapter (Matejka, 1958). For example, the knowledge that the desinence "a" occurs in a noun of class *N1* and subclass *I1* is sufficient to determine the interpretation of this desinence to be "GS". The affix "a" corresponds uniquely to the desinence "a"; the essential affix homography which it exhibits is resolved by virtue of the resolution of the underlying desinence homography. Internal homography cannot, of course, be resolved by these means. When "a" occurs in a noun of class *N1* and subclass *A1*, its interpretation is "GSAS". The table in the appendix shows in detail how the subclassification system of Sec. 6.3 serves to resolve desinence homography. The table used when subclassifying nouns during compilation (Table 32) was obtained from that of the appendix by selecting a minimum number of desinences whose differing interpretations define a distinction among subclasses.

Nonessential affix homography generally occurs when an affix is in one-to-many correspondence with desinences. For example, the affix "ат", when factored from certain verbal strings, corresponds to the desinence "ат", but, when factored from certain nominal forms as illustrated in Table 44, it corresponds to the desinence "#". This type of homography can be resolved if means can be found for associating a given instance of an affix with a unique desinence. Any remaining essential homography may then be resolved in the manner already described.

Providing for the resolution of nonessential affix homographs requires a careful study of the properties of the representations of Russian words and

Table 45. Nonessential affix homography in class *N1*.

Desinences	Potential nonessentially homographic affixes
#	в, ев, ов*, ам*, ем, им, ом*, ым, ям, ат, ет, ит, ут, ют, ят
а	
у	ему, ому
ом	
е	ете, ите
ы	
ов	
ам	
ами	
ах	

of the properties of the inverse inflection algorithm (Sherry, 1959). Very few words are actually affected, but such a study is essential to achieve a complete and reliable method of affix interpretation.

Potential instances of nonessential affix homography, as they affect nouns in class *N1*, are listed in Table 45. The desinences associated with the representations of nouns in class *N1* are shown in the first column. Affixes corresponding to each desinence are shown in the second column. In general, the affixes in the second column correspond to different desinences either in nouns of other classes or else in verbs or adjectives. Their factoring in nouns of class *N1* is an accident generally correlated with the production of multiple stems (Sec. 9.3B, Table 44). The starred affixes in the second column correspond to more than one desinence within the same paradigm.

Empirical verification of the accuracy of tables like Table 45 as well as a list of dictionary entries affected by affix homography may be

```
            SPOSOB-AX  NO1.00  1NO
             SCHET-AX  NO1.00  1NO
             TEKST-AX  NO1.00  1NO
         TZOLJATOR-AX  NO1.00  1NO
               TIP-AX  NO1.00  1NO
             TRIOD-AX  NO1.00  1NO
           FIL.TR-AX   NO1.00  1NO
               FON-AX  NO1.00  1NO
             FRONT-AX  NO1.00  1NO
            TSENTR-AX  NO1.00  1NO
              SHUM-AX  NO1.00  1NO
            EHKRAN-AX  NO1.00  1NO
      EHK SPERIMENT-AX NO1.00  1NO
         EHLEKTROD-AX  NO1.00  1NO
           EHFFEKT-AX  NO1.00  1NO
              IZVI-V   NO1.00  300
          INFINITI-V   NO1.00  300
            PERERY-V   NO1.00  300
           ABAMPER-E   NO1.00  600
           ABVOL.T-E   NO1.00  600
           ABKULON-E   NO1.00  600
            ABLITS-E   NO1.00  600
              ABOM-E   NO1.00  600
         ABONEMENT-E   NO1.00  600
             ABRIS-E   NO1.00  600
            ABSURD-E   NO1.00  600
       AV IOPEREXVAT-E NO1.00  600
       AVT OGETERODIN-E NO1.00 600
           AVTODIN-E   NO1.00  600
           AVTOZAL-E   NO1.00  600
        AVTOMATIZM-E   NO1.00  600
           AVTOMAT-E   NO1.00  600
          AVTOPILOT-E  NO1.00  600
      AVTO REGULJATOR-E NO1.00 600
             AVTOR-E   NO1.00  600
           AVTOSIN-E   NO1.00  600
```

Fig. 86. Affixes in class *N1*.

obtained by rearranging the file of inflected factored forms (File 12ISM-S, Sec. 7.7) obtained during compilation. A portion of this file arranged by affixes within classes is shown in Fig. 86. The forms "изви-в", "инфинити-в", and "переры-в" illustrate the phenomena of multiple stem generation and nonessential affix homography.

The resolution of nonessential affix homography presents no serious conceptual difficulties. For example, when "в" occurs as a factor in a representation of a noun of class *N1*, it may simply be treated as equivalent to "#". The co-occurrence of nonessential affix homography and multiple stems provides the key to the interpretation of the starred affixes. For example, *остров** will be represented in the dictionary file by an entry with

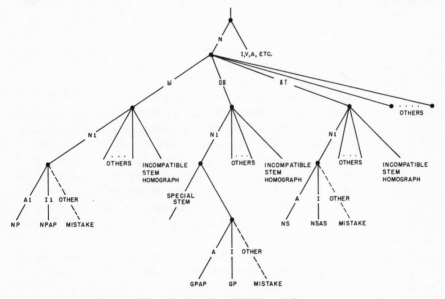

Fig. 87. The structure of an affix-interpreting program.

key "остр" obtained from "остров" alone, and by another entry with key "остров" obtained from all other members of the reduced paradigm, including "островов". A special mark in the entry for "остр" might be provided to supply the interpretation of "ов" directly in this case. When the other entry is referred to, interpretation is normal.

The outline of a program for automatic affix interpretation is sketched in Fig. 87. The effect of the program is equivalent to that of look-up in a suitable extension of the table of the appendix. Storing such a table in machine memory would be highly uneconomical, and it is better to use an algorithm with a treelike structure, similar to that used in the inverse-inflection algorithm described in Sec. 5.5.

The affix-interpreting program operates on items in the augmented-text file described in Sec. 8.3. An initial branching is determined by the first character in the organized word (Sec. 6.3), that is to say, in word 26 of the item (Fig. 56). Once it is established that an item is an entry for a noun, a further branching is made according to the affix in character positions 8–10 of word 3. The next branch is chosen according to the numerical portion of the morphological class mark in positions 2–6 of word 3.

At this stage, entries for which the combination of morphological class mark and affix leads to an empty cell in the table of the appendix will be rejected as incompatible stem homographs, since further branches are provided only for those classes in which the affix in question can legitimately occur.

Figure 87 shows how instances of the affixes "ы", "ов", and "ат" will be interpreted when occurring with stems of class *N1*. If the stem with which "ы" occurs has been subclassified *A1*, the interpretation of "ы" is "NP" and if the subclass is *I1*, the interpretation is "NPAP".

It may happen that, by mistake, some characters other than *A, I, 1, 2*, and so on will appear in the positions allocated for the purpose in the organized word. A branch labeled "mistake" should therefore be provided. It would be quite uneconomical, and not very dignified, to allow mistakes to be detected in this way. In practice a special program, similar in structure but simpler than the affix-interpreting program, is used to verify the validity of the code marks that will control branch selection (Coppinger *et al.*, 1959). Misclassification of an animate noun as inanimate cannot, of course, be detected in this way, but a mistake through which *2* is typed as *Z* during compilation, and the like, can easily be detected.

Because "ов" is starred in Table 45, the tree for this affix contains a branch which is selected if the entry under consideration has the special mark used in such cases. The interpretation of "ов" may then be supplied either by additional marks in the entry or by tying the "special stem" branch to an appropriate point in the tree for the affix "*#*". The interpretation of "ов" otherwise depends only on whether the noun is animate or inanimate, hence a distinction between *1* and *2* is not necessary in this case, as it is for "ы" or for "а".

The branch for "ат" ensures the proper treatment of such forms as "аппар-ат". The portion of the tree following the selection of the branch labeled *N1* is, in this case, identical to the similar portion of the tree for "*#*". Hence some economy in programming may be achieved by tying this branch to the corresponding point in the tree for "*#*". A similar comment applies to all other unstarred affixes listed opposite "*#*" in Table 45.

FIRST ENGLISH CORRESPONDENT	RUSSIAN WORD	TEXT SERIAL NUMBER	ORGANIZED WORD	INTERPRETATION OF AFFIX	FIRST SEMIORGANIZED WORD	THIRD SEMIORGANIZED WORD	CLASS MARK	DICTIONARY ENTRY SERIAL NUMBER
$SOURCE:	$SOURCE:							●
RADIO TECHNI CIAN	RADIOTEXNIK_ A	0001	NDA1H000 ###	-GA-------	AOA1		NO1.10.168720	●
RADIO ENGINE ERING		0002	ND1F100	N---------	AOA1		NO4.10.166750	●
AND	-1	0003			AO		IO1.00.000080	●
ELECTRONICS	EHLEKTRONIK_ A	0004	ND1F100	N---------	A1		NO4.10.217840	●
*.	*.	0005	*					
PUBLISHING H OUSE	IZDATEL.STV_ O	0006	ND1H000	N-A-------	AO		NO6.00.075380	●
ACADEMY	AKADEMI_I	0007	ND1F0	-G-D-PN-A-	AOA1		NO7.00.001550	●
	NAUK-	0008	***					
	SSSR-	0009	***					
$TITLE:	$TITLE:	0010	###					
MASS	MASS-	0011	ND12F000	------G---	AOA1		NO4.00.104550	●
*-	*-	0012	*					
SPEKTROMETRI CHESK-OE	SPEKTROMETRI CHESK-OE	0013	***					
DETERMINING	OPREDELFNI-E	0014	ND1H0	N-A-------	AOA1		N10.00.127720	●
COMPOUND	SOSTAV-A	0015	ND1H000	-G-------	AOA1		NO1.00.188990	●
WILL FORM		0016	VSR30000000		A-A1A2	B2	VO4.00.189000	●
REMAINING	OSTATOCHN-YX	0016	ND0000	-G-------	AOA1		AO2.00.129900	●
GAS	GAZ-OV	0017	ND12M0		A-A1	B1	NO1.00.039950	●
GASSY					A-A1		AO3.00.039960	●
GAS	-V	0018	AD0000		A1		VO3.00.039970	●
IN		0019	VNO0F30000		A-A1		IO1.00.000020	●
ELECTRONIC	EHLEKTRONN-Y X	0020	AD0100		A-A1		AO1.00.217850	●
DEVICE	PRIBOR-AX	0021	ND1H0	------P	A-A1		NO1.00.157740	●
WITH	S-	0022			AOA1		IO1.00.078910	●
PORIST-YM	PORIST-YM	0023	ND1H000 ***	INCOMPAT A	AOA1		NO1.00.006230	●
METAL	METALL-O	0024	*					
*-	*-	0025	AD0100		A1		AO2.00.144030	●
TAPE	PLENOCHN-YM	0026	ND1H0	----I------	AOA1		NO1.C00.086800	●
CATHODE	KATOD-OM	0027	###					
$AUTHORS:	$AUTHORS:	0028	***					
JU.G.PTUSHIN SK-IJ	JU.G.PTUSHIN SK-IJ	0029	**					
*.	*.	0030	***					
B.A.CHUJK-OV	B.A.CHUJK-OV	0031	###					
$FOREWORD:	$FOREWORD:	0032						
ATTACHED TO	PR-1	0033	ND1H000	INCOMPAT A	AO		IO1.00.154240	●
STICK		0034	AD0000		AOA1		NO1.00.154250	●
DIRECT		0035	ND12F000	-G-D-PN-A-	A-A1		AO3.00.154260	●
HELP	POMOSHCH-I	0036	*		AO		NO6.10.151710	●
MASS	MASS-	0037	ND1F0	------G---	AOA1		NO4.00.104550	●
*-	*-							
SPEKTROMETRI CHESK-OJ	SPEKTROMETRI CHESK-OJ	0038	AD0000	-G---N-A---	AO		NO4.10.106670	●
METHOD	METODIK-I	0039			A1A2	Q3	AO1.00.083540	●
INVESTIGATED	ISSLEDOVAN-				A-		AO1.00.087080	●
QUALITATIVE	KACHESTVENN- YJ	0040	ND1H000 ***	N-A-------	AOA1		NO1.00.188980	●
COMPOUND	SOSTA-V							

Fig. 88. Interpreted and condensed augmented-text file (nouns only are interpreted).

Words in classes other than *N1* will be treated in a similar way. For example, "радиотехник-ов", when referred to the entry for *радио-техника**, leads to the combination of "ов" with *N4.1*, for which no interpretation is given in the table, a fact reflected by the absence of a branch for *N4.1* at this stage in a complete tree. The path labeled "incompatible stem homograph" is therefore chosen, and the entry may be marked for elimination.

If there were no entry in the dictionary for *радиотехник**, "радио-техник-ов" would be treated as a safe-failing missing homograph. However, when an entry is present, one of the paths labeled "others" in Fig. 87 and corresponding to the morphological class *N1.I* will be followed. This path is similar to that for *N1*, and leads to the interpretation of "ов" as "GPAP".

Through the operation of the affix-interpreting program, the information in each dictionary entry in an alphabetic subdictionary file or augmented-text file is further augmented by marks denoting the interpretation of the affix. A program like that described in Sec. 9.10 is used to obtain a condensed file which may be printed in a format suitable either for postediting or for research purposes. A portion of such a condensed augmented-text file is shown in Fig. 88.

Russian words and their text identification serial numbers are shown in columns 2 and 3. When a text string is homographic, and therefore represented by more than one dictionary entry in the augmented-text file, the string and serial number are printed only once (see, for example, "радио-техника" (0002) and "состава" (0015)). The first English correspondent in each dictionary entry is shown in column 1. For dummy entries (Sec. 8.3), a punctuation mark, comment, or transliterated Russian word, as the case may be, appears instead of an English correspondent.

The interpretations of affixes associated with nominal entries are shown in column 5, which is one machine word wide. Character positions 1–6 are used to denote the singular, and character positions 7–12 to denote the plural. The characters "N", "G", "A", "D", "I", "P" (nominative, genitive, accusative, dative, instrumental, prepositional) appearing in positions 1, 2, 3, 4, 5, 6, or 7, 8, 9, 10, 11, 12, respectively, give the interpretation of the affix. For example "–G–D–PN–A– – –" shows that "помощи" (0033) represents either the genitive, dative, or prepositional singular, or else the nominative or accusative plural, thus exhibiting a high degree of internal homography. The interpretation of "а" in "радиотехника" (0002) is given by "–GA– – – – – – – – –" with the entry for *радиотехник** and by "N– – – – – – – – – – –" with the entry for *радиотехника**. The mark "INCOMPAT" denotes incompatible stem homographs.

The dictionary entry serial number, the morphological class mark, the organized word, and portions of the semiorganized words in each entry appear in the remaining columns. The marks "*", "***", and "###" in place of the organized word denote punctuation, words not in the dictionary, and "dollar-sign items", respectively.

A similar interpretation system is being prepared for adjectives and verbs. Our treatment of the problem of affix interpretation independently of dictionary look-up is a historical and technical accident. The development of methods of interpretation followed that of the compilation and operation processes described in Chapters 7 and 8. Moreover, the internal memory of the Univac is too small to contain look-up and interpretation programs simultaneously. There is no fundamental obstacle to combining dictionary search (Fig. 63) and affix interpretation into a single program, more efficient than the series of programs employed to obtain the experimental results described here.

10.2. English Inflection and Derivation

The description of affix interpretation given in the preceding section concludes the portion of the presentation that can be based on theoretical and experimental work accomplished at the time of writing. The questions discussed in the remainder of this chapter relate in part to certain lexical problems, but chiefly to the transition from lexical to syntactic problems. Most of these questions, though still unresolved, are now under active study.

One necessary step toward the automatic production of smooth translations is the choice not only of the appropriate English correspondent for a Russian word, but of a representation of this correspondent reflecting certain properties of the original. For example, a Russian plural should, in most instances, be replaced by an English plural. The proper choice of English representation is not wholly a lexical problem.

First, the interpretation of the affix of a Russian word in the manner of Sec. 10.1 may not in itself uniquely determine the proper English representation. Such is the case in the presence of internal homography, for example, when a given member of a Russian reduced paradigm represents both a singular and a plural form. Second, the functions of morphology and syntax are not identical in Russian and in English. What in Russian may be denoted by the use of a particular desinence may be denoted in English by a preposition. The English use of *of** where Russian uses the genitive case is one example. Third, the interpretation of a Russian desinence, even when unique, need not imply a unique function. For example, the Russian instrumental case sometimes functions in a way similar to the English preposition *by**. In some contexts, however, the instrumental plays a purely formal role, without parallel in English (for instance, in "между Ленинградом и Москвой"). Recognizing and properly treating such cases poses serious syntactic and semantic problems.

In spite of these problems, direct English inflection and also English derivation may be studied toward the end of producing algorithms that will generate proper strings, even if they cannot, by themselves, determine how these strings should be used. As pointed out in Sec. 5.6, such algorithms should be capable either of producing a specific English string on demand, or else of locating it in an existing table. The first kind of algorithm is most useful with a stem dictionary, and the second with a paradigm dictionary, but the first kind is more fundamental, since it may also be used in compiling a paradigm dictionary automatically (Sec. 10.3).

English inflection is so rudimentary that linguists have questioned the validity or usefulness of speaking of English as an inflected language. Nevertheless, it is convenient to think of English words as having reduced paradigms like those shown for *dog**, *man**, *to repair**, and *to do** in Table 46.

Table 46. English reduced paradigms.

"dog"	"man"	"repair"	"do"
"dogs"	"men"	"repairs"	"does"
		"repaired"	"did"
		"repairing"	"done"
			"doing"
a	b	c	d

Methods analogous to those used to obtain algorithms for direct Russian inflection (Chapters 5 and 6) may be applied to English. The direct inflection of English is somewhat simpler than that of Russian not only intrinsically but also because, when the English words under consideration appear as correspondents in Russian-English dictionary entries, they are already classified, at least roughly.

Figure 89 shows sections of an end-alphabetized list of the English correspondents appearing in the entries for stems arising from Russian classical canonical strings (those marked by "F"; see Sec. 7.7 and Fig. 53). This list was obtained from the dictionary file by a series of programs of which the last arranged the words first according to the alphabetic order of the first character of the organized word, then by the order of the terminal characters in the representations of the words (Frink, 1959). The marks in the second column indicate the rank of each English correspondent in the parent entry of the automatic dictionary file; for example, *photograph** is the first

Fig. 89. End-alphabetized English correspondents.

English	Code	Russian	ID	Morph	Cat	Ref
RADIOMETEOROGRAPH	%03	RADIOZOND-	018-0762	ND1M000	AOA1	N01.00 16761
TELECTROGRAPH	%01	TELEKTOGRAF-	018-2056	ND1M000	A1	N01.00 19637
SPECTROGRAPH	%01	SPEKTROGRAF-	018-1721	ND1M000	A1	N01.00 18985
RADIOSPECTROGRAPH	%01	RADIOSPEKTROGRAF-	018-0843	ND1M000	A1	N01.00 16850
PSYCHOSOMATOGRAPH	%01	PSIXOSOMATOGRAF-	018-0630	ND1M000	A1	N01.00 16552
PHOTOGRAPH	103	SNIMOK-	018-1655	ND1M000	A1	N01.30 18643
PHOTOGRAPH	%03	KARTOCHK-A	014-0290	ND1F000	AOA1	N04.30 08657
X-RAY PHOTOGRAPH	102	RENTGENOSNIMOK-	018-1186	ND1M000	A1	N01.30 17743
X-RAY PHOTOGRAPH	%01	RENTGENOGRAMM-A	018-1176	ND1F000	A1	N04.00 17725
TELEAUTOGRAPH	%03	BIL-DAPPARAT-AT	006-0013	ND1M000	A1	N01.00 00782
TELAUTOGRAPH	102	TELEAVTOGRAF-	018-2031	ND1M000	A1	N01.00 19594
RADIOELECTROGRAPH	%01	RADIOEHLEKTROMIOGRAF-	018-0893	ND1M000	A1	N01.00 16915
ELECTROMYOGRAPH	%02	EHLEKTROMIOGRAF-	030-1042	ND1M000	A1	N04.00 21778
AIRGRAPH	102	MIKROPOCHT-A	017-0281	ND12F000	AOA1A2	N04.00 21340
DASH	102	CHERTOCHK-A	030-0786	ND1F000	AOA1	
DASH	%03	TIR-E,	018-0400	ND1N000	AOA1	I01.00 19780
DASH	%03	PRIM-ES,	030-0098	ND1F000	AOA1	N06.00 15892
FLASH	105	VSPYSHK-A	009-0026	ND1F000	AOA1A2	N04.30 02841
FLASH	%02	OPLAVLENI-E	022-0156	ND1N000	A1	N10.00 12721
FLASH	%04	BLESK-	006-0214	ND1MOY0	AOA1	N01.10 00841
BACKLASH	103	ZATJAGIVANI-E	013-0039	ND1N000	A1	N10.00 07006
SPLASH	%02	VSPLESK-	009-0023	ND1M000	AOA1	N01.10 02792
CRASH	305	GROXOT-	009-0289	ND1M000	A1AO	N01.00 04438
MESH	204	OBVOD-	021-0123	ND1M000	A1	N01.00 12072
MESH	204	JACHEJK-A	004-0618	ND1F000	AOA1	N04.30 21653
DISH	203	OTRAZHATEL-*	023-0120	ND1M000	AOA1A2	N03.00 13552
FARADISH	103	LECHENI-E	016-0233	ND1N000	AOA1A2	N10.00 09979
BLEMISH	203	DEFEKT-	010-0233	ND1M000	AOA1A2	N01.00 04970
BLEMISH	304	PJATNISTOST-*	018-0685	ND1F000	A1	N06.00 16666
FINISH	103	ZAKONCHENNOST-*,	012-0052	ND1F100	AO	N06.00 06306
VARNISH	102	LAK-	016-0167	ND1M000	A1	N01.10 09849
INSULATING VARNISH	204	IZOLAK-	013-0338	ND1M000	AOA1	N01.10 07748
FLOURISH	%02	ZAVITOK-	011-0276	ND1M000	AO	N01.30 05873
TIDAL MARSH	102	VATT-	001-0105	ND1MO	AOA1	N01.00 01171
BUSH	203	KUST-	016-0148	ND1M000	A1	N01.00 09824
BUSH	102	VKLADYSH-	007-0054	ND2M000	AOA1	N03.10 01740
MUSH	102	KASH-A	014-0314	ND4FOY0	A1	N04.10 08709
MUSH	%03	GRJAZ-*	009-0296	ND1F000	A1	N06.00 04462
MUSH	%02	XRIP-	030-0696	ND1M300	AOA1A2	N01.00 21147
MUSH	104	SUBGARMONIK-A	01A-1923	ND1F000	A1	N04.10 19345
PUSH	105	TOLCHOK-	030-0113	ND1F000	A1	N01.30 19813
PUSH	%02	KNOPK-A	015-0093	ND1F000	A1AO	N04.31 08921
PEAK PUSH	203	GRUSH-A	001-0294	ND1MO	AOA1	N01.30 04458
RUSH	%01	BROSOK-	001-0084	ND1MO	AOA1	N04.10 01008
BRUSH	%01	SHCHETK-A	030-0943	ND1F000	A1	N04.31 21621
BRUSH	%05	POBEGUSHK-A	018-0335	ND1F000	AO	N04.30 14434
BRUSH	204	KIS-T*	015-0051	ND1F000	A1	N06.00 08816
DEATH	305	GIBEL-*	009-0186	ND1M000	A1AO	N06.00 04143
SHEATH	406	GIL*Z-A	009-0191	ND1F000	A1AO	N04.00 04160
SHEATH	203	KOZHUX-*	015-0112	ND1F000	AOA1	N01.10 08970
LATH	203	PLANK-A	018-0293	ND1F000	A1	N04.31 14380
LATH	%02	REJK-A	018-1152	ND1F000	AO	N04.30 17665
PATH	103	PU-T*	045-0242	ND000000	A1AO	N99.99 16555
PATH	%02	TRAEKTORI-JA	006-0178	ND1F000	A1AO	N07.00 19887
PATH	%02	DOROZHK-A	011-0120	ND1F000	A1AO	N04.30 05505
BREADTH	302	SHIRIN-A	004-0601	ND2F000	A1	N04.00 21450
WIDTH	102	SHIRIN-A	004-0601	ND12F000	AO	N04.00 21450
LENGTH	%01	DLIN-A	011-0013	ND1N100	A1AO	N04.00 05198
STRENGTH	%02	SOPROTIVLENI-E	004-0486	ND12F100	AOA1	N10.00 18831
STRENGTH	304	SIL-A	018-1509	ND12F000	A1AO	N04.00 18370

correspondent out of three for *снимок**, and the last out of three for *карточка**, as denoted by "103" and "%03" respectively. The list is useful in completing a system of classification for English analogous to that for Russian described in Sec. 6.1.

One problem requiring special attention is that created by the presence of phrases in the list. However, the proper inflection of *radio technician**, *publishing house**, or *to put together** follows from a knowledge of the inflection of *technician**, *house**, and *to put**, and of the position of the representations of these words in reducible representations of the phrases.

Some provision for English derivation may be necessary for the effective handling of the adverb/adjective homography as described in Sec. 9.7B. The basic problem lies in distinguishing adjectives with representations that may be transformed into representations of adverbs merely by the addition of "ly" from those in which more complicated algorithms are required, or which must be handled as individual cases.

10.3. Paradigm Dictionaries

It is no accident that the experimental work described in this book has been based on the use of a canonical (stem) dictionary. When research began, no available machine had an internal storage capacity or a tape-motion speed adequate for storing entries for every inflected form of many thousand words; this still holds at the time of writing. Also, the prospect of compiling a large paradigm dictionary manually seemed appalling, especially because there seemed to be no adequate way of controlling the accuracy of the product.

Accordingly, the development of semiautomatic methods for compiling canonical dictionaries was undertaken in the hope that the compilation task could be reduced to a routine one susceptible of careful control of mistakes, and that a canonical dictionary would be the foundation for effective research on syntactic and semantic problems, as well as immediately useful in the production of rough translations. The reader is now in a position to judge for himself the extent to which these hopes were fulfilled.

The chief advantage of a canonical dictionary over a paradigm dictionary lies in its greater compactness. Its major disadvantages arise from the creation of stem homographs and from the related need for affix interpretation following look-up. The advantage is reduced as the cost of storage devices is decreased, and as the speed of access to these devices is increased. In one sense, the pain of dealing with stem homographs and with affix interpretation is transient, since it is reduced and can be ignored once the methods described in Sec. 10.1 are fully developed. This pain is also made more bearable by the prospect that the techniques developed for compiling canonical dictionaries and for interpreting affixes can, if it should prove desirable to do so, be applied to the semiautomatic compilation of paradigm dictionaries.

The chief difference between paradigm and canonical dictionaries arises from the fundamental choice, already noted several times, between high memory requirements and simple algorithms on the one hand, and smaller memory requirements but more complex algorithms on the other. With a canonical dictionary, affix interpretation is essential for every entry that appears in an augmented text, and for every text. With a paradigm dictionary, affix interpretation may be done once and for all during compilation. If most information is concentrated in the entry for the classical canonical form, and all others are reduced to the status of cross-reference entries, the storage requirements for a paradigm dictionary may be sharply reduced, of course at the expense of increased algorithmic complexity. As usual, a whole spectrum of compromises is possible.

With hindsight, the process of compiling a canonical dictionary may be simplified, and a system for compiling paradigm dictionaries may be designed to take account of these simplifications. The starting point of compilation is, again, a list of classical canonical forms. However, not only a morphological class mark but all grammatical marks and a set of canonical English correspondents are assigned manually prior to transcription. The direct inflection of Russian may be accompanied by automatic interpretation of the generated forms, and by automatic modification of the English correspondents to match the interpretation. For nouns and adjectives, there is no doubt that all steps following transcription can be made fully automatic. For verbs, this possibility is quite likely, but less clear-cut. It seems, however, that if the generation of some "nonexistent" strings (Sec. 9.8) is tolerated almost all verbs will lend themselves to fully automatic compilation.

If and when it should prove worth while to produce paradigm dictionaries, existing canonical dictionaries would not be a loss. Since entries with stems factored from classical canonical strings are identified by a special mark (Sec. 7.7 and Fig. 53), these entries, automatically extracted from the stem dictionary file, provide a starting point for the automatic portion of the compilation process for a paradigm dictionary.

10.4. The Analysis of Contexts

Word-by-word translation is a linear approximation to more sophisticated transformations capable of mapping elements of the domain of translation into elements of the range. It is generally recognized that these more sophisticated transformations will be functions, not of isolated words, but of words and their *contexts*.

Broadly speaking, the context may include every aspect of the culture of which a text is a product that may be felt relevant. Jumping from word-by-word analysis to an analysis of contextual influence in the broadest sense is not likely to be fruitful; initially restricting analysis to linguistic context, and to a small neighborhood of a word at that, is more likely to lead to results both linguistically significant and concrete. This is not to say that all broader considerations can be wholly ignored, any more than syntactic problems could be wholly ignored in the development of lexical processes. They must, however, remain in the background as a guide, not as an object of immediate, active consideration.

Several of the intermediate files produced by the operation of an automatic dictionary provide the raw material for producing lists of great value in the study of contexts (Greenberg, 1959; Jones, 1958). Given a file containing items A, B, C, D, . . . in sequence, a simple program can produce a *master context file* with items such as ABCDE, BCDEF, CDEFG, DEFGH, . . . each consisting of an item of the original file surrounded by neighboring items, two on each side in this example. Contexts of certain kinds, for instance, all contexts of a given word or all contexts in which a noun is preceded by an adjective, may be selected from the master context file to form smaller subfiles. The master file, or any subfile, may be ordered in a variety of ways considered useful for various purposes.

One useful arrangement is obtained by sorting the entire master file, using the center word of each context as a primary key. The resulting file, suitably edited, provides a concordance to the text. A section of a concordance is shown in Fig. 90. The center of each group of contexts is identified in the left-hand column, and the contexts are displayed in the middle column. The numbers at the right are the text serial numbers of the context centers, which provide for cross reference to the original text. The concordance has long been a favorite tool of philologists and clergymen, among others. Recently, it has come into experimental use as a means for the automatic compilation of indexes to publications (Luhn and James, 1958); in this case, the contexts of words of habitually high frequency (prepositions, conjunctions, and the like) and those of words of very low frequency in the particular text are rejected. The remaining contexts are those of words likely to be significant in the particular text. As a further refinement, the use of inverse inflection could bring together under one heading all members of the same reduced paradigm.

The format of Fig. 90 is not very well suited for syntactic studies, for which a columnar layout has greater advantages. Figure 91 shows a set of contexts of the preposition $npu*$. The center of each context was used as the primary sorting key in the preparation of this list. The immediate right-hand neighbors

MUL.TIVIBRATOR
```
. ZAPERTYJ MUL.TIVIBRATOR S SAMOSTOJATEL.NYM                        3076
POKAZAN VTOROJ MUL.TIVIBRATOR TSEPOCHKI S                           2807
ZAPUSKAJUSHCHIJ PERVYJ MUL.TIVIBRATOR TSEPOCHKI .                   3114
. TRETIJ MUL.TIVIBRATOR TSEPOCHKI .                                 2917
```
MUL.TIVIBRATORA
```
VOZVRASHCHENIE VTOROGO MUL.TIVIBRATORA V ISXODNOE                   2904
ZAPERTOGO MUL.TIVIBRATORA MUL.TIVIBRATORA ISXODNOE SOSTOJANIE       2845
VOZVRASHCHENIE ZAPERTOGO MUL.TIVIBRATORA MUL.TIVIBRATORA ISXODNOE   2844
PERVOGO UPRAVLJAJUSHCHEGO MUL.TIVIBRATORA PROISXODJAT .             3188
ZAPUSK PERVOGO MUL.TIVIBRATORA TSEPOCHKI I                          2997
PREDYDUSHCHEGO PERVOGO MUL.TIVIBRATORA TSEPOCHKI OTPIRAET           2825
ZAPUSK TRET.EGO MUL.TIVIBRATORA TSEPOCHKI PROISXODIT                2941
PRAVYJ TRIOD MUL.TIVIBRATORA . CHEM                                 2834
PRAVOGO TRIODA MUL.TIVIBRATORA . VOZVRASHCHENIE                     2900
PERVOGO UPRAVLJAJUSHCHEGO MUL.TIVIBRATORA . TSEP.                   2962
```
MUL.TIVIBRATOROV
```
SAMOSTOJATEL.NOE VOZVRASHCHENIE MUL.TIVIBRATOROV V ISXODNOE         2886
VSEX DESJATI MUL.TIVIBRATOROV V ISXODNOE                            3003
OT UPRAVLJAEMYX MUL.TIVIBRATOROV DO TEX                             3202
TSEPOCHKI ZAPERTYX MUL.TIVIBRATOROV I UPRAVLJAJUSHCHEJ              2796
VSEX DESJATI MUL.TIVIBRATOROV ODNIM SOVOENNYM                       2872
DVUX ZAPERTYX MUL.TIVIBRATOROV S ODINAKOVYM                         3139
ZAPUSKA UPRAVLJAJUSHCHIX MUL.TIVIBRATOROV SODERZHITSJA TOL.KO       2967
TSEPOCHKA UPRAVLJAMYX MUL.TIVIBRATOROV . POSLE                      3213
JACHEJKAX TSEPOCHKI MUL.TIVIBRATOROV . $10-                         2977
SXEMU UPRAVLJAJUSHCHIX MUL.TIVIBRATOROV . SAMOSTOJATEL.NOE          2882
IZ UPRAVLJAJUSHCHIX MUL.TIVIBRATOROV . UPRAVLJAJUSHCHIE             3129
```
MUL.TIVIBRATORY
```
CHETNYE UPRAVLJAEMYE MUL.TIVIBRATORY ODNOVREMENNO .                 3181
. UPRAVLJAJUSHCHIE MUL.TIVIBRATORY PREDSTAVLJAJUT SOBOJ             3132
TSEPOCHKI . MUL.TIVIBRATORY SRABATYVAJUT TAKIM                      2920
```
MUL.TIVI.RATORA
```
PRAVYJ TRIOD MUL.TIVI.RATORA . POLOZHITELNYJ                        2818
```

Fig. 90. Concordance.

VERXNJAJA		GRANITSA	PRI	VYBORE	SOPROTIVLENIJA	
OSTSILLOGRAFA		*.	PRI	PODACHE	SINUSOIDAL.NOGO	0293
EHFFEKTIVNOST.		DOSTIGAETSJA	PRI	ISPOL.ZOVANII	NAIBOL.SHEJ	
NE		IZMENJAJUTSJA	PRI	REGULIROVANII	DLITEL.NOSTI	
EMKOSTI	$C SUB 2		PRI	KONSTRUIROVANII	PRIBORA	
IMENNO		*.	PRI	KONSTRUIROVANII	SINXRONNOGO	
DETEKTORA		*.	PRI	KONSTRUIROVANII	SINXRONNOGO	
JAVLJAJUSHCHIJSJA		TSEL.JU	PRI	KONSTRUIROVANII	PRIBORA	0294
PEREXODNYE		PROTSESSY	PRI	VOSPROIZVEDENII	SIGNALA	
SHUM		*.	PRI	OSUSHCHESTVLENII	SINXRONNOGO	
*.		ISPOL.ZOVANNYX	PRI	NAKOPLENII	*.	
JAVLJAETSJA		KRITERIEM	PRI	SRAVNENII	*.	
VOL.TMETROM		*2	PRI	USREDNENII	PO	
CHASTOTAM		*.	PRI	PRIMENENII	POLUCHENNYX	0295
FLUKTUATSII		OTSCHETA	PRI	IZMERENII	VSEX	
EHLEKTRONNOJ		APPARATURY	PRI	IZMERENII	MALYX	
MEDLENNO		RASTET	PRI	UVELICHENII	$DELTA T.	
USREDNENNYX		ORDINAT	PRI	IZUCHENII	NEBOL.SHOJ	
KONDENSATORE	$EQUATION 5		PRI	PODKLJUCHENII	SLEDUJUSHCHEGO	
OBNARUZHIT.		OTRAZHENIE	PRI	OTNOSHENII	SIGNAL	0296
NAPRIMER		*.	PRI	OTNOSHENII	SIGNAL	
*.		I	PRI	UMEN.SHENII	POLOSY	
PRIBORA		POLUCHAETSJA	PRI	PRAKTICHESKI	ISPOL.ZUEMYX	
VVEDENIE		SHUMA	PRI	NEIZMENNOJ	AMPLITUDE	
NA		KONDENSATORE	PRI	KAZHDOM	PODKLJUCHENII	
CHASTOT		SNIZU	PRI	PREDVARITEL.NOM	USILENII	0297
NAIBOL.SHEJ		VOZMOZHNOJ	PRI	ZADANNYX	PARAMETRAX	
TEORETICHESKI		VOZMOZHNOMU	PRI	ZADANNYX	VREMENI	
NA		VYXODE	PRI	RAZLICHNYX	AMPLITUDAX	
NE		MENJAETSJA	PRI	ZNACHITEL.NYX	KOLEBANIJAX	
NABLJUDAT.		SIGNAL	PRI	OTNOSHENIJAX	SIGNAL	
DANNYMI		DOLZHEN	PRI	$DELTA T APPROX. 88 MU	SEK.	0298
PEREXODNYX		PROTSESSOV	PRI	$DELTA T MUCH GREATER TH	PRENEBREZHIMO	
MALO		*.	PRI	$DELTA T MUCH LESS THAN	POSTOJANNAJA	
VREMENI	$EQUATION 6		PRI	$DELTA T RC SUB 2	POSTOJANNAJA	
VREMENI		RAVNA	PRIBLIZHENNO	$1.44 RC SUB 2.	POSTOJANNAJA	
*.		POSTROENNYJ	PRIBOR	BYL	PODROBNO	
RASCHETA		*.	PRIBOR	POZVOLJAET	TAKIM	0299
K		VXODU	PRIBORA	CHEREZ	SOPROTIVLENIE	
PRINTSIP		DEJSTVIJA	PRIBORA	I	EGO	
USTANOVLENIJA		POKAZANIJ	PRIBORA	SREDNIJ	KVADRAT	
USTANOVLENIJA		POKAZANIJ	PRIBORA	$T	RAVNA	
ISPYTANIE		POSTROENNOGO	PRIBORA	$TRXT:	REALIZATSIJA	
OSNOVNYE		PARAMETRA	PRIBORA	*	SHIRINA	0300
POLNOJ		SXEME	PRIBORA	*2	RIS.	
DLJA		RAZRABOTKI	PRIBORA	*.	UDOVLETVORJAJUSHCHEGO	
OSNOVNYE		KASKADY	PRIBORA	*.	OSUSHCHESTVLJAJUSHCHIE	
KONKRETNOJ		SXEMY	PRIBORA	*.	V	
SHIRINA		POLOSY	PRIBORA	*.	VYPOLNJAJUSHCHEGO	
KONSTRUKTIVNOGO		OFORMLENIJA	PRIBORA	*.	I	0301
V		SXEME	PRIBORA	*.	VOZNIKNOVENIE	
PRI		KONSTRUIROVANII	PRIBORA	*.	ODNOJ	
PRI		KONSTRUIROVANII	PRIBORA	*.	CHTOBY	
K		KORPUSU	PRIBORA	*.	V	
EHFFEKTIVNOST.		PRIMENENIJA	PRIBORA	*.	MAKSIMAL.NAJA	
PRIMENENIJA		TAKOGO	PRIBORA	DLJA	NABLJUDENIJA	0302
POLOSA		PROPUSKANIJA	PRIBORA	DLJA	SINXRONNOGO	
USTANOVLENIJA		POKAZANIJ	PRIBORA	POLUCHAETSJA	PRI	
POKAZANIJ		KAZHDOGO	PRIBORA	XARAKTERIZUETSJA	SHIRINOJ	

Fig. 91. Contexts of *npu**.

of the centers are end-alphabetized. The remaining three words served as subsidiary keys, but their effect is hardly noticeable in so short a list. The left-justification of a column indicates that the words in it were used as an ordinary key, while right-justification denotes end-alphabetization.

Two major groups of contexts of *при** are apparent: those where it is followed directly by a noun, and those where an adjective intervenes. It is well known, of course, that *при** governs the prepositional case. The dative/prepositional ambiguity in the interpretation of "амплитуде" ("неизменной амплитуде", Fig. 91) may therefore, if desirable, be resolved in this context.

From a purely structuralist viewpoint, the function of context lists would be to aid in the *discovery* of the governing properties of *при**. It is more convenient to assume these properties as a working hypothesis and to use context lists to check for exceptions. For this purpose, context lists incorporating some of the information given in Fig. 88 would be quite useful. One arrangement could be obtained by using the interpretations of affixes as primary keys. Like ambiguities would be brought together, the contexts in which well-known classical properties can serve to resolve the ambiguities would be readily identifiable, and attention could be focused on exceptional situations. Contexts of nouns exhibiting internal homography involving genitives are shown in Fig. 92. Extensive lists of this type are being used to study the resolution of internal homography as a function of context.

One question of great importance in the analysis of technical texts is the syntactic role of equations. Figure 93 shows how context lists may be used in approaching this question. Contexts a few words long may also be used in studying the macroscopic features of sentence organization. For example, the conjunction of comma and *что**, illustrated in Fig. 94, marks a boundary between major components of a sentence.

Numerous other possibilities suggest themselves. As one example, class contexts, rather than word contexts, may be studied on lists produced from interpreted augmented texts, in which only the functional class marks (N, A, V, and so on) and the affix interpretations appear. The full exploitation of these possibilities, exciting for all realms of linguistics, is still in the future.

10.5. Trial Translating

An experimental system suitable for advanced research on automatic translation is sketched in Fig 95. The automatic dictionary is a fundamental part of this system; it is extended by affix-interpreting and English inflecting programs described in Secs. 10.1 and 10.2 respectively. The *trial translator* (Giuliano, 1958) is visualized in a research instrument that will enable the testing of translation algorithms on large samples of Russian text. The *formula finder* (Giuliano, 1959*b*) is conceived as a process for the automatic derivation of certain classes of translation algorithms.

While several rules for producing smooth Russian-English automatic translations have been proposed in the literature (see the bibliographic notes, Sec. 10.7), published experimental results have been conspicuously absent—discounting several newspaper reports that have never been adequately substantiated in the technical literature. Until very recently, the reported use of automatic machines has been confined to the applications of computer programs tailored *ad hoc* to the processing of particular sentences or of carefully selected texts. Since few if any of the algorithms proposed for Russian-English translation have been tested on large bodies of Russian text, it has been exceedingly difficult, if not impossible, to evaluate them objectively.

Trial translating is the process of applying experimental translation algorithms to representative Russian texts, examining the results, and evaluating the algorithms. For such a process to be practical and meaningful, the algorithms must be applied by a machine. Giuliano has proposed an automatic programming system designed to put the computer readily at the disposal of linguists, Slavic scholars, and other individuals not usually trained in computer programming. Only the essential features of this system, called the trial translator, will be mentioned here.

The inputs to the trial translator are a set of experimental syntactic and semantic algorithms expressed in a notation based on that of the propositional calculus. Each of the algorithms is expressed in the form: "if the logical condition P is satisfied, then apply the transformation Q". P is an expression compounded of the logical variables that determine when the transformation Q is to be applied to a text. If k is a serial number defining the position of an entry in the augmented text, typical variables might be $N(k)$ standing for "text word k is a noun", $GP(k)$ standing for "text word k is in the genitive plural", $PREP(k-1)$ standing for "the text word preceding word k is a preposition", and so on. A logical expression P is constructed by connecting variables with the connective functors of the propositional calculus: "$\cdot$" standing for "and", "v" standing for "or", and "$\sim$" standing for "not". Typical transformations might be $PERM(k,k+1)$ standing for "permute the translations of text words k and $k+1$", INS(of the, k) standing for "insert *of the* before the translation of k", and so on. We might then consider the sample algorithm: "If a noun in the genitive plural is not immediately preceded by a preposition, then insert *of the* before its translation". This can simply be abbreviated as $\sim PREP(k-1) \cdot N(k) \cdot GP(k) \to INS$(of the, k). The algorithm is obviously too simple to be valid, and is included here only to illustrate the use of the notation.

```
- - - - - - - - - - - -   - - - - - - - - - - - -   - - - - - - - - - - - -   - - - - - - - - - - - -   - - - - - - - - - - - -

FORM-Y                    EHT-OGO                   URAVNENI-JA               S-                        TAKOV-YM
                                                                 -G----N-A---
BEND                      THIS                      EQUATION                  WITH                      SUCH
FORM                                                EQUALIZATION              AND                       THE SAME
                                                                              FROM
                                                                              SINCE
FORM-Y                                                                        IN                                  00Q-1018
+++++(HOMOGRAPH OF PREV)

- - - - - - - - - - - -   - - - - - - - - - - - -   - - - - - - - - - - - -   - - - - - - - - - - - -   - - - - - - - - - - - -

XIMICHESK-IX              JAVLENI-J                 IZMENENI-JA               SOSTAV-A                  SOSTAV-A
                                   -------G---                   -G----N-A---          -G----------
CHEMICAL                  APPEARANCE                ALTERATION                COMPOUND                  WILL FORM
                          PHENOMENON                CHANGE                    COMPOSITION               WILL BE PUT
                          EFFECT                    MODIFICATION              CONSTITUTION              TOGETHER
                                                                              STRUCTURE                 WILL BE FORMULATED
                                                                              STAFF                     WILL BE CONSTITUTED
                                                                                                        +++++(HOMOGRAPH OF PREV.)

- - - - - - - - - - - -   - - - - - - - - - - - -   - - - - - - - - - - - -   - - - - - - - - - - - -   - - - - - - - - - - - -

P-O                       POVOD-U                   TEORI-I                   OBOBSHCHENN-YX            KRITICHESK-IX
                                   ---D--------                  -G-D-PN-A---
ON                        OCCASION                  THEORY                    GENERALIZED               CRITICAL
ALONG                     GROUND                                              UNIFIED                   CENSORIOUS
BY                        CAUSE                                                                         CARPING
THROUGH
IN                                                                                                                00Q-1053

- - - - - - - - - - - -   - - - - - - - - - - - -   - - - - - - - - - - - -   - - - - - - - - - - - -   - - - - - - - - - - - -

RASCHET-U                 KONSTANT-                 RAVNOVESI-JA              N-A                       PREDEL-AX
          ---D--------             -------G---                   -G----N-A---                                  ----------P
ACCOUNT                   CONSTANT                  EQUILIBRIUM               ON                        LIMIT
CALCULATION                                         BALANCE                   UPON                      BOUND
COMPUTATION                                         EQUIPOSE                  IN                        END
ESTIMATE                                                                      AT
SETTLING                                                                      TO

- - - - - - - - - - - -   - - - - - - - - - - - -   - - - - - - - - - - - -   - - - - - - - - - - - -   - - - - - - - - - - - -

TAKZH-E                   -O                        VOZMOZHNOST-I             SUSHCHESTVOVANI-JA        DVUX-
                                                                 -G-D-PN-A---
TOO                       OF                        POSSIBILITY               SUSHCHESTVOVANI-JA        TWO
EITHER                    ABOUT                     OPPORTUNITY                                         OTHER
                          ON                        MEANS                     ###                       FEW

                                                                                                                  00Q-1081
```

Fig. 92. Internal homography involving genitives.

$EQUATION 1	$T, OTKUDA $EQUATION V SLUCHAE	0776
$EQUATION 2	, OTKUDA $EQUATION GDE $F	0797
$EQUATION 4	, OTKUDA $EQUATION EHTO ZHE	1908
$EQUATION 5	NA KONDENSATORE $EQUATION PRI PODKLJUCHENII	2136
$EQUATION 6	POSTOJANNOJ VREMENI $EQUATION PRI $DELTA	2174
$EQUATION 7	SOOTVETSTVUJUSHCHEJ $DELTA $EQUATION DLITEL,NOSTI PODKLJUCHENIJA	2239
$EQUATION 8	VOSPROIZVEDENIJA SIGNALA $EQUATION EHTO SOOTNOSHENIE	2287

Fig. 93. Equations in context.

ISPYTANIJA POKAZALI , CHTO VVEDENIE	2608
NAIBOLEE SUSHCHESTVENNO , CHTO VYVEDENNAJA	1184
SUSHCHESTVENNO OTMETIT, , CHTO DLITEL,NOST,	0376
. SUSHCHESTVENNO , CHTO DLJA	1045
$(5) POKAZYVAET , CHTO ZATUXANIE	2162
SINXRONNYX DETEKTOROV , CHTO I	1025
NEUSREDNENNYJ SIGNAL , CHTO I	0640
OSUSHCHESTVLJAETSJA GAL,VANICHESKI , CHTO ISKLJUCHAET	2337
TOGDA OKAZYVAETSJA , CHTO NAPRJAZHENIE	1877
ESLI ZAMETIT, , CHTO POSTOJANNAJA	1921
$(2) POKAZYVAET , CHTO PRIMENENIE	1011
MOZHNO UTVEPZHDAT, , CHTO SINXRONNYJ	2733
V VIDU , CHTO SPEKTRAL,NAJA	1150
LEGKO POKAZAT, , CHTO ,	2050

Fig. 94. The conjunction of comma and *что**.

The operation of the trial translator is based on the automatic association of algorithms with dictionary entries, the automatic specification of the truth values of logical variables, and the automatic evaluation of logical formulas. The system applies the given algorithms to English inflected augmented texts (Fig. 95). Its outputs are improved translations resulting from the application of the given rules to the given texts. Linguists, psychologists,

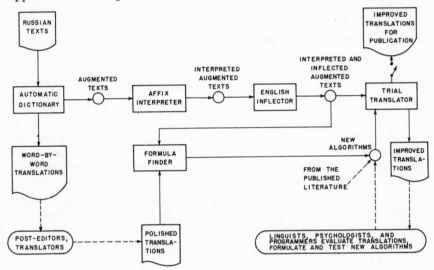

Fig. 95. An experimental system for research on automatic translation.

and computer specialists examine these translations and suggest modified algorithms that are, in turn, coded in the abbreviated notation and applied by the trial translator. The same man-machine cycle can be repeated until a satisfactory set of algorithms is determined, or until it is obvious that some major change must be made in the machine system.

10.6. Formula Finding

The first formulas to be tested while experimenting with the trial translator may be either drawn from the existing literature or else obtained in the course of experiments with the products of the automatic dictionary (Fig. 92). It may also prove possible to find algorithms automatically, by means of a system called *formula finder*, now being investigated (Giuliano, 1959b).

The inputs to the formula finder are an augmented Russian text, and a postedited translation of the same text. The English translation must first be transcribed onto magnetic tape by manual or automatic means. When given proper clues by a linguist, the system is expected to produce algorithms that can be used to transform the augmented text into the edited version.

The clues that must be given to the formula finder are: (1) a list of logical variables that might conceivably determine a certain transformation, and (2) a statement of the transformation being investigated. The variables and transformation are assumed to be stated in the notation used as input to the trial translator (Sec. 10.5). The formula finder compares the augmented text and the postedited text. Whenever a product of the indicated transformation is found in the postedited text and certain auxiliary conditions are satisfied, the formula finder examines the truth-value configuration of the given variables in the augmented text. After examining all instances of the transformation, the formula finder can ascertain whether the indicated variables can be combined into a logical formula that implies the given transformation. The output of the formula finder can be.

1. A logical formula that always implies the given transformation, thus defining a translation algorithm valid for the given corpus of text; the formula finder will reduce this formula to its simplest logical form and eliminate vacuous variables not actually needed in the algorithm;

2. A statement of the "closest" logical formula in case other variables besides those originally given are required to determine the transformation; the statement will be accompanied by indications of the quality of the approximate formula, and by clues that will help linguists to suggest additional variables for testing;

3. A machine-coded statement of the exact or approximate algorithm, ready to be tested by the trial translator.

Since natural languages are open systems, an algorithm-finding process can never quite be finished. Nevertheless, there is hope that the processes of trial translating and formula finding will eventually lead to an acceptably stable and elegant set of algorithms and that, in the interim, the quality of the trial translations will steadily improve.

10.7. Bibliographic Notes

A survey of early efforts in the field of automatic language translation, including statements of many problems that are still unsolved, may be found in Locke and Booth (1955). Booth, Brandwood, and Cleave (1958) discuss the application of information-processing machines to linguistic problems, with special emphasis on automatic translation from German to English.

The journal *Mechanical Translation*, published at the Massachusetts Institute of Technology since 1954 under the editorship of Locke and Yngve, is the major source of articles in English. Each issue of this journal includes a bibliography with short abstracts that will guide the interested reader to scattered international journals. The journal *Babel* published in Bonn by the International Federation of Translators occasionally publishes papers on automatic translation. Soviet work on automatic translation has been reported chiefly in the journal *Voprosy Jazykoznanija*. A comprehensive bibliography of Russian publications may be found in Oettinger (1959*b*). Papers by Mel'chuk (1958) and by Kulagina (1958) of which earlier informal versions appeared in the *Bjulleten' Ob"edinenija po Problemam Mashinnogo Perevoda*, and one by Moloshnaja (1958), appear in Ljapunov (1958).

Reports of research on automatic translation have been made by the University of Washington (1958) and by the Ramo-Wooldridge Corporation (1958). A series of research memoranda, of which the latest available at the time of writing (Edmundson *et al.*, 1958) carries a list of the earlier ones, has been prepared at the Rand Corporation. The first of a series entitled "Occasional Papers on Machine Translation" was prepared at Georgetown University by Brown (1959). Bar Hillel (1955) gives a critical review of the state of automatic translation in the United States and Great Britain at the beginning of 1959.

Descriptions of work on translation between natural and artificial languages or among artificial ones may be found in some of the references on automatic coding cited in Sec. 3.5. The National Science Foundation issues semiannual reports on current research in scientific documentation (see, for example, National Science Foundation, 1958), which are a good source of current information and of bibliographic leads, not only in the field of automatic language translation, but also regarding related research on the general problems of information storage and retrieval.

APPENDIX

The Interpretation of Desinences in Nouns

No.	Desi-nence	N1 I1	N1 A1	N1 I2	N1 A2	N1.1 I1	N1.1 A1	N1.1 I2	N1.1 A2
1	#	NsAs	Ns	NsAs	Ns	NsAs	Ns	NsAs	Ns
2	а	Gs	GsAs	GsNpAp	GsAsNp	Gs	GsAs	GsNpAp	GsAsNp
3	у	Ds	Ds	Ds	Ds	Ds	Ds	Ds	Ds
4	е	Ps	Ps	Ps	Ps	Ps	Ps	Ps	Ps
5	ом	Is	Is	Is	Is	Is	Is	Is	Is
6	ов	Gp	GpAp	Gp	GpAp	Gp	GpAp	Gp	GpAp
7	ам	Dp	Dp	Dp	Dp	Dp	Dp	Dp	Dp
8	ах	Pp	Pp	Pp	Pp	Pp	Pp	Pp	Pp
9	ами	Ip	Ip	Ip	Ip	Ip	Ip	Ip	Ip
10	ы	NpAp	Np						
11	и					NpAp	Np		
12	я								
13	ю								
14	й								
15	ев								
16	ем								
17	ях								
18	ям								
19	ями								
20	ь								
21	ей								
22	ой								
23	о								
24	ью								
25	ьми								

APPENDIX (*continued*)

No.	Desi-nence	N1.2 I1	N1.2 A1	N1.2 I2	N1.2 A2	N1.3 I1	N1.3 A1	N1.4 I1	N1.4 A1
					Class				
1	#	NsAs	Ns	NsAs	Ns	NsAs	Ns	NsAs	Ns
2	a	Gs	GsAs	GsNpAp	GsAsNp	Gs	GsAs	Gs	GsAs
3	y	Ds	Ds	Ds	Ds	Ds	Ds	Ds	Ds
4	e	Ps	Ps	Ps	Ps	Ps	Ps	Ps	Ps
5	ом	Is	Is	Is	Is	Is	Is		
6	ов	Gp	GpAp	Gp	GpAp	Gp	GpAp		
7	ам	Dp	Dp	Dp	Dp	Dp	Dp	Dp	Dp
8	ах	Pp	Pp	Pp	Pp	Pp	Pp	Pp	Pp
9	ами	Ip	Ip	Ip	Ip	Ip	Ip	Ip	Ip
10	ы	NpAp	Np					NpAp	Np
11	и					NpAp	Np		
12	я								
13	ю								
14	й								
15	ев							Gp	GpAp
16	ем							Is	Is
17	ях								
18	ям								
19	ями								
20	ь								
21	ей								
22	ой								
23	о								
24	ью								
25	ьми								

APPENDIX (*continued*)

		Class							
No.	Desi-nence	*N2* *I1*	*N2* *A1*	*N2* *I2*	*N2* *A2*	*N2* *I3*	*N2* *A3*	*N2.1* *I1*	*N2.1* *A1*
1	#								
2	а								
3	у								
4	е	Ps	Ps			Ps	Ps	Ps	Ps
5	ом								
6	ов								
7	ам								
8	ах								
9	ами								
10	ы								
11	и	NpAp	Np	PsNpAp	PsNp			NpAp	Np
12	я	Gs	GsAs	Gs	GsAs	GsNpAp	GsAsNp	Gs	GsAs
13	ю	Ds	Ds	Ds	Ds	Ds	Ds	Ds	Ds
14	й	NsAs	Ns	NsAs	Ns	NsAs	Ns	NsAs	Ns
15	ев	Gp	GpAp	Gp	GpAp	Gp	GpAp	Gp	GpAp
16	ем	Is	Is	Is	Is	Is	Is	Is	Is
17	ях	Pp	Pp	Pp	Pp	Pp	Pp	Pp	Pp
18	ям	Dp	Dp	Dp	Dp	Dp	Dp	Dp	Dp
19	ями	Ip	Ip	Ip	Ip	Ip	Ip	Ip	Ip
20	ь								
21	ей								
22	ой								
23	о								
24	ью								
25	ьми								

APPENDIX (*continued*)

No.	Desinence	N3 I1	N3 A1	N3 I2	N3 A2	N3.05 I1	N3.05 A1
1	#						
2	а						
3	у						
4	е	Ps	Ps	Ps	Ps	Ps	Ps
5	ом						
6	ов						
7	ам						
8	ах						
9	ами						
10	ы						
11	и	NpAp	Np			NpAp	Np
12	я	Gs	GsAs	GsNpAp	GsAsNp	Gs	GsAs
13	ю	Ds	Ds	Ds	Ds	Ds	Ds
14	й						
15	ев						
16	ем	Is	Is	Is	Is	Is	Is
17	ях	Pp	Pp	Pp	Pp	Pp	Pp
18	ям	Dp	Dp	Dp	Dp	Dp	Dp
19	ями	Ip	Ip	Ip	Ip	Ip	Ip
20	ь	NsAs	Ns	NsAs	Ns	NsAs	Ns
21	ей	Gp	GpAp	Gp	GpAp	Gp	GpAp
22	ой						
23	о						
24	ью						
25	ьми						

APPENDIX (*continued*)

		Class					
No.	Desi-nence	*N3.1* *I1*	*N3.1* *A1*	*N3.1* *I2*	*N3.1* *A2*	*N3.1* *I3*	*N3.1* *A3*
1	#	NsAs	Ns	NsAs	Ns	NsAs	Ns
2	a	Gs	GsAs	Gs	GsAs	GsNpAp	GsAsNp
3	у	Ds	Ds	Ds	Ds	Ds	Ds
4	е	Ps	Ps	Ps	Ps	Ps	Ps
5	ом	Is	Is				
6	ов						
7	ам	Dp	Dp	Dp	Dp	Dp	Dp
8	ах	Pp	Pp	Pp	Pp	Pp	Pp
9	ами	Ip	Ip	Ip	Ip	Ip	Ip
10	ы						
11	и	NpAp	Np	NpAp	Np		
12	я						
13	ю						
14	й						
15	ев						
16	ем			Is	Is	Is	Is
17	ях						
18	ям						
19	ями						
20	ь						
21	ей	Gp	GpAp	Gp	GpAp	Gp	GpAp
22	ой						
23	о						
24	ью						
25	ьми						

APPENDIX (*continued*)

No.	Desi-nence	N4 I1	N4 A1	N4 I2	N4 A2	N4.05 I1	N4.05 A1	N4.06 I1	N4.06 A1
						Class			
1	#	Gp	GpAp	Gp	GpAp	Gp	GpAp	Gp	GpAp
2	а	Ns	Ns	Ns	Ns	Ns	Ns	Ns	Ns
3	у	As	As	As	As	As	As	As	As
4	е	DsPs	DsPs	DsPs	DsPs	DsPs	DsPs	DsPs	DsPs
5	ом								
6	ов								
7	ам	Dp	Dp	Dp	Dp	Dp	Dp	Dp	Dp
8	ах	Pp	Pp	Pp	Pp	Pp	Pp	Pp	Pp
9	ами	Ip	Ip	Ip	Ip	Ip	Ip	Ip	Ip
10	ы	GsNpAp	GsNp	GsNpAp	GsNp	GsNpAp	GsNp	GsNpAp	GsNp
11	и								
12	я								
13	ю								
14	й								
15	ев								
16	ем								
17	ях								
18	ям								
19	ями								
20	ь								
21	ей	Is	Is						
22	ой			Is	Is	Is	Is	Is	Is
23	о								
24	ью								
25	ьми								

APPENDIX (*continued*)

No.	Desi-nence	Class N4.1 I1	N4.1 A1	N4.1 I2	N4.1 A2	N4.1 I3	N4.1 A3	N4.1 I4	N4.1 A4
1	#	Gp	GpAp					Gp	GpAp
2	а	Ns	Ns	Ns	Ns	Ns	Ns	Ns	Ns
3	у	As	As	As	As	As	As	As	As
4	е	DsPs	DsPs	DsPs	DsPs	DsPs	DsPs	DsPs	DsPs
5	ом								
6	ов								
7	ам	Dp	Dp	Dp	Dp	Dp	Dp	Dp	Dp
8	ах	Pp	Pp	Pp	Pp	Pp	Pp	Pp	Pp
9	ами	Ip	Ip	Ip	Ip	Ip	Ip	Ip	Ip
10	ы								
11	и	GsNpAp	GsNp	GsNpAp	GsNp	GsNpAp	GsNp	GsNpAp	GsNp
12	я								
13	ю								
14	й								
15	ев								
16	ем								
17	ях								
18	ям								
19	ями								
20	ь								
21	ей			IsGp	IsGpAp	Gp	GpAp	Is	Is
22	ой	Is	Is			Is	Is		
23	о								
24	ью								
25	ьми								

APPENDIX (*continued*)

No.	Desi-nence	N4.3 I1	N4.3 A1	N4.31 I1	N4.31 A1	N5 I1	N5 A1	N5 I2	N5 A2
1	#	Gp	GpAp	Gp	GpAp				
2	а	Ns	Ns	Ns	Ns				
3	у	As	As	As	As				
4	е	DsPs	DsPs	DsPs	DsPs	DsPs	DsPs	DsPs	DsPs
5	ом								
6	ов								
7	ам	Dp	Dp	Dp	Dp				
8	ах	Pp	Pp	Pp	Pp				
9	ами	Ip	Ip	Ip	Ip				
10	ы								
11	и	GsNpAp	GsNp	GsNpAp	GsNp	GsNpAp	GsNp	GsNpAp	GsNp
12	я					Ns	Ns	Ns	Ns
13	ю					As	As	As	As
14	й								
15	ев								
16	ем								
17	ях					Pp	Pp	Pp	Pp
18	ям					Dp	Dp	Dp	Dp
19	ями					Ip	Ip	Ip	Ip
20	ь							Gp	GpAp
21	ей					IsGp	IsGpAp	Is	Is
22	ой	Is	Is	Is	Is				
23	о								
24	ью								
25	ьми								

APPENDIX (*continued*)

		Class							
No.	Desi-nence	*N5.05* *I1*	*N5.05* *A1*	*N5.1* *I1*	*N5.1* *A1*	*N5.1* *I2*	*N5.1* *A2*	*N5.15* *I1*	*N5.15* *A1*
1	#			Gp	GpAp				
2	а								
3	у								
4	е	DsPs	DsPs	DsPs	DsPs	DsPs	DsPs	DsPs	DsPs
5	ом								
6	ов								
7	ам								
8	ах								
9	ами								
10	ы								
11	и	GsNpAp	GsNp	GsNpAp	GsNp	GsNpAp	GsNp	GsNpAp	GsNp
12	я	Ns	Ns	Ns	Ns	Ns	Ns	Ns	Ns
13	ю	As	As	As	As	As	As	As	As
14	й	Gp	GpAp						
15	ев								
16	ем								
17	ях	Pp	Pp	Pp	Pp	Pp	Pp	Pp	Pp
18	ям	Dp	Dp	Dp	Dp	Dp	Dp	Dp	Dp
19	ями	Ip	Ip	Ip	Ip	Ip	Ip	Ip	Ip
20	ь					Gp	GpAp	Gp	GpAp
21	ей	Is	Is	Is	Is	Is	Is	Is	Is
22	ой								
23	о								
24	ью								
25	ьми								

APPENDIX (*continued*)

No.	Desi-nence	N5.2 I1	N5.2 A1	N5.3 I1	N5.3 A1	N6 I1	N6 A1	N6 I2	N6 A2
1	#								
2	а								
3	у								
4	е	DsPs	DsPs	DsPs	DsPs				
5	ом								
6	ов								
7	ам								
8	ах								
9	ами								
10	ы								
11	и	GsNpAp	GsNp	GsNpAp	GsNp	GsDsPs NpAp	GsDsPs Np	GsDsPs NpAp	GsDsPs Np
12	я	Ns	Ns	Ns	Ns				
13	ю	As	As	As	As				
14	й	Gp	GpAp	Gp	GpAp				
15	ев								
16	ем								
17	ях	Pp	Pp	Pp	Pp	Pp	Pp	Pp	Pp
18	ям	Dp	Dp	Dp	Dp	Dp	Dp	Dp	Dp
19	ями	Ip	Ip	Ip	Ip	Ip	Ip		
20	ь					NsAs	NsAs	NsAs	NsAs
21	ей	Is	Is	Is	Is	Gp	GpAp	Gp	GpAp
22	ой								
23	о								
24	ью					Is	Is	Is	Is
25	ьми							Ip	Ip

APPENDIX (*continued*)

		Class						
No.	Desi-nence	*N6.1* *Il*	*N6.1* *Al*	*N7* *Il*	*N7* *Al*	*N8* *Il*	*N8* *Al*	*N8.1* *Il*	*N8.1* *Al*
1	#					Gp	GpAp	Gp	GpAp
2	а					GsNpAp	GsNp	GsNpAp	GsNp
3	у					Ds	Ds	Ds	Ds
4	е					Ps	Ps	Ps	Ps
5	ом					Is	Is	Is	Is
6	ов								
7	ам	Dp	Dp			Dp	Dp	Dp	Dp
8	ах	Pp	Pp			Pp	Pp	Pp	Pp
9	ами	Ip	Ip			Ip	Ip	Ip	Ip
10	ы								
11	и	{ GsDsPs NpAp	GsDsPs Np	GsDsPs NpAp	GsDsPs Np				
12	я			Ns	Ns				
13	ю			As	As				
14	й			Gp	GpAp				
15	ев								
16	ем								
17	ях			Pp	Pp				
18	ям			Dp	Dp				
19	ями			Ip	Ip				
20	ь	NsAs	NsAs						
21	ей	Gp	GpAp	Is	Is				
22	ой								
23	о					NsAs	NsAs	NsAs	NsAs
24	ью	Is	Is						
25	ьми								

APPENDIX (*continued*)

No.	Desi-nence	Class							
		N8.15 *I1*	*N8.15* *A1*	*N8.3* *I1*	*N8.3* *A1*	*N8.3* *I2*	*N8.3* *A2*	*N8.3* *I3*	*N8.3* *A3*
1	#	Gp	GpAp					Gp	GpAp
2	a	GsNpAp	GsNp	Gs	Gs	GsNpAp	GsNp	Gs	Gs
3	у	Ds	Ds	Ds	Ds	Ds	Ds	Ds	Ds
4	е	Ps	Ps	Ps	Ps	Ps	Ps	Ps	Ps
5	ом	Is	Is	Is	Is	Is	Is	Is	Is
6	ов			Gp	GpAp	Gp	GpAp		
7	ам	Dp	Dp	Dp	Dp	Dp	Dp	Dp	Dp
8	ах	Pp	Pp	Pp	Pp	Pp	Pp	Pp	Pp
9	ами	Ip	Ip	Ip	Ip	Ip	Ip	Ip	Ip
10	ы								
11	и			NpAp	Np			NpAp	Np
12	я								
13	ю								
14	й								
15	ев								
16	ем								
17	ях								
18	ям								
19	ями								
20	ь								
21	ей								
22	ой								
23	о	NsAs	NsAs	NsAs	NsAs	NsAs	NsAs	NsAs	NsAs
24	ью								
25	ьми								

APPENDIX (*continued*)

No.	Desi-nence	N8.4 I1	N8.4 A1	N9 I1	N9 A1	N9 I2	N9 A2	N9.2 I1	N9.2 A1
						Class			
1	#	Gp	GpAp			Gp	GpAp	Gp	GpAp
2	а	Gs	Gs	GsNpAp	GsNp	GsNpAp	GsNp	GsNpAp	GsNp
3	у	Ds	Ds	Ds	Ds	Ds	Ds	Ds	Ds
4	е	Ps	Ps	NsAsPs	NsAsPs	NsAsPs	NsAsPs	NsAsPs	NsAsPs
5	ом	Is	Is						
6	ов								
7	ам	Dp	Dp	Dp	Dp	Dp	Dp	Dp	Dp
8	ах	Pp	Pp	Pp	Pp	Pp	Pp	Pp	Pp
9	ами	Ip	Ip	Ip	Ip	Ip	Ip	Ip	Ip
10	ы								
11	и	NpAp	Np						
12	я								
13	ю								
14	й								
15	ев			Gp	GpAp				
16	ем			Is	Is	Is	Is	Is	Is
17	ях								
18	ям								
19	ями								
20	ь								
21	ей								
22	ой								
23	о	NsAs	NsAs						
24	ью								
25	ьми								

APPENDIX (*continued*)

No.	Desi-nence	N10 II	N10 A1	N11 II	N11 A1	N11.1 II	N11.1 A1	N11.2 II	N11.2 A1
		Class							
1	#								
2	a								
3	y								
4	e	NsAs	NsAs	NsAsPs	NsAsPs	NsAsPs	NsAsPs	NsAsPs	NsAsPs
5	ом								
6	ов								
7	ам								
8	ах								
9	ами								
10	ы								
11	и	Ps	Ps						
12	я	GsNpAp	GsNp	GsNpAp	GsNp	GsNpAp	GsNp	GsNpAp	GsNp
13	ю	Ds	Ds	Ds	Ds	Ds	Ds	Ds	Ds
14	й	Gp	GpAp	Gp	GpAp	Gp	GpAp		
15	ев							Gp	GpAp
16	ем	Is	Is	Is	Is	Is	Is	Is	Is
17	ях	Pp	Pp	Pp	Pp	Pp	Pp	Pp	Pp
18	ям	Dp	Dp	Dp	Dp	Dp	Dp	Dp	Dp
19	ями	Ip	Ip	Ip	Ip	Ip	Ip	Ip	Ip
20	ь								
21	ей								
22	ой								
23	о								
24	ью								
25	ьми								

APPENDIX (*continued*)

		Class			
No.	Desi-nence	*N12 I1*	*N12 A1*	*N13 I1*	*N13 A1*
1	#				
2	а				
3	у				
4	е			NsAsPs	NsAsPs
5	ом				
6	ов				
7	ам				
8	ах				
9	ами				
10	ы				
11	и				
12	я	NsAs	Ns	GsNpAp	GsNp
13	ю			Ds	Ds
14	й				
15	ев				
16	ем			Is	Is
17	ях			Pp	Pp
18	ям			Dp	Dp
19	ями			Ip	Ip
20	ь				
21	ей			Gp	GpAp
22	ой				
23	о				
24	ью				
25	ьми				

BIBLIOGRAPHY

Throughout this Bibliography the following abbreviations will be used, with an asterisk preceding the entry:

*AF-46,49,50—Design and operation of digital calculating machinery, Progress Reports by the Staff of the Harvard Computation Laboratory to the United States Air Force, Cambridge, Massachusetts.

*NSF-2—Mathematical linguistics and automatic translation, Report to National Science Foundation, Harvard Computation Laboratory, Cambridge, Massachusetts.

*Seminar papers—L. G. Jones and A. G. Oettinger, Papers presented at the seminar in mathematical linguistics, Harvard University (on deposit at Widener Library, Harvard University, Cambridge, Massachusetts).

Abaev, V. I. (1957), "O podache omonimov v slovare" (On the treatment of homonyms in dictionaries), Voprosy jazykoznanija, No. 3, pp. 31–43.

Alt, F. L. (1958), Electronic digital computers, Academic Press, New York.

American Institute of Physics (1957), Soviet physics doklady, Vol. 1, No. 4, pp. 499–500.

American Mathematical Society (1953), Russian-English vocabulary with a grammatical sketch, American Mathematical Society, Providence, Rhode Island.

Andree, R. V. (1958), Selections from modern abstract algebra, Holt, New York.

Apostel, L., B. Mandelbrot, and A. Morf (1957), Logique, langage, et théorie de l'information, Presses Universitaires de France, Paris.

Ashenhurst, R. L. (1954), "Sorting and arranging," Theory of switching, Report No. BL-7, Sec. I, Harvard Computation Laboratory, Cambridge, Massachusetts.

*—— (1956), "The symbolic representation of structure," Seminar papers, Vol. II.

Association for Computing Machinery (1957), Proceedings of the symposium "New computers—a report from the manufacturers," Los Angeles, California.

Bar Hillel, Y. (1959), Report on the state of machine translation in the United States and Great Britain, Technical Report No. 1, Hebrew University, Jerusalem.

*Barnes, V. L. (1958), "An index routine," Seminar papers, Vol. IV.

Baumann, D. M. (1958), "A high-scanning-rate storage device for computer applications," Journal of the Association for Computing Machinery, Vol. 5, pp. 76–88.

Bazilevskij, Ju. Ja., editor (1958), Voprosy teorii matematicheskix mashin (On the theory of mathematical machines), Gosudarstvennoe Izdatel'stvo Fiziko-Matematicheskoj Literatury, Moscow.

Beberman, M. (1958), An emerging program of secondary school mathematics, Harvard University Press, Cambridge, Massachusetts.

Begun, S. J. (1955), "Magnetic memory device for business machines," Electrical engineering, Vol. 74 (June 1955), pp. 466–468.

Belevitch, V. (1956), Langage des machines et langage humain, Office de Publicité, Brussels.

Bennett, B. J., *et al.* (1955), "Electronics in financial accounting," *Proceedings of the Eastern Joint Computer Conference*, pp. 26–32.

*Berson, J. S. (1958), "The calculus of propositions in automatic translation," *Seminar papers*, Vol. IV.

Bielfeldt, H. H. (1958), *Rückläufiges Wörterbuch der Russischen Sprache der Gegenwart*, Akademie-Verlag, Berlin.

Birkhoff, G., and S. MacLane (1950), *A survey of modern algebra*, Macmillan, New York.

Bloch, B., and G. L. Trager (1942), *Outline of linguistic analysis*, Special Publications of the Linguistic Society of America, Waverly Press, Baltimore, Maryland.

Bloomfield, L. (1933), *Language*, Holt, New York.

Booth, A. D., and K. H. V. Booth (1956), *Automatic digital calculators* (2nd ed.), Butterworths Scientific Publications, London.

———— L. Brandwood, and J. P. Cleave (1958), *Mechanical resolution of linguistic problems*, Butterworths Scientific Publications, London.

*Bossert, W. (1958), "The problem of homographs in the automatic dictionary," *Seminar papers*, Vol. IV.

Bowden, B. V. (1953), *Faster than thought*, Pitman, London.

Brower, R. A., editor (1959), *On translation*, Harvard University Press, Cambridge, Massachusetts.

Brown, A. F. R. (1959), *Manual for a "simulated linguistic computer"—A system for direct coding of machine translation*, Occasional Papers on Machine Translation No. 1, Georgetown University, Institute of Languages and Linguistics, Machine Translation Research Center, Washington, D.C.

Buchler, J., editor (1955), *Philosophical writings of Peirce*, Dover, New York.

Bull, W. E., C. Africa, and D. Teichroew (1955), "Some problems of the 'word'," in Locke and Booth (1955).

Burge, W. H. (1958), "Sorting, trees, and measures of order," *Information and Control*, Vol. 1, No. 3, pp. 181–197.

Burks, A. W., D. W. Warren, and J. B. Wright (1954), "An analysis of a logical machine using parenthesis-free notation," *Mathematical Tables and Other Aids to Computation*, Vol. VIII, Nos. 45–48, pp. 53–57.

Caldwell, S. H. (1958), *Switching circuits and logical design*, Wiley, New York.

Carroll, J. B. (1953), *The study of language*, Harvard University Press, Cambridge, Massachusetts.

Carroll, J. M. (1956), "Trends in computer input/output devices," *Electronics*, Vol. 29 (September 1956), pp. 142–149.

Cary, E. (1956), *La traduction dans le monde moderne*, Georg, Geneva.

Chaundy, T. W., P. R. Barrett, and C. Batey (1954), *The printing of mathematics*, Oxford University Press, London.

Cherry, C. (1957), *On human communication*, Technology Press, Cambridge, Massachusetts, and Wiley, New York.

Chomsky, N. (1956), "Three models for the description of languages," *IRE Transactions*, Vol. IT-2, No. 3, pp. 113–124.

———— (1957), *Syntactic structures*, Mouton, The Hague.

Church, A. (1956), *Introduction to mathematical logic*, Vol. 1, Princeton University Press, Princeton, New Jersey.

Churchill, W. (1901), *The crisis*, Macmillan, New York.

Citroen, I. J. (1959), "The translation of texts dealing with applied science," *Babel*, Vol. V, No. 1, pp. 30–33.

*Cohn, M. (1958), "On the design of an affix splitting device," *AF-50*, Sec. IV.

*Coppinger, L. L., J. W. Hannan, D. Isenberg, C. J. Rogers, and S. von Susich (1959), "Correcting and updating the Harvard automatic dictionary," *NSF-2*, Sec. VIII.

Curry, H. B. (1953), "Mathematics, syntactics and logic," *Mind*, Vol. 62, pp. 172–183.

Daum, E., and W. Schenk (1954), *Die Russischen Verben*, VEB Bibliographisches Institut, Leipzig.

David, E. E., Jr. (1958), "Artificial auditory recognition in telephony," *IBM Journal of Research and Development*, Vol. 2, No. 4, pp. 294–309.

Davidson, J. T., and R. L. Fortune (1955), "Automatic translation of printed code to impulses acceptable to computing equipment," *Proceedings of the Western Joint Computer Conference*, pp. 29–33.

Davies, D. W. (1956), "Sorting of data on an electronic computer," *Proceedings of the Institute of Electrical Engineers 103*, Part B, Supplement 1–3, pp. 87–93.

Davis, K. H., R. Biddulph, and S. Balashek (1952), "Automatic recognition of spoken digits," *Journal of the Acoustical Society of America*, Vol. 24, pp. 637–642.

Dean, N. J. (1957), "The variable word and record length problem and the combined record approach on electronic data processing systems," *Proceedings of the Western Joint Computer Conference*, pp. 214–218.

Demuth, H. B. (1956), *Electronic data sorting*, Stanford Research Institute, Menlo Park, California.

Department of the Army (1956), *English-Russian, Russian-English electronics dictionary*, TM 30–545.

Dewey, G. (1923), *Relativ frequency of English speech sounds*, Harvard University Press, Cambridge, Massachusetts.

Dimond, T. L. (1957), "Devices for reading handwritten characters," *Proceedings of the Eastern Joint Computer Conference*, pp. 232–237.

Dostert, L. E. (1955), "The Georgetown–I.B.M. experiment," in Locke and Booth (1955).

Dudley, H. (1955), "Fundamentals of speech synthesis," *Journal of the Audio Engineering Society*, Vol. 3, pp. 170–185.

Eckert, J. P., Jr. (1953), "A survey of digital computer memory systems," *Proceedings of the Institute of Radio Engineers 41*, pp. 1393–1406.

Edmundson, H. P., K. E. Harper, D. G. Hays, and A. M. Koutsoudas (1958), "Studies in machine translation—9: Bibliography of Russian scientific articles," Report RM 2069 (October 16, 1958), The Rand Corporation, Santa Monica, California.

Eisler, G. (1956), "Requirements for a rapid access data file," *Proceedings of the Western Joint Computer Conference*, pp. 39–42.

Eldredge, K. R., F. J. Kamphoefner, and P. H. Wendt (1956), "Automatic input for business data processing systems," *Proceedings of the Eastern Joint Computer Conference*, pp. 69–73.

Esch, R., and P. Calingaert (1957), *Univac I central computer programming*, Harvard Computation Laboratory, Cambridge, Massachusetts.

Fairthorne, R. A. (1958), "Algebraic representation of storage and retrieval languages," *Preprints of Papers for the International Conference on Scientific Information*, National Academy of Sciences—National Research Council, Washington, D.C., Area VI, pp. 43–56.

Fedorov, A. V. (1953), *Vvedenie v teoriju perevoda* (Introduction to the theory of translation), Biblioteka filologa, Moscow (second revised edition appeared in 1958).

*Foust, W. (1957), "Inflected form generators," *AF-49*, Sec. VI.

*—— (1958), "Inflectors," *AF-50*, Sec. VI.

Franklin Institute (1957), *Automatic coding*, Monograph No. 3, Lancaster, Pennsylvania.

Friend, E. H. (1956), "Sorting on electronic computer systems," *Journal of the Association for Computing Machinery*, Vol. 3, pp. 134–168.

Fries, C. C. (1952), *The structure of English*, Harcourt, Brace, New York.

*Frink, O. (1959), "Programs for correcting the Harvard automatic dictionary and for syntactic study (Conhadic, Checkhadic, Texthadic, and Freqhadic)," *NSF-2*, Sec. V.

Fry, D. B., and P. Denes (1956), "Experiments in mechanical speech recognition," in C. Cherry, editor, *Information theory*, Butterworths Scientific Publications, London, pp. 206–212.

Gibbons, J. (1957), "How input/output units affect data-processor performance," *Control Engineering*, Vol. 4, No. 7 (July 1957), pp. 97–102.

Gill, S. (1958), "Parallel programming," *The Computer Journal*, Vol. 1, No. 1, pp. 2–10.

*Giuliano, V. E. (1957a), "Programming an automatic dictionary," *AF-49*, Sec. I.

*—— (1957b), "Compilation of an automatic dictionary," *AF-49*, Sec. V.

—— (1958), "The trial translator, an automatic programming system for the experimental machine translation of Russian to English," *Proceedings of the Eastern Joint Computer Conference*, pp. 138–144.

—— (1959a), "An experimental study of automatic language translation," doctoral thesis, Harvard University, Cambridge, Massachusetts.

*—— (1959b), "A formula finder for the automatic synthesis of translation algorithms," *NSF-2*, Sec. IX.

—— and A. G. Oettinger (1959), "Research in automatic translation at the Harvard Computation Laboratory," preprint of paper to appear in *Proceedings of an International Conference on Information Processing* (Paris, June 1959).

*Glantz, R. S. (1959), "Russian-English semantic equivalence classes," *Seminar papers*, Vol. V.

Gleason, H. A. (1955), *An introduction to descriptive linguistics*, Holt, New York.

Gold, B. (1959), "Machine recognition of hand-sent Morse code," *IRE Transactions*, Vol. IT-5, No. 1 (March 1959), pp. 17–24.

Goodman, N. (1952), "On likeness of meaning," in L. Linsky, editor, *Semantics and the philosophy of language*, University of Illinois Press, Urbana, Illinois.

Gorn, S. (1957), "Standardized programming methods and universal coding," *Journal of the Association for Computing Machinery*, Vol. 4, pp. 254–273.

Gotlieb, C. C., and J. N. P. Hume (1958), *High-speed data processing*, McGraw-Hill, New York.

Greanias, E. C., and Y. M. Hill (1957), "Considerations in the design of character recognition devices," *IRE National Convention Record*, Part 4, pp. 119–126.

*Greenberg, M. (1959), "A method for contextual analysis of Russian word classes," *Seminar papers*, Vol. V.

Guiraud, P. (1954), *Bibliographie critique de la statistique linguistique*, Editions Spectrum, Utrecht/Antwerpen.

Hamming, R. W. (1950), "Error detecting and error correcting codes," *Bell System Technical Journal*, Vol. 29, pp. 147–160.

Hamp, E. P. (1957), *A glossary of American technical linguistic usage*, Editions Spectrum, Utrecht/Antwerpen.

Harris, Z. S. (1951), *Methods in structural linguistics*, University of Chicago Press, Chicago, Illinois.

—— (1954), "Transfer grammar," *International Journal of American Linguistics*, Vol. 20, No. 4, pp. 259–270.

—— (1957), "Co-occurrence and Transformation in Linguistic Structure," *Language*, Vol. 33, No. 3, pp. 283–340.

Hayes, R. M., and J. Wiener (1957), "Magnacard—A new concept in data handling," *IRE Wescon Convention Record*, Part 4, pp. 205–209.

Henle, P. (1949), "Mysticism and semantics," *Philosophy and Phenomenological Research*, Vol. IX, No. 3, pp. 416–422.

Herdan, G. (1956), *Language as choice and chance*, Noordhoff, Groningen, Holland.

Hildebrandt, P., and H. Isbitz (1959), "Radix exchange—An internal sorting method for digital computers," *Journal of the Association for Computing Machinery*, Vol. 6, pp. 156–163.

Higonnet, R., and R. Gréa (1955), *Étude logique des circuits électriques*, Berger-Levrault, Paris.

Hockett, C. F. (1958), *A course in modern linguistics*, Macmillan, New York.

Hollander, G. L. (1956), "Quasi-random access memory systems," *Proceedings of the Eastern Joint Computer Conference*, pp. 128–135.

Hoover, C. W., Jr., R. E. Staehler, and R. W. Ketchledge (1958), "Fundamental concepts in the design of the flying spot store," *Bell System Technical Journal*, Vol. 37, pp. 1161–1194.

Hosken, J. C. (1955), "Evaluation of sorting methods," *Proceedings of the Eastern Joint Computer Conference*, pp. 39–55.

Hunt, F. V. (1954), *Electroacoustics*, Harvard University Press, Cambridge, Massachusetts.

Isaac, E. J., and R. C. Singleton (1956), "Sorting by address calculation," *Journal of the Association for Computing Machinery*, Vol. 3, pp. 169–174.

Iverson, K. E. (1955), "Immediate access storage devices for business records," *Progress Report No. 1* by the Staff of the Computation Laboratory to the American Gas Association and Edison Electric Institute, Sec. III, Harvard University, Cambridge, Massachusetts.

Jakobson, R. (1959), "On linguistic aspects of translation," in Brower (1959).

Janov, Ju. I. (1958), "O logicheskix sxemax algoritmov" (On logical schemata of algorithms), in Ljapunov (1958).

Jeenel, J. (1959), *Programming for digital computers*, McGraw-Hill, New York.

*Jones, P. E., Jr. (1958), "A method of contextual analysis applicable to linguistic research," *Seminar papers*, Vol. IV.

*—— (1959), "The continuous dictionary run," *NSF-2*, Sec. I.

Karp, R. M. (1959), "Some applications of logical syntax to digital computer programming," doctoral thesis, Harvard University, Cambridge, Massachusetts.

Kemeny, J. G., H. Mirkil, J. L. Snell, and G. L. Thompson (1959), *Finite mathematical structures*, Prentice-Hall, Englewood Cliffs, New Jersey.

—— J. L. Snell, and G. L. Thompson (1957), *Introduction to finite mathematics*, Prentice-Hall, Englewood Cliffs, New Jersey.

King, G. W. (1955), "A new approach to information storage," *Control Engineering*, Vol. 2, No. 8 (August 1955), pp. 48–53.

Klee, P. (1948), *On modern art*, Faber and Faber, London.

Kleene, S. C. (1952), *Introduction to metamathematics*, Van Nostrand, New York.

Kolmogorov, A. N. (1956), *Sessija Akademii Nauk SSSR po nauchnym problemam avtomatizatsii proizvodstva, 15–20 oktjabrja 1956g.*, *Plenarnye zasedanija* (Session of the Academy of Sciences of the USSR on the scientific problems of automation of production, October 15–20, 1956, Plenary sessions), Academy of Sciences of the USSR, Moscow, p. 161.

Korolev, L. N. (1957), "Kodirovanie i svertyvanie kodov" (Coding and code compression), *Doklady, AN SSSR*, Vol. 113, No. 4, pp. 746–747. (An English translation appears in the *Journal of the Association for Computing Machinery*, Vol. 5, No. 4 (1958), pp. 328–330.)

Koutsoudas, A., and A. Humecky (1957), "Ambiguity of syntactic function resolved by linear context," *Word*, Vol. 13, No. 3, pp. 403–414.

Kulagina, O. S. (1958), "Ob odnom sposobe opredelenija grammaticheskix ponjatij na baze teorii mnozhestv" (On a method for defining linguistic concepts based on set theory), in Ljapunov (1958).

—— and I. A. Mel'chuk (1956), "Mashinnyj perevod s frantsuzkogo jazyka na russkij" (Machine translation from French to Russian), *Voprosy Jazykoznanija*, No. 5, pp. 111–121.

*Landau, H. J. (1955), "An automatic glossary and problems of case translation," *Seminar papers*, Vol. I.

Langer, Susanne (1948), *Philosophy in a new key* (1st ed.), Penguin Books, New York.

Lebedev, S. A. (1956), *Elektronnye vychislitel'nye mashiny* (Electronic computers), Nauchno-populjarnaja serija, Izdatel'stvo Akademii Nauk SSSR, Moscow.

Ljapunov, A. A., editor (1958), *Problemy kibernetiki* (Problems of cybernetics), No. 1, Gosudarstvennoe Izdatel'stvo Fiziko-Matematicheskoj Literatury, Moscow.

Locke, W. N., and A. D. Booth, editors (1955), *Machine translation of languages*, Technology Press, Cambridge, Massachusetts, and Wiley, New York.

Luhn, H. P., and P. James (1958), "Bibliography and index: Literature on information retrieval and machine translation," Service Bureau Corporation, New York.

Łukasiewicz, J. (1951), *Aristotle's syllogistic from the standpoint of modern formal logic*, Clarendon Press, Oxford (second enlarged edition appeared in 1957).

Lynch, I., "On Russian verbal voice—the -sja verbs", doctoral thesis, Radcliffe College, Cambridge, Massachusetts.

MacDonald, D. N. (1956), "Datafile—A new tool for extensive file storage," *Proceedings of the Eastern Joint Computer Conference*, pp. 124–128.

*Magassy, K. (1956), "An automatic method of inflection for Russian," *AF–46*, Sec. V.

*—— (1957), "An automatic method of inflection for Russian (II)," *AF–49*, Sec. III.

*Matejka, L. (1957), "Selection and classification of the vocabulary for the automatic dictionary," *AF–49*, Sec. IV.

*—— (1958), "Grammatical specifications in the Russian-English automatic dictionary," *AF–50*, Sec. V.

McCracken, D. D. (1957), *Digital computer programming*, Wiley, New York.

Mel'chuk, I. A. (1958), "O mashinnom perevode s vengerskogo jazyka na russkij" (On machine translation from Hungarian to Russian), in Ljapunov (1958).

Menger, K. (1957), *The basic concepts of mathematics*, The Bookstore, Illinois Institute of Technology, Chicago, Illinois.

Miehle, W. (1957), "Burroughs truth function evaluator," *Journal of the Association for Computing Machinery*, Vol. 4, No. 2, pp. 189–192.

Miller, G. A., and J. G. Beebe-Center (1956), "Some psychological methods for evaluating the quality of translations," *Mechanical Translation*, Vol. 3, No. 3, pp. 73–80.

Moloshnaja, T. N. (1958), "Voprosy razlichenija omonimov pri mashinnom perevode s anglijskogo jazyka na russkij" (On discerning homonyms in machine translation of English to Russian), in Ljapunov (1958).

National Bureau of Standards (1957), "Fosdic II—Reads microfilmed punched cards," *N.B.S. Technical News Bulletin*, Vol. 41, pp. 72–74.

National Science Foundation (1958), *Current research and development in scientific documentation*, No. 3, Science Information Service, National Science Foundation, Washington, D.C.

Northcutt, S., and J. Kilbane (1958), "Letter counting routine," Harvard Systems, unpublished report.

Noyes, T., and W. E. Dickinson (1956), "Engineering design of a magnetic disk random-access memory," *Proceedings of the Western Joint Computer Conference*, pp. 42–44.

*Oettinger, A. G. (1953), "A study for the design of an automatic dictionary", *AF-26*, Sec. VIII.

—— (1954), "A study for the design of an automatic dictionary," doctoral thesis, Harvard University, Cambridge, Massachusetts.

—— (1955a), "Chart representations of data processing systems," *Progress Report No. 1* by The Staff of the Computation Laboratory to The American Gas Association and Edison Electric Institute, Sec. V, Harvard University, Cambridge, Massachusetts.

—— (1955b); "The manual use of automatic records," *Proceedings of the Eastern Joint Computer Conference*, pp. 33–39.

—— (1956a), "Chart representations of data processing systems—(II) The abstract data processing system," *Progress Report No. 3* by The Staff of the Computation Laboratory to The American Gas Association and Edison Electric Institute, Sec. V, Harvard University, Cambridge, Massachusetts.

—— (1956b), "A new basic approach to automatic data processing," *Progress Report No. 3* by The Staff of the Computation Laboratory to The American Gas Association and Edison Electric Institute, Sec. III, Harvard University Cambridge, Massachusetts.

—— (1957a), "Account identification for automatic data processing," *Journal of the Association for Computing Machinery*, Vol. 4, pp. 245–253.

—— (1957b), "Linguistics and mathematics," in E. Pulgram, editor, *Studies presented to Joshua Whatmough*, Mouton, The Hague.

—— (1958), "An input device for the Harvard automatic dictionary," *Mechanical Translation*, Vol. 5, No. 1, pp. 2–7.

—— (1959a), "Automatic translation," in Brower (1959).

—— (1959b), "A survey of Soviet work on automatic translation," *Mechanical Translation*, Vol. 5, No. 3, pp. 101–110.

—— W. Foust, V. E. Giuliano, K. Magassy, and L. Matejka (1958), "Linguistic and machine methods for compiling and updating the Harvard automatic dictionary," *Preprints of Papers for the International Conference on Scientific Information*, National Academy of Sciences—National Research Council, Washington, D.C., Area V, pp. 137–159.

Ostrowski, A. (1954), "On two problems in abstract algebra connected with Horner's rule," *Studies in Mathematics and Mechanics presented to R. von Mises*, Academic Press, New York.

Oswald, V. A., Jr., and R. H. Lawson (1953), "An idioglossary for mechanical translation," *Modern Language Forum*, Vol. XXXVIII, Nos. 3–4, pp. 1–11.

Parkinson, Professor C. Northcote (1957), *Parkinson's law*, Houghton-Mifflin, Boston.

Perlis, A. J., and K. Samelson, editors (1959), "Report on the algorithmic language ALGOL," *Numerische Mathematik*, Vol. 1, pp. 41–60.

Peterson, W. W. (1957), "Addressing for random-access storage," *IBM Journal of Research and Development*, Vol. 1, pp. 130–146.

Postley, J. A. (1955), "File reference," Report P-691 (August 18, 1955), Rand Corporation, Santa Monica, California.

Quine, W. V. (1950), *Methods of logic*, Holt, New York.

—— (1953), *From a logical point of view*, Harvard University Press, Cambridge, Massachusetts.

—— (1955), *Mathematical logic* (revised edition, second printing), Harvard University Press, Cambridge, Massachusetts.

Ramo-Wooldridge Corporation (1958), *Experimental machine translation of Russian to English, a project progress report*, (M20-8U13), Ramo-Wooldridge Corp., Los Angeles, California.

Razumovskij, S. N. (1957), "K voprosu ob avtomatizatsii programmirovanija zadach perevoda s odnogo jazyka na drugoj" (On the question of automatizing the programming of the process of translation from one language into another), *Doklady*, AN SSSR, Vol. 113, No. 4, pp. 760–761.

Reichenbach, H. (1947), *Elements of symbolic logic*, Macmillan, New York.

Rhodes, I. (1959), "A new approach to the mechanical translation of Russian," National Bureau of Standards, Washington, D.C., unpublished report.

Sapir, E. (1921), *Language*, Harcourt, Brace, New York.

Savory, T. H. (1957), *The art of translation*, Cape, London.

Segal, L. (1953), *Russian-English dictionary*, Lund, Humphries, London.

Shannon, C. E., and W. Weaver (1949), *The mathematical theory of communication*, University of Illinois Press, Urbana, Illinois.

Shaw, J. C., A. Newell, H. A. Simon, and T. O. Ellis (1958), "A command structure for complex information processing," Report P-1277, The Rand Corporation, Santa Monica, California.

Shepard, D. H., P. F. Bargh, and C. C. Heasly, Jr. (1957), "A reliable character sensing system for documents prepared on conventional business devices," *IRE Wescon Convention Record*, Part 4, pp. 111–120.

—— and C. C. Heasly, Jr. (1955), "Photoelectric reader feeds business machines," *Electronics*, Vol. 28 (May 1955), pp. 134–138.

*Sherry, M. E. (1958), "Analysis of case and number of Russian nouns for automatic translation," *Seminar papers*, Vol. IV.

*—— (1959), "Automatic interpretation of Russian affixes on a word-by-word basis," *NSF-2*, Sec. VI.

Smith, H. M. (1955), "The Typotron: A novel character display storage tube," *IRE Convention Record*, Part 4, pp. 129–134.

Smirnitskij, A. I. (1949), *Russko-anglijskij slovar'* (Russian-English dictionary), OGIZ, Moscow. New 1958 edition, with O. S. Axmanova.

Staff of the Computation Laboratory (1951), *Synthesis of electronic computing and control circuits*, Annals of the Computation Laboratory of Harvard University, Vol. 27, Harvard University Press, Cambridge, Massachusetts.

Unbegaun, B. O. (1957), *Russian grammar*, Clarendon Press, Oxford.

U.S. Department of Commerce (1955), *The ISCC-NBS method of designating colors and a dictionary of color names*, National Bureau of Standards Circular 553, Washington, D.C.

University of Illinois (1959), *High school mathematics*, University of Illinois Committee on School Mathematics, Urbana, Illinois, mimeographed report.

University of Washington (1958), *Linguistic and engineering studies in automatic language translation of scientific Russian into English*, Department of Far Eastern and Slavic Languages and Literature and Department of Electrical Engineering, University of Washington, Seattle, Washington.

Ushakov, D. N. (1935), *Tolkovyj slovar' russkogo jazyka* (Explanatory dictionary of the Russian language), Vols. I-IV, OGIZ, Moscow.

van der Pol, B. (1956), "An iterative translation test," in C. Cherry, editor, *Information Theory*, Third London Symposium 1955; Butterworths Scientific Publications, London.

Vasmer, M., R. Greve, and B. Kroesche (1957), *Russisches rückläufiges Wörterbuch*, Harrassowitz, Wiesbaden.

Webster's new collegiate dictionary (1953), Merriam, Springfield, Massachusetts.

Weik, M. H. (1957), *A second survey of domestic electronic digital computing systems*, Report No. 1010, Ballistic Research Laboratories, Aberdeen Proving Ground, Maryland.

—— (1958), *A minimum "ones" binary code for English text*, Ballistic Research Laboratories Technical Note No. 1215, Aberdeen Proving Ground, Maryland.

Welsh, H. F., and V. J. Porter (1956), "A large-capacity drum-file memory system," *Proceedings of the Eastern Joint Computer Conference*, pp. 136–139.

Whatmough, J. (1956a), *Language*, St. Martin's Press, New York.

—— (1956b), *Poetic, scientific, and other forms of discourse*, University of California Press, Berkeley and Los Angeles, California.

Whorf, B. L. (1956), *Language, thought, and reality*, Technology Press, Cambridge, Massachusetts, and Wiley, New York.

Wilkes, M. V. (1956), *Automatic digital computers*, Methuen, London.

—— D. J. Wheeler, and S. Gill (1957), *The preparation of programs for an electronic digital computer* (2nd ed.), Addison-Wesley, Cambridge, Massachusetts.

Worth, D. S. (1959), " 'Linear contexts,' linguistics and machine translation," *Word*, Vol. 15, No. 1, pp. 183–191.

Yngve, V. H. (1957), "A framework for syntactic translation," *Mechanical Translation*, Vol. 4, No. 3, pp. 59–65.

—— (1958), "A programming language for mechanical translation," *Mechanical Translation*, Vol. 5, No. 1, pp. 25–41.

Young, J. W., Jr., and H. K. Kent (1958), "Abstract formulation of data processing problems," *Journal of Industrial Engineering*, Vol. IX, No. 6, pp. 471–479.

Yule, G. U. (1944), *The statistical study of literary vocabulary*, Cambridge University Press, Cambridge.

Zipf, G. K. (1935), *The psycho-biology of language*, Houghton-Mifflin, Boston.

—— (1949), *Human behavior, and the principle of least effort*, Addison-Wesley, Cambridge, Massachusetts.

INDEX

Abaev, V. I., 132
Access to storage, 27, 29, 32, 321, 322
Access ratio, 29, 31, 322
Address, 2, 25, 31; absolute, 96; modification, 68; origin, 96; relative, 95f
Affix, 137f, 143, 150, 218, 307, 311
Affix interpretation, 328f, 340
Africa, C., 228
Algorithm, 57f, 61, 84, 98, 114, 117, 118
Alphabet, 55, 107; Cyrillic, 72; natural, 72; Roman, 104
Alphabetic subdictionary, 274, 276, 307, 336
American Institute of Physics, 287
American Mathematical Society, 232, 253
Apostel, L., 282, 322
Arranging, 31
Ashenhurst, R. L., 58
Augmented text, 265, 271f, 323, 336, 341, 349

Bar Hillel, Y., 125
Barnes, V. L., 267
Barrett, P. R., 63
Batey, C., 63
Beebe-Center, J. G., 114, 288
Berson, J. S., 102
Bielfeldt, H. H., 142
Binary-coded decimal system, 6
Binary coding system, 7
Binary partition access, 32
Bistable device, 5, 65
Bit, 8
Bloch, B., 110
Block, 8, 17, 71, 220, 226
Blockette, 8, 17, 71, 220
Booth, A. D., 125, 151
Branch, 148
Brandwood, L., 151
Brower, R. A., 111, 113f
Bull, W. E., 228
Burks, A. W., 101

Caldwell, S. H., 143
Calingaert, P., 38, 67
Canonical dictionary, 135, 152, 155, 255, 308, 316, 338, 340f
Canonical form of words, 134f, 234, 303
Cary, E., 111, 112f

Character, 8, 49, 121
Chaundy, T. W., 63
Chomsky, N., 117f
Churchill, W., 229
Cipher, 49
Classification: detecting and correcting mistakes in, 236f; functional, 163; morphological, 154f, 234; of adjectival forms, 169; of nominal forms, 163f; of verbal forms, 170f
Cleave, J. P., 151
Code: conversion, 71f; grammatical, 248f; Morse, 4, 105
Coding of machine programs, 84f
Cohn, M., 151
Compilation of a dictionary, 216f, 302f, 340; detecting and correcting mistakes in, 323f
Concordance, 343
Condensation, 244, 248
Connectors in flow charts, 88
Content of *m*, 10
Context, 253, 341f
Control unit, 3
Coordinates of word ends, 237, 244
Coordinate access, 32, 322
Coppinger, L. L., 323, 334
Correspondence between sets, 45, 47f, 107
Curry, H. B., 40f, 44

Daum, E., 170, 173
Delayed access, 27
Department of the Army, 232, 253
Derivation, 127f, 158, 319, 337f
Desinence, 135, 137, 139, 151, 307
Dewey, G., 228f
Dictionary: by-products of, 275f; file structure, 321f; keys, 127f, 248, 311; look-up, 138, 271f, 337; operation, 265f, 302f
Domain of a transformation, 110
Dostert, L. E., 63

End-alphabetic order, 139, 141, 241, 338, 345
English correspondents, 248f
Entry number, 252, 324
Equivalence under translation, 110

377